GENETICS, GENOMICS AND BREEDING OF GRAPES

Genetics, Genomics and Breeding of Crop Plants

Series Editor

Chittaranjan Kole
Department of Genetics and Biochemistry
Clemson University
Clemson, SC
USA

Books in this Series:

Published or in Press:

- Jinguo Hu, Gerald Seiler & Chittaranjan Kole: *Sunflower*
- Kristin D. Bilyeu, Milind B. Ratnaparkhe & Chittaranjan Kole: *Soybean*
- Robert Henry & Chittaranjan Kole: *Sugarcane*
- Kevin Folta & Chittaranjan Kole: *Berries*
- Jan Sadowsky & Chittaranjan Kole: *Vegetable Brassicas*
- James M. Bradeen & Chittaranjan Kole: *Potato*
- C.P. Joshi, Stephen DiFazio & Chittaranjan Kole: *Poplar*
- Anne-Françoise Adam-Blondon, José M. Martínez-Zapater & Chittaranjan Kole: *Grapes*
- Christophe Plomion, Jean Bousquet & Chittaranjan Kole: *Conifers*
- Dave Edwards, Jacqueline Batley, Isobel Parkin & Chittaranjan Kole: *Oilseed Brassicas*
- Marcelino Pérez de la Vega, Ana María Torres, José Ignacio Cubero & Chittaranjan Kole: *Cool Season Grain Legumes*
- Yi-Hong Wang, Tusar Kanti Behera & Chittaranjan Kole: *Cucurbit*

GENETICS, GENOMICS AND BREEDING OF GRAPES

Editors

Anne-Françoise Adam-Blondon
INRA
UMR INRA UEVE URL CNRS Genomique végétale (URGV)
Evry Cedex, France

José M. Martínez-Zapater
Instituto de Ciencias de la Vid y del Vino
(CSIC, UR, Gobierno de La Rioja)
Logroño
Spain

Chittaranjan Kole
Department of Genetics and Biochemistry
Clemson University
Clemson, SC
USA

CRC Press
Taylor & Francis Group
an informa business
www.crcpress.com

6000 Broken Sound Parkway, NW
Suite 300, Boca Raton, FL 33487
270 Madison Avenue
New York, NY 10016
2 Park Square, Milton Park
Abingdon, Oxon OX14 4RN, UK

Science Publishers
Jersey, British Isles
Enfield, New Hampshire

Published by Science Publishers, an imprint of Edenbridge Ltd.
- St. Helier, Jersey, British Channel Islands
- P.O. Box 699, Enfield, NH 03748, USA

E-mail: info@scipub.net Website: www.scipub.net

Marketed and distributed by:

CRC Press
Taylor & Francis Group
an informa business
www.crcpress.com

6000 Broken Sound Parkway, NW
Suite 300, Boca Raton, FL 33487
270 Madison Avenue
New York, NY 10016
2 Park Square, Milton Park
Abingdon, Oxon OX14 4RN, UK

ISBN 978-1-57808-717-4

Library of Congress Cataloging-in-Publication Data

Genetics, genomics and breeding of grapes / editors, Anne-Françoise
Adam-Blondon, José-Miguel Martinez-Zapater, Chittaranjan Kole. -- 1st
ed.
p. cm. -- (Genetics, genomics and breeding of crop plants)
Includes bibliographical references and index.
ISBN 978-1-57808-717-4 (hardcover)
1. Grapes--Genetics. 2. Grapes--Genome mapping. 3. Grapes--Breeding. I. Adam-Blondon, Anne-Françoise. II. Martínez-Zapater, José M. III. Kole, Chittaranjan. IV. Series:
Genetics, genomics and breeding of crop plants.
SB398.3.G46 2011
634.8'8233--dc22

2011000002

Printed in the United States of America

Preface to the Series

Genetics, genomics and breeding has emerged as three overlapping and complimentary disciplines for comprehensive and fine-scale analysis of plant genomes and their precise and rapid improvement. While genetics and plant breeding have contributed enormously towards several new concepts and strategies for elucidation of plant genes and genomes as well as development of a huge number of crop varieties with desirable traits, genomics has depicted the chemical nature of genes, gene products and genomes and also provided additional resources for crop improvement.

In today's world, teaching, research, funding, regulation and utilization of plant genetics, genomics and breeding essentially require thorough understanding of their components including classical, biochemical, cytological and molecular genetics; and traditional, molecular, transgenic and genomics-assisted breeding. There are several book volumes and reviews available that cover individually or in combination of a few of these components for the major plants or plant groups; and also on the concepts and strategies for these individual components with examples drawn mainly from the major plants. Therefore, we planned to fill an existing gap with individual book volumes dedicated to the leading crop and model plants with comprehensive deliberations on all the classical, advanced and modern concepts of depiction and improvement of genomes. The success stories and limitations in the different plant species, crop or model, must vary; however, we have tried to include a more or less general outline of the contents of the chapters of the volumes to maintain uniformity as far as possible.

Often genetics, genomics and plant breeding and particularly their complimentary and supplementary disciplines are studied and practiced by people who do not have, and reasonably so, the basic understanding of biology of the plants for which they are contributing. A general description of the plants and their botany would surely instill more interest among them on the plant species they are working for and therefore we presented lucid details on the economic and/or academic importance of the plant(s); historical information on geographical origin and distribution; botanical origin and evolution; available germplasms and gene pools, and genetic and cytogenetic stocks as genetic, genomic and breeding resources; and basic information on taxonomy, habit, habitat, morphology, karyotype, ploidy level and genome size, etc.

Classical genetics and traditional breeding have contributed enormously even by employing the phenotype-to-genotype approach. We included detailed descriptions on these classical efforts such as genetic mapping using morphological, cytological and isozyme markers; and achievements of conventional breeding for desirable and against undesirable traits. Employment of the in vitro culture techniques such as micro- and megaspore culture, and somatic mutation and hybridization, has also been enumerated. In addition, an assessment of the achievements and limitations of the basic genetics and conventional breeding efforts has been presented.

It is a hard truth that in many instances we depend too much on a few advanced technologies, we are trained in, for creating and using novel or alien genes but forget the infinite wealth of desirable genes in the indigenous cultivars and wild allied species besides the available germplasms in national and international institutes or centers. Exploring as broad as possible natural genetic diversity not only provides information on availability of target donor genes but also on genetically divergent genotypes, botanical varieties, subspecies, species and even genera to be used as potential parents in crosses to realize optimum genetic polymorphism required for mapping and breeding. Genetic divergence has been evaluated using the available tools at a particular point of time. We included discussions on phenotype-based strategies employing morphological markers, genotype-based strategies employing molecular markers; the statistical procedures utilized; their utilities for evaluation of genetic divergence among genotypes, local landraces, species and genera; and also on the effects of breeding pedigrees and geographical locations on the degree of genetic diversity.

Association mapping using molecular markers is a recent strategy to utilize the natural genetic variability to detect marker-trait association and to validate the genomic locations of genes, particularly those controlling the quantitative traits. Association mapping has been employed effectively in genetic studies in human and other animal models and those have inspired the plant scientists to take advantage of this tool. We included examples of its use and implication in some of the volumes that devote to the plants for which this technique has been successfully employed for assessment of the degree of linkage disequilibrium related to a particular gene or genome, and for germplasm enhancement.

Genetic linkage mapping using molecular markers have been discussed in many books, reviews and book series. However, in this series, genetic mapping has been discussed at length with more elaborations and examples on diverse markers including the anonymous type 2 markers such as RFLPs, RAPDs, AFLPs, etc. and the gene-specific type 1 markers such as EST-SSRs, SNPs, etc.; various mapping populations including F_2, backcross, recombinant inbred, doubled haploid, near-isogenic and pseudotestcross; computer software including MapMaker, JoinMap, etc. used; and different types of genetic maps including preliminary, high-resolution, high-density, saturated, reference, consensus and integrated developed so far.

Mapping of simply inherited traits and quantitative traits controlled by oligogenes and polygenes, respectively has been deliberated in the earlier literature crop-wise or crop group-wise. However, more detailed information on mapping or tagging oligogenes by linkage mapping or bulked segregant analysis, mapping polygenes by QTL analysis, and different computer software employed such as MapMaker, JoinMap, QTL Cartographer, Map Manager, etc. for these purposes have been discussed at more depth in the present volumes.

The strategies and achievements of marker-assisted or molecular breeding have been discussed in a few books and reviews earlier. However, those mostly deliberated on the general aspects with examples drawn mainly from major plants. In this series, we included comprehensive descriptions on the use of molecular markers for germplasm characterization, detection and maintenance of distinctiveness, uniformity and stability of genotypes, introgression and pyramiding of genes. We have also included elucidations on the strategies and achievements of transgenic breeding for developing genotypes particularly with resistance to herbicide, biotic and abiotic stresses; for biofuel production, biopharming, phytoremediation; and also for producing resources for functional genomics.

A number of desirable genes and QTLs have been cloned in plants since 1992 and 2000, respectively using different strategies, mainly positional cloning and transposon tagging. We included enumeration of these and other strategies for isolation of genes and QTLs, testing of their expression and their effective utilization in the relevant volumes.

Physical maps and integrated physical-genetic maps are now available in most of the leading crop and model plants owing mainly to the BAC, YAC, EST and cDNA libraries. Similar libraries and other required genomic resources have also been developed for the remaining crops. We have devoted a section on the library development and sequencing of these resources; detection, validation and utilization of gene-based molecular markers; and impact of new generation sequencing technologies on structural genomics.

As mentioned earlier, whole genome sequencing has been completed in one model plant (Arabidopsis) and seven economic plants (rice, poplar, peach, papaya, grapes, soybean and sorghum) and is progressing in an array of model and economic plants. Advent of massively parallel DNA sequencing using 454-pyrosequencing, Solexa Genome Analyzer, SOLiD system, Heliscope and SMRT have facilitated whole genome sequencing in many other plants more rapidly, cheaply and precisely. We have included extensive coverage on the level (national or international) of collaboration and the strategies and status of whole genome sequencing in plants for which sequencing efforts have been completed or are progressing currently. We have also included critical assessment of the impact of these genome initiatives in the respective volumes.

Comparative genome mapping based on molecular markers and map positions of genes and QTLs practiced during the last two decades of the last century provided answers to many basic questions related to evolution, origin and phylogenetic relationship of close plant taxa. Enrichment of genomic resources has reinforced the study of genome homology and synteny of genes among plants not only in the same family but also of taxonomically distant families. Comparative genomics is not only delivering answers to the questions of academic interest but also providing many candidate genes for plant genetic improvement.

The 'central dogma' enunciated in 1958 provided a simple picture of gene function—gene to mRNA to transcripts to proteins (enzymes) to metabolites. The enormous amount of information generated on characterization of transcripts, proteins and metabolites now have led to the emergence of individual disciplines including functional genomics, transcriptomics, proteomics and metabolomics. Although all of them ultimately strengthen the analysis and improvement of a genome, they deserve individual deliberations for each plant species. For example, microarrays, SAGE, MPSS for transcriptome analysis; and 2D gel electrophoresis, MALDI, NMR, MS for proteomics and metabolomics studies require elaboration. Besides transcriptome, proteome or metabolome QTL mapping and application of transcriptomics, proteomics and metabolomics in genomics-assisted breeding are frontier fields now. We included discussions on them in the relevant volumes.

The databases for storage, search and utilization on the genomes, genes, gene products and their sequences are growing enormously in each second and they require robust bioinformatics tools plant-wise and purpose-wise. We included a section on databases on the gene and genomes, gene expression, comparative genomes, molecular marker and genetic maps, protein and metabolomes, and their integration.

Notwithstanding the progress made so far, each crop or model plant species requires more pragmatic retrospect. For the model plants we need to answer how much they have been utilized to answer the basic questions of genetics and genomics as compared to other wild and domesticated species. For the economic plants we need to answer as to whether they have been genetically tailored perfectly for expanded geographical regions and current requirements for green fuel, plant-based bioproducts and for improvements of ecology and environment. These futuristic explanations have been addressed finally in the volumes.

We are aware of exclusions of some plants for which we have comprehensive compilations on genetics, genomics and breeding in hard copy or digital format and also some other plants which will have enough achievements to claim for individual book volume only in distant future. However, we feel satisfied that we could present comprehensive deliberations on genetics, genomics and breeding of 30 model and economic plants, and their groups in a few cases, in this series. I personally feel also happy that I could work with many internationally celebrated scientists who edited the book volumes on the leading plants and plant groups and included chapters authored by many scientists reputed globally for their contributions on the concerned plant or plant group.

We paid serious attention to reviewing, revising and updating of the manuscripts of all the chapters of this book series, but some technical and formatting mistakes will remain for sure. As the series editor, I take complete responsibility for all these mistakes and will look forward to the readers for corrections of these mistakes and also for their suggestions for further improvement of the volumes and the series so that future editions can serve better the purposes of the students, scientists, industries, and the society of this and future generations.

Science publishers, Inc. has been serving the requirements of science and society for a long time with publications of books devoted to advanced concepts, strategies, tools, methodologies and achievements of various science disciplines. Myself as the editor and also on behalf of the volume editors, chapter authors and the ultimate beneficiaries of the volumes take this opportunity to acknowledge the publisher for presenting these books that could be useful for teaching, research and extension of genetics, genomics and breeding.

Chittaranjan Kole

Preface to the Volume

Grapevine (*Vitis vinifera* L.) is one of the most economically important perennial fruit crop species in the world with more than 7.4 million hectares planted in temperate and tropical climatic regions. Grapes are not only produced for fruit, juice or raisins but are the basis of high added value products, mainly wine and spirits. Contrary to other crops where innovation is primarily based on cultivar improvement, the notable innovation suffered by viticulture in the last decades has been based on management techniques, whereas wine grape cultivars have changed slowly or not at all. Most vineyards in the world are planted with cultivars that have been perpetuated for centuries by vegetative propagation.

The original wild species from which grapevines were domesticated can still be found in remnant populations in temperate Eurasian regions. This species is dioecious, with male and female plants, while most current cultivars are hermaphrodite plants. Still, as other fruit tree species, grapevine cultivars are highly heterozygous which require vegetative propagation to maintain their phenotypic features. While breeding has been very successful in the generation of new table grape cultivars in the last decades, wine grape cultivars have not evolved in the same way. The need to be attached to a few elite cultivars recognized world-wide and the added difficulty of introducing wine quality evaluation and selection in breeding programs have resulted in very few new bred wine cultivars being incorporated to the wine market in the last century.

Two decades ago, the application of molecular markers in the construction of genetic maps and the analyses of quantitative traits opened the way to understanding the genetic basis of relevant agronomic and quality traits in grapevine and related species. More recently, the completion of its genome sequence has speeded up the development of genomics and is generating an explosion in grapevine research. The application of these fast evolving technologies will greatly increase our understanding of grapevine biology and should pave the way to support the development of new wine and table grape cultivars more adapted to the needs of a sustainable viticulture in a changing environment. In addition, these genomics can also help to improve viticulture techniques by providing information on the plant status and its relationships with other organisms in the vineyard.

This volume presents in 14 Chapters the recent advances in grapevine genetics and genomics. The book begins with basic information on the species and the crop in the introductory Chapter 1 and dedicates two chapters to describe the existent natural variation in the *Vitis* genus (Chapter 2) as well as the origin and utilization of somatic variation in grapevine (Chapter 3). Natural genetic variation is the raw material for breeding but it is also a basic tool to understand gene biological function. Genetic analyses and molecular breeding are considered in the following four chapters including association genetics (Chapter 4), genetic maps (Chapter 5), genetic analysis of complex traits (Chapter 6) and molecular breeding (Chapter 7). The connection between genetic and physical maps is considered in Chapter 8 which demonstrates the feasibility of positional cloning in grapevine. The following chapters review the available information on the grapevine genome sequence (Chapter 9), the application of global gene expression analytical strategies (Chapters 10 and 11), the development of proteomics and metabolomics technologies for grapevine (Chapter 12) and the availability of databases and bioinformatics tools (Chapter 13). The book ends with a chapter on future prospects of genome research in viticulture (Chapter 14). We greatly appreciate the effort and time devoted to this volume by all contributors. We expect that this new knowledge will help coping with future viticulture challenges.

Finally, we would like to dedicate this book to the memory of Dr Alain Bouquet, who left us suddenly in May 2009, while we were completing the preparation of the book. His entire career (nearly 40 years) in the French National Institute for Agronomical Research (INRA), first at Versailles (1971–1972), then in Bordeaux (1973–1983) and finally in Montpellier (1984–2009) was devoted to grapevine genetic improvement, mainly pathogen resistance but also other important traits, such as seedlessness. Alain Bouquet was passionate about the challenge to increase and exploit genetic variation in grapevine for breeding purposes. He shared his wide knowledge on grapevine and its cultivation by writing the introductory chapter. He also co-authored Chapter 3 and participated, through the creation of innovative genetic materials, in the results described in many other Chapters such as 2, 5, 6, 7 and 8. Alain was a very enthusiastic, committed and extremely generous scientist whose vision and creativity has paved the way for many of the breakthroughs in grapevine genetics that are now being achieved by grape researchers throughout the world. He will be sadly missed but his legacy lives on.

Anne Françoise Adam Blondon
José Miguel Martínez Zapater
Chittaranjan Kole

Contents

List of Contributors

Anne-Françoise Adam-Blondon
UMR INRA UEVE ERL CNRS, Génomique végétale (URGV), 2 rue Gaston Crémieux, BP5708, 91 057 Evry cedex, France.
Tel: (33) 160874534
Fax: (33) 160874549
Email: *adam@evry.inra fr*

Claire Anderson
School of Forest Resources and Conservation, University of Florida, Gainsville, FL 32611.
Tel: (1) 3522738122
Fax: (1) 3523921707
Email: *anderson@ufl.edu*

Michel Boulay
60, rue de Vaux, 77000 Livry-sur-Seine, France, ex. Responsable du Développement Viticole, Moët et Chandon, 20 Avenue de Champagne, 51200 Epernay cedex.

Alain Bouquet
UMR INRA1334 CIRAD Montpellier SupAgro AGAP 2, place P. Viala, 34060 Montpellier Cedex, France.

Jean-Michel Boursiquot
UMT Géno-Vigne, 2, place P. Viala, 34060 Montpellier Cedex, France.
Tel: (33) 499612258
Fax: (33) 499612064
Email: *boursiquot@supagro.inra.fr*

Aurélie Canaguier
UMR INRA UEVE ERL CNRS, Génomique végétale (URGV), 2 rue Gaston Crémieux, BP5708, 91 057 Evry cedex, France.
Tel: (33) 160874534
Fax: (33) 160874549
Email: *canaguier@evry.inra.fr*

Nathalie Choisne
Unité de Recherches en Génomique-info (URGI), INRA, Route de Saint-Cyr, 78026 Versailles, France.
Tel: (33) 130833000
Fax: (33) 130833458
Email: *nchoisne@versailles.inra.fr*

Guido Cipriani
Dipartmento de Scienze Agrarie e Ambientali, University of Udine, via delle scienze 208, 33100 Udine, Italy.
Tel: (39) 432629911
Fax: (39) 432603887
Email: *guido.cipriani@uniud.it*

John C. Cushman
Department of Biochemistry and Molecular Biology, MS 200, University of Nevada, Reno, Nevada 89557-0200, USA.
Tel: (1) 7757841419
Fax: (1) 7757841650
Email: *jcushman@unr.edu*

Grant Cramer
1664 North Virginia Street, Department of Biochemistry and Molecular Biology, University of Nevada, Reno, NY 89557, USA.
Tel: (1) 7757844204
Fax: (1) 7757841650
Email: *cramer@unr.edu*

Serge Delrot
Lab Ecophysiology and Grape Functional Genomics, Institute of Vine and Wine Sciences, University of Bordeaux, INRA, 210 Chemin de, Leysotte, CS 2008, 33882 Villenave d'Omon, France.
Tel: (33) 55757900
Fax: (33) 557575923
Email: *serge.delrot@bordeaux.inra.fr*

Julie Dickerson
Electrical and Computer Engineering, Iowa State University, Ames, IA, 50011-3060, USA.
Tel: (1) 51529947705
Fax: (1) 5152948432
Email: *julied@iastate.edu*

Gabriele Di Gaspero
Dipartmento de Scienze Agrarie e Ambientali, University of Udine, via delle scienze 208, 33100 Udine, Italy, Istituto di Genomica Applicata, Parco Scientifico e Tecnologico, Via Jacopo Linussio 51, 33100 Udine, Italy.
Tel: (39) 432629911
Fax: (39) 432603887
Email: *gabriele.digaspero@uniud.it*

Ian B. Dry
CSIRO Plant Industry, Waite Campus, Hartley Grove, Urrbrae, SA 5064, Australia.
Tel: (61) 883038632
Fax: (61) 883038601
Email: *ian.dry@csiro.au*

Rudolf Eibach
JKI Institute for Grapevine Breeding, Geilweilerhof, D-76833 Siebeldingen, Germany.
Tel: (49) 634541118
Fax: (49) 6345919050
Email: *rudolf.eibach@jki.bund.de*

Lucie Fernandez
Instituto de Ciencias de la Vid y del Vino (ICVV), (CSIC, Universidad de La Rioja, Gobierno de La Rioja), C/Madre de Dios 51, 26006 Logroño, Spain.
Tel: (34) 941299694
Fax: (34) 941299608
Email: *lucie.fernandez@icvv.es*

Stella Grando
IASMA Research and Innovation Centre, Fondazione Edmund Mach-Genomics and Crop Biology Area, Via Mach, 1 38010 San Michele all'Adige (Trento), Italy.
Tel: (39) 461 615197
Fax: (39) 461 650956
Email: *stella.grando@iasma.it*

Jérôme Grimplet
Instituto de Ciencias de la Vid y del Vino (CSIC,UR, Gobierno de La Rioja), C/Madre de Dios 51, 26006 Logroño, Spain.
Tel: (34) 941299695
Fax: (34) 941299608
Email: *jerome.grimplet@icvv.es*

Ludger Hausmann
JKI Institute for Grapevine Breeding, Geilweilerhof, D-76833 Siebeldingen, Germany.
Tel: (49) 63451117
Fax: (49) 634591050
Email: *ludger.hausmann@jki.bund.de*

Olivier Jaillon
CEA, DSV, Institut de Génomique, Genoscope, 2 rue Gaston Crémieux, CP5706, 91057 Evry, France.
Tel: (33) 160872500
Fax: (33) 160872514
Email: *ojaillon@cns.genoscope.fr*

Julie Jusseaume
UMR INRA UEVE ERL CNRS, Génomique végétale (URGV), 2 rue Gaston Crémieux, BP5708, 91 057 Evry cedex, France.
Tel: (33) 160874534
Fax: (33) 160874549

Christian Kappel
Lab Ecophysiology and Grape Functional Genomics, Institute of Vine and Wine Sciences, University of Bordeaux, INRA, 210 Chemin de, Leysotte, CS 2008, 33882 Villenave d'Omon, France.
Tel: (33) 55757900
Fax: (33) 557575923
Email: *christian.kappel@bordeaux.inra.fr*

Thierry Lacombe
UMR INRA1334 CIRAD Montpellier SupAgro AGAP 2, place P. Viala, 34060 Montpellier Cedex, France.
Tel: (33) 499612253
Fax: (33) 499612064
Email: *lacombe@supagro.inra.fr*

Arnaud Lemainque
Centre National de Génotypage, CNG-IG CEA, 2 rue Gastron Crémieux CP 5721, 91057 Evry Cedex, France.
Tel: (33) 160878344
Fax: (33) 160878485

Steven Lund
241-2205 East Mall, Wine Research Centre, Faculty of Land and Food Systems, University of British Columbia, Vancouver, BC, Canada, V6T 1Z4.
Tel: (1) 6048225708
Fax: (1) 6048222016
Email: *steven.lund@ubc.ca*

José M. Martinez-Zapater
Instituto de Ciencias de la Vid y del Vino (ICVV), (CSIC, Universidad de La Rioja, Gobierno de La Rioja), C/Madre de Dios 51, 26006 Logroño, Spain.
Tel: (34) 941299698
Fax: (34) 941299608
Email: *zapater@icvv.es*

Christopher L. Owens
USDA-ARS Grape Genetics Research Unit, Cornell University, Geneva, NY, USA.
Tel: (1) 315782437
Fax: (1) 315782339
Email: *chris.owens@ars.usda.gov*

Frédérique Pelsy
INRA, UMR SVQV INRA-ULP Strasbourg, 28 rue de Herrlisheim, 68021 Colmar Cedex.
Tel: (33) 389224946
Fax: (33) 389224933
Email: *pelsy@colmar.inra.fr*

Jean-Pierre Péros
UMR INRA1334 CIRAD Montpellier SupAgro AGAP 2, place P. Viala, 34060 Montpellier Cedex, France.
Tel: (33) 499612026
Fax: (33) 499612064
Email: *peros@supagro.inra.fr*

Charles Romieu
UMR INRA1334 CIRAD Montpellier SupAgro AGAP 2, place P. Viala, 34060 Montpellier Cedex, France.
Tel: (33) 499612888
Fax: (33) 499612064
Email: *romieu@supagro.inra.fr*

Johan Tassin
Centre National de Génotypage, CNG-IG CEA, 2 rue Gastron Crémieux CP 5721, 91057 Evry Cedex, France.
Tel: (33) 160878344
Fax: (33) 160878485
Email: *tassin@cng.fr*

Raffaele Testolin
Dipartiment di Scienze Agrarie e Ambientali, University of Udine, via delle scienze 208, 33100 Udine, Italy.

Vincent Thareau
IBP, UMR CNRS Université Paris-Sud, bat 630, 91405 Orsay cedex, France.
Tel: (33)169153330
Fax: (33) 1691153330
Email: *vincent.thareau@u-psud.fr*

Patrice This
UMR INRA1334 CIRAD Montpellier SupAgro AGAP 2, place P. Viala, 34060 Montpellier Cedex, France.
Tel: (33) 499612292
Fax: (33) 499612064
Email: *this@supagro.inra.fr*

Richard L. Tillett
Department of Biochemistry and Molecular Biology, MS 200, University of Nevada, Reno, Nevada 89557-0200, USA.
Tel: (1) 7757841918
Fax: (1) 7757841650

Reinhard Töpfer
JKI Institute for Grapevine Breeding, Geilweilerhof, D-76833 Siebeldingen, Germany.
Tel: (49) 634541114
Fax: (49) 6345919050
Email: *reinhard.toepfer@jki.bund.de*

Laurent Torregrosa
UMR INRA1334 CIRAD Montpellier SupAgro AGAP 2, place P. Viala, 34060 Montpellier Cedex, France.
Tel: (33) 499612455
Fax: (33) 499612064
Email: *laurent.torregrosa@supagro.inra.fr*

Michela Troggio
IASMA Research and Innovation Centre, Fondazione Edmund Mach, Via E. Mach 1, 38010, S. Michele all'Adige, Trento, Italy.
Tel: (39) 0461615132
Fax: (39) 0461650956
Email: *michela.troggia@iasma.it*

Riccardo Velasco
IASMA Research and Innovation Centre, Fondazione Edmund Mach, Via E. Mach 1, 38010, S. Michele all'Adige, Trento, Italy.
Tel: (39) 0461615257
Fax: (39) 0461650956
Email: *riccardo.velasco@iasma.it*

Silvia Vezzulli
IASMA Research and Innovation Centre, Fondazione Edmund Mach, Via E. Mach 1, 38010, S. Michele all'Adige, Trento, Italy.
Tel: (39) 0461615536
Fax: (39) 0461650956
Email: *silvia.vezzulli@iasma.it*

Leocir Welter
Federal University of Pampa, Rua Luiz Joaquim de Sa Britto, 97650-000, Itaqui, Rio Grande do Sul, Brazil.
Tel: (55) 34331669213
Fax: (55) 34331669
Email: *leocirwelter@unipampa.edu.br*

Eva Zyprian
Julius Kühn Institut, Federal Centre for Cultivated Plants, Institute for Grapevine Breeding Geilweilerhof, 76833 Siebeldingen, Germany.
Tel: (49) 63451126
Fax: (49) 634591050
Email: *eva.zyprian@jki.bund.de*

Andrey Zharkikh
Myriad Genetics Inc., 320 Wakara Way, Salt Lake City, UT 84108, USA.
Tel: (1) 8015843600
Fax: (1) 8015843640
Email: *zharkikh@myriad.com*

Abbreviations

2DGE	Two-dimensional gel electrophoresis (2DGE)
ABA	Abscisic acid
AFLP	Amplified fragment length polymorphism
AGI	Arabidopsis Genome Initiative
ANR	Anthocyanidin reductase
ArMV	Arabis Mosaic Virus
AROS	Array-Ready Oligo Set
BAC	Bacterial artificial chromosome
BC	Back-Cross or Before Christ depending on the context
BES	BAC end sequences
bp	base pair
BSA	Bulked segregant analysis
CaOMT	Caffeoyl methyl transferase
CAPS	Cleaved amplified polymorphic sequence
cDNA	Complementary DNA
cDNA-AFLP	cDNA-amplified fragment length polymorphism
ChIP	Chromatine immuno precipitation
CHS	Chalcone synthase
CLS	Carpel less
cM	centi Morgan
cpDNA	Chloroplastic DNA
cv	Cultivar
DFCI	Dana-Fraber Cancer Institute
DFR	di hydroflavonol reductase
DNA	Deoxyribonucleic acid
DXS	1-deoxy-D-xylulose 5-phosphate synthase
EB	Ethidium bromide
EBI	European Bioinformatics Institute
EMBL	European Molecular Biology Laboratory
EMS	Ethylmethane sulfonate
e-QTL	Expression quantitative trait locus
ESI	Electrospray ionization
EST	Expressed Sequence Tag
EU	European Union

F3′5′H	Flavonoid 3′ 5′ hydroxylase
F3′H	Flavonoid 3′ hydroxylase
F3H	Flavanone 3 hydroxylase
FAME	Fatty acid methyl ester
flb	Fleshless berry
flcDNAs	full length cDNA
FT-ICR	Fourier transform ion cyclotron resonance
GAI1	Gibberellic Acid Insensitive1
GCB	Grapevine Corky Bark
GC	Gas chromatography
GEO	Gene Expression Omnibus
GFkV	Grapevine Fleck Virus
GFLV	Grapevine Fanleaf Virus
GIS	Geographical Information Systems
GLRaV	Grapevine leafroll associated virus
GM/GMO	Genetically modified (organism)
GO	Gene Ontology
GPP	Geranyl *diphosphate*
GPS	Global Positioning System
GRIP	Grape Ripening-Induced Protein
GS	Genome Sequencer
GSS	Genome Survey Sequences
GST	Glutathione S-transferase
GVM	Grapevine Vein Mosaic
GVN	Grapevine Vein Necrosis
H_0	Observed heterozygosity
HC	Hydrogen Cyanamide
HMA/HTA	Homozygous assembly/heterozygous assembly
HPLC	High Performance Liquid Chromatography
IBD	Identity-by-descent
ICAT	Isotope-coded affinity tags
IGGP	International Grapevine Genome Program
IGS	Intergenic spacer
Indel	Insertion/deletion
IPCC	Intergovernmental Panel on Climatic Change
IPP	Isopentenyl diphosphate
IRGSP	International Rice Genome Sequencing Project
ITMS	Ion trap mass spectrometry
iTRAQ	Isobaric tagging for relative and absolute quantitation
ITS	Internally transcribed spacer
kbp	Kilobase pair
KSG	Kober Stem Grooving
LAR	Leucoanthocyanidin reductase

LC	Liquid chromatograhy
LD	Linkage disequilibrium
LG	Linkage Group
LN33SG	LN33 Stem Grooving
LRR	Leucine-rich repeat
LTR	Long Terminal Repeat
MABC	Marker assisted back-crossing
MALDI	Matrix-assisted laser desorption/ionization
MAMP	Microbe-associated molecular pattern
MAS	Marker assisted selection
Mbp	Megabase pair
MIAME	Minimum information about a microarray experiment
MINSEQE	Minimum information about a high-throughput seQuencing experiment
miRNA	Micro RNA
MPSS	Massively parallel signature sequencing
MPW	Multiple perianth whorls
mRNA	messenger RNA
MS	Mass Spectrometry
MSAP	Methylation-sensitive amplification polymorphism
NBS	Nucleotide-binding site
NCBI	National Center for Biotechnology Information
NGS	Next-generation sequencing
NIL	Near isogenic line
NMR	Nuclear magnetic resonance
OIV	International Organization of Vine and Wine
PA	Proanthocyanidin
PCR	Polymerase Chain Reaction
PD	Pierce's disease
Pdr	Pierce's disease resistance
PEG	Polyethylene Glycol
PI	Identification power
PIC	Polymorphism information content
PIP	Plasma membrane intrinsic proteins
PMF	Peptide mass fingerprinting
PMN	The Plant Metabolic Network
PO	Plant Ontology
PPO	Polyphenol oxidase
PR	Pathogenesis-related
PTP	Picotiter plate
qRT-PCR	Quantitative real time PCR
QTL	Quantitative trait locus
R gene	Resistance gene

RAPD	Random Amplified Polymorphic DNA
REN	Resistance to *Erisyphe necator*
RFLP	Restriction Fragment Length Polymorphism
RGA	Resistance gene analog
RIL	Recombinant inbreed lines
RNA	Ribonucleic acid
ROS	Reactive Oxygen Species
RPKM	Reads per exon kilobase per million
Rpv	Resistance to *Plasmopora viticola*
RRM	Reiterated reproductive meristems
rRNA	Ribosomal RNA
RSP	Rupestris Stem Pitting
RT-PCR	Real-Time PCR
RuBisCO	Ribulose-1,5-bisphosphate carboxylase/oxygenase
Run	Resistance to *Uncinula necator*
SAGE	Serial Analysis of Gene Expression
SAM	Shoot apical meristem
SAMPL	Selective Amplification of Polymorphic Loci
SBS	Sequencing-by-synthesis
SCAR	Sequence-characterized amplified region
SdI	Seed development inhibitor
SMRT	single-molecule real-time
snoRNA	Small nucleolar RNA
SNP	Single Nucleotide Polymorphism
SOLiD	Sequencing by Oligo Ligation and Detection
S-SAP	Sequence-Specific Amplified Polymorphism
SSH	Subtractive Suppression Hybridization
SSR	Simple Sequence Repeat
STMS	Sequence Tagged Microsatellite Site
STS	Sequence Tagged Sites
TAIR	The Arabidopsis Information Resource
TC	Tentative consensus
TDF	Transcript-derived fragments
TE	Transposable elements
TF	Transcription factors
TIGR	The Institute for Genomics Research
TIP	Tonoplast intrinsic proteins
TIR	Toll/Interleukin-1 receptor
TOF	Time-of-flight
TPA	Third Party Annotation
TPM	Transcripts Per Million
tRNA	Transfer RNA
tSMS™	True single-molecule sequencing

UFC	Unfused carpels (UFC)
UFGT	UDP glucose-flavonoid 3-O-glucosyl transferase
UTR	Untranslated region
VMC	Vitis Microsatellite Consortium
VRP	*Vitis riparia* downy mildew resistance
V-SDR	Vitis-Segregation Disorder Region
VvAROS	Vitis vinifera Array-Ready Oligo Set
VvGI	Grape Gene Index (TIGR)
WGD	Whole genome duplication
WGS	Whole Genome Shotgun
WT	Wild type
XET	Xyloglucan endotransglycosylase
Xf	Xylella fastidiosa
XiR1	Xiphinema index resistance 1
XTH	Xyloglucan endotransglucosylases
ZMW	Zero-mode-waveguide

1

Grapevines and Viticulture

Alain Bouquet

ABSTRACT

Grapevine is an important crop in terms of economic value and for social activities, used mainly for wine and spirit production but also for fresh fruit, raisins, fruit juices, jams, etc. It is cultivated throughout the world, from temperate to tropical climates. In the past decades, while innovative viticulture management technologies were developed, the panel of grown varieties was by contrast greatly reduced. Grapevine breeding was poorly encouraged over the last century. This is a trend that is currently changing in order to cope with the need to reduce the use of phytochemicals and to adapt to environmental changes and also thanks to newly developed genomic tools. The origin of viticulture and its spread around the Mediterranean basin is associated with the development of wine making and linked to myths and religion. Grapevine is part of the Vitaceae family, which is a basic branch of the Eurosids I. Their genome was long suspected to have a polyploid origin, which was recently confirmed after sequencing the *Vitis vinifera* genome. Domestication of *V. vinifera* was followed by the development of genetically related groups of varieties adapted to local environments and, during the centuries, somatic variations was also selected and conserved. Since the spread in Europe of the Phylloxera and several fungal diseases, other species of the genus *Vitis*, that are resistant to those pests and pathogens, are being used for genetic improvement.

1.1 Introduction

In the recent history of grapevine cultivation, contrary to many other crops where production innovation is primarily in the variety, innovation is mainly focused on management techniques. In cereal crops new cultivars are introduced rapidly in response to changes in biological or economic

---

UMR INRA Montpellier SupAgro IRD Diversité et Adaptation des Plantes Cultivées, 2 Place P. Viala, 34060 Montpellier Cedex, France.
Corresponding author: *laurent.torregrosa@supagro.inra.fr*

factors of production. In contrast, during the last decades, the cultivars of wine grape have changed slowly or not at all. Most of the world's vineyards are planted with varieties, which have been perpetuated for centuries by vegetative propagation. The reasons are many and involve a complex mixture of plant and human factors. Nowadays, viticulture is confronted with great pressure for changes in the methods of production. The present generation may be too presumptuous in believing that the success of traditional cultivars, based on today's context, will continue for ever on a such unchanged basis to face the new challenges for the future. However, in the scope of the "sustainable viticulture" format which is progressively imposed to viticulture and the challenges linked to climatic changes, it is likely that the improvement of cultivation techniques will not be inadequate. Thus, high priority must be given henceforth to plant breeding. Given the difficulties, it is reasonable to question whether these goals are attainable. The advance of knowledge in the biology of grapevines and the recent progresses in genomics encourages an optimistic response.

1.2 Economic Importance of Viticulture

Grapevine (*Vitis vinifera* L.) is one of the most worldwide-grown perennial fruit crops. It is mainly grown at latitudes from 50°N to 30°N and 40°S to 30°S, that approximate to the 10°C and 20°C isotherms (Mullins et al. 1992). Within these zones there are climatic variations caused by mountains, large masses of land or water and oceanic currents, which greatly affect the distribution of vineyards in the different continents. The grapevine flourishes in climates with warm dry summers and cool wet winters. Countries of the Mediterranean basin are the world's main producers of grapes, but Mediterranean climates are also found in the coastal areas of southern Australia, South Africa, California and Chile. Oceanic climates may be suitable for grape growing provided that summer temperatures are not too low and summer rainfall is not too high. In the humid tropics, table grapes can be grown with careful disease management. Finally, it is worth noting that *V. vinifera* is a highly adaptable species and that grapevines can be grown in a very wide range of environments. In many cases, the limitations imposed by climate can be countered by technical inputs, and climate is only one factor among others which determines the success of grape-growing in a given area.

1.2.1 Wine Production

Though a part of the production of grapevines is eaten fresh (table grapes) or dried (raisins), its main use is wine making. Total world grape production was estimated in 2007 at 65 million tons, from which 65% was used for a wine production estimated to be 271 million hectoliters, 28% consumed as table grapes and 7% processed into raisins. According the OIV statistics (*http://www.oiv.int*) the worldwide surface area of vineyards was estimated in 2007 to be 7.9 million hectares distributed between Europe (58.0%), Asia (21.5%), America (13.0%), Africa (5.0%) and Oceania (2.5%).

The 10 highest wine producing countries were France (48.4 Mhl), Italy (48.0 Mhl), Spain (34.7 Mhl), USA, Argentina, China, Germany, South Africa, Australia and Chile. The main consumers of wine are the wine-producing countries, but during the last 50 years, there has been a continuous drop in consumption in many of these countries. The reduction of wine consumption in the three leading wine-producing countries has been offset to some extent by an increase of wine consumption in northern Europe, the USA and other countries. In 2007, 33.7% (91.4 Mhl) of the world wine production was exported, with Italy, Spain and France responsible for half of all exports (49.3 Mhl) (see also Chapter 14).

The gap between worldwide wine production (271 Mhl in 2007) and consumption (241 Mhl) can be considered only partly as a surplus because, in some countries, large quantities of wine are distilled to produce fortifying spirits and high-proof spirits, e.g., cognac. Since the "French paradox" was brought to the fore (Renaud and De Lorgeril 1992), there are epidemiological and biological evidence that moderate consumption of wine protects against cardiovascular diseases (German and Walzem 2000). These health benefits are based on the presence in the wine of various phytonutrients, e.g., resveratrol, flavonoids, flavans, flavonols and other phenolic compounds. Recently, melatonine, a compound with powerful antioxidant activity, was detected in berries, and its role was debatable (Iriti et al. 2006).

1.2.2 Table Grapes Production

Contrary to wine production, it is difficult to precisely state the world production of table grapes. Summarizing 2005 OIV statistics, Asia represents more than 55% of the 18.3 million tons (Mt) produced, with China as a leading country (3.70 Mt), followed by Iran (1.83 Mt), Turkey (1.55 Mt) and India (1.40 Mt). In the Americas table grapes are produced mainly in the USA (0.86 Mt), Chile (0.81 Mt), and Brazil (0.68 Mt). In Europe, table grapes are produced mainly in Italy (1.27 Mt) and Spain (0.31 Mt). In Africa table grapes are produced mainly in Egypt (1.18 Mt). Despite the perishability of the fruit and the high cost of transportation, table grapes support a significant international trade of nearly 3 million tons with imports mainly into the USA, Germany and the UK and exports mainly from Chile and Italy. Annual consumption of table grapes is low and does not exceed 10 kg per capita in most producing countries. In Europe and North America table grapes represent less than 5% of consumption of fresh fruits. Table grape cultivars are usually seeded, but growing of seedless grapes is increasing worldwide, as they are generally preferred for fresh consumption. The size of the berries is smaller than in normally seeded grapes, but can be improved by treatments with gibberellic acid (Weaver 1958) or by genetic selection (Ledbetter and Ramming 1989). Fresh grapes have important nutritional qualities and provide about 700 kcal/kg with high levels of vitamins (Morgan et al. 1934). Contrary to wine, the health benefits of their consumption does not have the negative impact of excessive alcohol consumption.

1.2.3 Raisin Production

The world's largest producers of raisins are Turkey (3.6 Mt), USA (3.3 Mt) and Iran (2.1 Mt). They are produced generally from seedless varieties such as "Black Corinth" or "Sultana" (syn. Thompson seedless). Raisins are typically sun-dried, but may also be water-dipped, or dehydrated. Raisins are high-energy food, rich in sugars, and providing 3,400 kcal/kg. As for all dried fruits, raisins have low vitamin content.

1.2.4 Other Uses

Grape berries have many other uses, such as canned fruit after pasteurization. The fresh juice of pressed grapes can be bottled for consumption, converted to jellied products, such as the "Concord" jelly from the USA or concentrated into syrups such as the "Pecmez" from Turkey. The health benefits of grapevine juice and derived products become more and more documented (Anselm et al. 2007). In addition there are many miscellaneous uses of grape products including the production of ethanol, vinegar, and seed oil. Grape seed oil is notable for its high contents in tocopherols, phytosterols and polyunsaturated fatty acids (Thomas et al. 2005). Grape seeds are also used in cosmeceuticals and skincare products. Finally the residues of wine making (marcs) can be processed for the industrial production of tartaric acid and anthocyanin pigments and the waste can be further utilized for fertilizer production.

1.3 Academic Importance of Grapevine and Viticulture

1.3.1 Development of Innovative Technologies and Disincentives to Varietal Innovation

In viticulture the main response to biological constraints or economic changes has been to optimize the growing of traditional cultivars by applying progressively higher technical inputs (Mullins et al. 1992). The innovations include cultural practices, e.g., rootstocks adapted to different types of soil, pruning methods such as "minimal" pruning, trellis and training systems adapted to optimal production of high-quality grapes in a given vineyard environment, and irrigation techniques such as the "partial root zone drying". There were many innovations in chemical-based intensive grape growing, e.g., use of fertilizers, herbicides, pesticides or growth regulators, such as gibberellic acid in table grapes production or hydrogen cyanamid to improve bud breaking in hot winter climates. Innovations were also made in the mechanization of pruning and harvesting and finally in post-harvest technology and wine making. The most recent innovation is precision (site-specific) viticulture that helps grape growers to optimize vineyard performances through the use of enabling technologies, including the global positioning system (GPS) and geographical information systems (GIS), coupled with tools for measuring and monitoring vineyards at high spatial resolution.

It is worth noting that among the oldest innovations, some have been remarkable. The introduction in 1868 of Phylloxera in Europe led to the grafting of *V. vinifera* grape varieties onto *Vitis* hybrids resistant to the aphid that was the first and most spectacular example of a successful biological struggle against a major parasite (Pouget 1990). Moreover, it should be kept in mind that the discoveries in 1853 and 1887 of the efficiency of sulfur and copper against the grapevine powdery mildew (Erisiphe necator) and downy mildew (Plasmopara viticola) were key steps in the development of phytopathology.

The breeding of grapevines present considerable difficulties. The biological cycle is long as in many other woody perennial fruit crops. Although cultivated grapes bear self-fertile flowers, they do not breed true from seed. In their wild state, they are dioecious outcrossers resulting in a strict allogamy that maintained a high level of heterozygosity. Domestication has resulted in a radical change in the floral biology of grapevines that passed from dioecy to hermaphroditism. As the morphology of perfect-flowered flowers leads to preferential self-pollination, the selection of plants in open-pollinated progenies was probably accompanied by an unconscious selection that maintained a level of heterozygoty compatible with good development. In grapevines, few traits of cultural importance are simply inherited. Most are controlled by large numbers of genes with minor effect. The perpetuation of the most effective genetic structures for centuries was made by vegetative propagation. Grapevine cultivars thus represent highly complex gene combinations and their wines have unique characteristics that enjoy a high level of consumer acceptance and have become firmly entrenched by custom or by law in European countries.

But the range of these "traditional" cultivars is not unchanging and has evolved considerably throughout history, particularly during the 18th and 19th centuries. This varietal shifting has been a means of adapting to changes of viticulture practices imposed either by nature or by man (Doazan 2000). During the 20th century, however, there was a slowing down of this evolution, and a dramatic reduction of the varietal range in commercial vineyards due to the development of a wine trade based on a few cultivars of international renown, among other causes. For instance, in 2004, five cultivars, i.e., Cabernet-Sauvignon, Chardonnay, Merlot, Sauvignon and Syrah represented 53% of the plantings in France, whereas more than 200 cultivars are certified for planting in France (Bouquet 2008).

1.3.2 Changes needed to Face New Challenges

This reduction of the varietal range resulted in part from the widespread ignorance about the genetic diversity that ensured the success of viticulture in the past. The long-term effects of this "genetic erosion" have not been carefully considered with regard to the challenges that viticulture will have to face in the future. The most important will obviously be the influence of climatic changes and global warming on the geographical distribution of vineyards and the steadiness of wine quality (Jones et al. 2005) but also on the host-pathogen interactions (Salinari et al. 2006). Moreover, the cost of production in modern intensive grape growing is high and includes a significant fossil energy consumption. In addition, control of pests and diseases presents serious risks associated with over-reliance on chemical methods. Development of resistances to fungicides is a recurrent problem in grape growing (Délye et al. 1997; Chen et al. 2007) but the main concerns are the occurrence of toxic residues in foodstuffs and contamination of the environment. Increasingly stringent legislation on food safety standards and environmental protection is also a cause of concern. Chemical control must be reduced and completed by alternative strategies with high priority given to breeding for genetic resistances (Bisson et al. 2002).

During the first half of the 20th century, the selection of winegrape cultivars resistant to pests and diseases was considerably developed in France by hybridization between *Vitis vinifera* and many American wild species. The resultant hybrids have been generally unacceptable to growers and winemakers due to the low quality of the wine, and their adoption has, therefore, been very limited throughout most viticultural regions of the world apart from eastern and mid-western USA (Pollefeys and Bousquet 2003). In France, these hybrids occupied one third of the grapevine acreage in 1958 and were practically all discarded. Such a complete failure led to a drastic change of breeding strategy from the 1970s with the only resort to one American species, *V. rotundifolia*, never used until then in crosses with *V. vinifera* (Bouquet 1980a). Nevertheless this should not draw attention away from the work accomplished in other countries, particularly in

Germany, to create varieties resistant to parasites from the old French Hybrids. The success of cultivars such as "Regent" (Eibach and Töpfer 2003) shows that there is great promise in genetic improvement for an environmentally friendly sustainable viticulture.

1.3.3 Grapevine as a Model Plant?

Progress in grape breeding will be obviously assisted by the two high quality sequences of grapevine genome that have been recently published (Jaillon et al. 2007; Velasco et al. 2007). Grapevine was the fourth one of the first flowering plants and the first fruiting perennial crop whose genome was completely sequenced (see Chapter 9). Progress was made easier by the relatively low size of the genome of *Vitis vinifera* (487–500 Millions pairs of bases with $2n = 38$ chromosomes). Public access to the grapevine genome sequences will assist in the identification of genes underlying the cultural and quality characteristics of this species, and opens considerable opportunities for molecular breeding (see Chapter 7). In parallel, important Expressed Sequence Tag (EST) sequencing projects have provided a basic resource for the study of the molecular basis of traits, and also for an efficient transfer of knowledge from model species to grapevine (da Silva et al. 2005; see Chapters 10 and 11). Recently, to boost genomic and genetic studies, new innovative lines of grapevines have been developed (Chaib et al. 2010).

A high number of genes related to disease resistance have been identified and mapped on genetic or physical maps (for review see Di Gaspero et al. 2007; Moroldo et al. 2008; Chapters 8, Chapter 6 Section 6.10.1). Progress is also expected in understanding of the molecular interactions between plants and abiotic factors such as water stress or salinity (Cramer et al. 2007). As many genes belonging to the metabolic pathways of terpenes and tannins have been identified, it may become possible to trace the diversity of wine flavors down to the genome level. More generally, the availability of the grapevine genome sequence will improve the understanding of the genetic and molecular control of the morphological, physiological and biochemical events which occur during berry growth and could shed supplementary light on the regulation of fruit development in Angiosperms.

Indeed, Angiosperms produce a great diversity of fruits, from dry single-seeded caryopses as in rice, capsules as in poplar or siliques as in *Arabidopsis*, to fleshy fruits such as berries in grapevine or tomato. Most fleshy fruits exhibit common characteristics and in the past few years, considerable attention has been focused on the molecular events that control ripening in tomato, the model of climacteric fruits. Even susceptible to ethylene, grapevine is a non-climacteric fruit such as citrus, cherry, pineapple, strawberry and few other species. There is no increase in respiratory CO_2 at the onset of ripening in grapevine and no production of ethylene. Berries do not continue to ripen when there are detached from the plant. However, recent studies lead one to assume some role of ethylene or ethylene signaling in ripening (Chervin et al. 2004; Tesniere et al. 2004). The onset of ripening in non-climacteric fruits is often less marked than in climacteric, but in grape berry, it shows very distinctive features.

Berry growth follows a characteristic double sigmoid pattern divided in three stages (Coombe 1992). Stage I is characterized by a rapid increase in size of green and hard berries, which accumulate organic acids and tannins. Growth of the pericarp is due firstly to cell division then cell expansion, and its duration is typically 40–60 days. Progress has been made recently in the knowledge of the physiological, biochemical and molecular events occurring during this key stage (Fernandez et al. 2007). Stage II is characterized by slow growth of the pericarp and maturation of the seeds. The berry remains a hard green organ and acid content reaches its maximum. This lag phase lasts 7–40 days and determines whether a cultivar is early- or late-maturing. The onset of stage III, called "véraison", is marked by softening of the tissues followed by color change. The resumption of rapid growth is due solely to cell expansion in relation to sugar uploading. Titratable acidity decreases and a massive accumulation of hexose sugars and water occurs. Stage III lasts approximately 35–55 days. The transition between stages II and III, which occurs within 24–48 hours, includes many physiological changes heavily dependent on gene expression revealed by transcriptome (Terrier et al. 2005), proteome (Deytieux et al. 2007) and metabolome analysis (Deluc et al. 2007).

Grapevine reproductive development displays special features when compared with herbaceous annual systems such as *Arabidopsis*, or woody plants, such as poplar. For proleptic axes, the reproductive cycle of grapevines is completed over two consecutive growing seasons separated by a dormancy period. Unlike *Arabidopsis*, in which vegetative meristems are switched to a reproductive mode of growth after a flowering stimulus, in grapevine both vegetative and reproductive meristematic structures are formed separately on the same shoot. Another major characteristic feature is the presence of tendril which is a modified reproductive structure. How the grapevine plant controls these processes is not completely understood, but considerable progress has been made (Boss et al. 2003). The increasing concern about the effects of climate change is needed to complete the understanding of the key interactions between environmental factors and genetic mechanisms controlling the formation of inflorescences, flowers and berries at the molecular level (Carmona et al. 2008).

1.4 Origin of Grapevine and History of Viticulture

The domestication of grapevine (*Vitis vinifera* L.) is closely linked to the discovery of wine, but it is unclear which process predated the other. Many archaeological findings suggest that these processes occurred 5500–5000 BC in the Near-East (McGovern 2003) probably in the region known as Transcaucasia, which supports till now a large genetic diversity of wild grapes. But paleobotanical findings in Spain dating 3000 BC (Nuñez and Walker 1989) raised the outstanding question about occurrence of secondary domestication centers in the western Mediterranean (Arroyo-García et al. 2006; Imazio et al. 2006). Moreover, the genetic relationships between wild and cultivated grapes are still unresolved (This et al. 2006).

The mythology associated with wine plays a prominent role in the history of western civilization for growing of grapes (Johnson 1989). From the primo-domestication sites in Transcaucasia, viticulture spreads to neighboring regions such as lower Mesopotamia (the "Fertile Crescent") then the Nile delta and the eastern Mediterranean. During the first millennium BC, the Phoenicians and Greeks introduced viticulture into the western Mediterranean (McGovern 2003). Grapes were probably brought to China 2000 years ago during the Han Dynasty but grape growing in Japan was not seen before the 12th century. Under the influence of Rome, grapevines extended throughout Europe from the western shores to the Rhine and Danube. The Romans were the first to give names to grape varieties (Bouquet 1982) but it is difficult to relate them to modern cultivars (Vuillamoz et al. 2007). With the fall of the Roman Empire grape growing and wine trade went into decline, but with the spread of Christianity, the mystical significance of wine and the need for sacramental purposes ensured the survival of vineyards and led to the establishment of a new trade in wine in Europe (Johnson 1989). By the end of the Middle Ages, wine drinking was a firmly established social custom in most of Europe, and viticulture grew steadily from the 16th to the 19th centuries despite a series of calamities such as the "phylloxera plague" (Pouget 1990).

By the 6th century, Islam prohibited the consumption of wine and this led to the development and growing of table grape varieties in the Middle East. The expansion of Islam led to their introduction and cultivation into North Africa and Spain, from where they spread to the New World. The planting of grapevines in Mexico was ordered by Cortés in 1525, and by 1550 the growing of grapes had spread to Peru, Chile and Argentina. In 1616, Dutch settlers introduced viticulture in South Africa. In 1697, the first grapes were introduced in California and the plantings expanded with the establishment of numerous Spanish missions. But the development of commercial viticulture in California during the second half of the 19th century was mainly due to European immigration. In Australia, German Lutherans developed grape growing in the Barossa Valley in the 1850s. Similarly, French settlers planted grapevines in New Zealand early in the 19th century.

1.5 Taxonomy, Origin, Evolution and Diversity of the Grapevines

1.5.1 The Family of Vitaceae

The cultivated vine (*Vitis vinifera* L.) belongs to the Vitaceae. This family is mainly distributed in inter-tropical climates and consists of perennial plants that take the form of woody or herbaceous lianas (tree-climbing plants) with leaf-opposed tendrils. It shows considerable morphological variability and adaptation to varied environments. The systematics of the Vitaceae is based on the classification of Planchon (1887) who identified 10 genera, with several subgenera. Subsequent revisions enlarged the family to 17 living genera (Table 1-1) and two fossil genera (*Cissites* and *Paleovitis*) from which only the genus *Vitis*, with its two subgenera *Vitis* (formerly *Euvitis*) and *Muscadinia*, produce edible fruits, with one species (*V. vinifera*)

Table 1-1 Genetic diversity and geographical distribution of the Vitaceae (Af. = Africa; Am. = America; As. = Asia; Aust. = Australia; Eur. = Eurasia; adapted from Galet 1988).

Genus	Number of species	Examples of species
Cissus	367	*C. quadrangularis* **(Af.)**; *C. discolor* **(As.)**; *C. gongylodes* **(Am.)**; *C. antartica* **(Aust.)**
Cyphostemma	258	*C. juttae* **(Af.)**
Tetrastigma	132	*T. lanceolarium* **(As.)**
Ampelocissus	92	*A. latifolia* **(As.)**; *A. acapulcensis* **(Am.)**; *A. abyssinica* **(Af.)**
Vitis subg. Vitis	68	*V. vinifera* **(Eur.)**; *V. labrusca* **(Am.)**; *V. coignetiae* **(As.)**
Vitis subg. Muscadinia	3	*V. rotundifolia* **(Am.)**
Cayratia	65	*C. pedata* **(As.)**; *C. ibuensis* **(Af.)**; *C. trifolia* **(Aust.)**
Ampelopsis	31	*A. heterophylla* **(As.)**; *A. cordata* **(Am.)**
Pterisanthes	21	*P. heterantha* **(As.)**
Parthenocissus	18	*P. tricuspidata* **(As.)**; *P. quinquefolia* **(Am.)**
Rhoicissus	11	*R. rhomboidea* **(Af.)**
Yua	3	*Y. austro-orientalis* **(As.)**
Clematicissus	1	*C. angustissima* **(Aust.)**
Landukia	1	*L. landuk* **(As.)**
Puria	1	*P. trilobata* **(As.)**
Nothocissus	1	*N. spicifera* **(As.)**
Acareosperma	1	*A. spireanum* **(As.)**
Pterocissus	1	*P. mirabilis* **(Am.)**

grown worldwide and another (*V. rotundifolia*) cultivated on a very limited scale in the US (Olien 1990).

Karyological analysis presents some difficulties in the vines, due to the high number of chromosomes ($2n = 38$ for *V. vinifera*) and their small size, from 1 to 2.4 µm (Raj and Seethaiah 1969; Haas and Alleweldt 2000). According to cytotaxonomy studies (Shetty 1959; Lavie 1970), the family ancestor may belong to the *Cissus* genus, which is typically inter-tropical, distributed throughout Africa and Asia, and consists of plants with tetramerous flowers and a basic chromosome karyotype of $2n = 24$. Some species of *Cissus* are important as ornamentals, notably *Cissus discolor* or *Cissus antartica*. The genera *Cyphostemma* and *Tetrastigma* are spread respectively in Africa and Asia. They have morphological and anatomical similarities and a common karyotype ($2n = 22$) with many cases of polyploidy. The genera *Cyphostemma* and *Cayratia* were formerly included in *Cissus* as subgenera by Planchon. The genera *Ampelocissus, Vitis, Ampelopsis* and *Parthenocissus* consist of plants with pentamerous flowers that are distributed in the temperate climates of America, Asia and Europe, except *Ampelocissus* that is typically found in tropical warm areas including Africa. They are characterized by a karyotype of $2n = 40$, except the genus *Vitis* ($2n = 38$) or at least its subgenus *Vitis,* the other subgenus *Muscadinia* showing the basal karyotype $2n = 40$ (Branas 1932). Taking into account its anatomical, morphological and karyological characteristics, the subgenus *Muscadinia* could be considered as a true genus as proposed a century ago by Small (1903).

1.5.2 Phylogeny of the Vitaceae

The phylogenetic position of Vitaceae and relationship of the grape family to core eudicots has been controversial for many years. Until recently, Vitaceae were included with the Rhamnaceae and Leeaceae within the Rhamnales order in the subclass Rosidae of Eudicots (Cronquist 1981). The studies of several chloroplastic DNA marker genes confirmed that the Vitaceae and Leeaceae (non-climbing vines) are closely related, but distant from the Rhamnaceae and located near the base of the large clade of Eurosids I from which they became separated at the end of the Cretaceous period about one hundred million years ago (Savolainen et al. 2000; Soltis et al. 2000; Ingrouille et al. 2002; Soejima and Wen 2006). The ancestral character of *Cissus* was confirmed (Fig. 1-1) and also the closeness of *Cyphostemma* and *Tetrastigma*, forming a large clade including also the genus *Cayratia* (Rossetto et al. 2002). The complete sequencing of the grapevine chloroplast genome provided strong support for the position of the Vitaceae as the earliest diverging lineage of rosids (Jansen et al. 2006). Finally, results obtained with the nuclear *GAI1* gene sequence supported most of the previous results

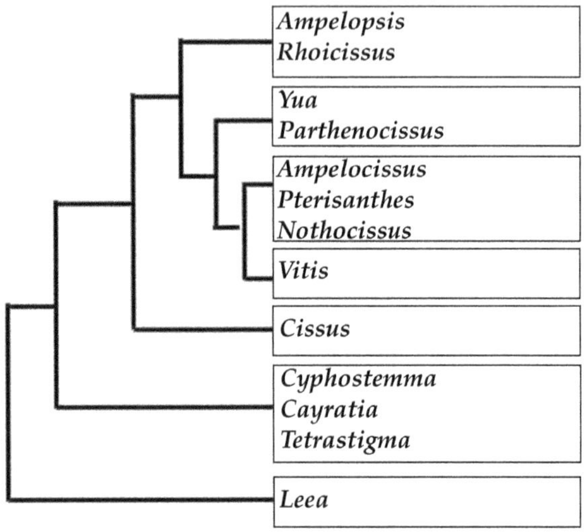

Figure 1-1 Phylogenetic relationships among *Vitaceae* using the chloroplastic markers: *trnL-F*, *atpb-rbcl* spacer and *rps16* intron (adapted from Soejima and Wen 2006).

obtained with chloroplast genes (Wen et al. 2007). All these studies have explained the relationships among several major lineages of Vitaceae and they have also emphasized the importance of taxon sampling for obtaining accurate phylogenies.

The phylogenetic distance between *Vitis* and *Ampelopsis* explains the unsuccessful attempts or controversial results in hybridization of these two genera (Millardet 1901; Cebrii 1950). Crosses between *Vitis* and *Parthenocissus* have not been reported. Therefore, the use of "Virginia creepers" and related species as genetic resources for grapevine breeding depends on the development of somatic hybridization methods. Some success has been obtained in the regeneration of protoplasts from somatic embryos (Reustle et al. 1995). However, this technique is a long way from being mastered in vines, despite a lot of work in protoplast technology (Papadakis et al. 2001). The phylogenetic proximity of the subgenera *Vitis* and *Muscadinia* revealed by the chloroplastic gene markers (Ingrouille et al. 2002; Soejima and Wen 2006) and the nuclear gene sequence *GAI1* (Wen et al. 2007) explains why the karyotypic differences have not been an insuperable obstacle in crossings between *V. vinifera* and *V. rotundifolia* (Wylie 1871; Detjen 1919; Patel and Olmo 1955; Bouquet 1980a).

1.5.3 Role of Transposable Elements in the Evolution of Grapevines

Among the Vitaceae with 2n = 40 or 38 chromosomes, the Ampelopsis genus appears to be ancestral, the genera *Parthenocissus* and *Ampelocissus* intermediary, and the *Vitis* genus more recent (Fig. 1-1), according the results of Ingrouille et al. (2002) and Soejima and Wen (2006). These conclusions can be compared with the results of Lodhi and Reisch (1995), who observed that the nuclear genome of *Ampelopsis brevipedunculata* was made up of 666 Mpb, and significantly larger than the genome of *Parthenocissus tricuspidata* (516 Mpb) or *Vitis* (475 Mpb). Throughout a diverse range of organisms, the genome size appears to be correlated with the number of copies of transposable elements. When active, these mobile genetic elements can lead to the expansion of the genome, disruption of wild type gene function resulting in altered phenotypes, and chromosomal rearrangements (Bennetzen 2000). These changes can contribute to species evolution as well as crop evolution and domestication (Naito et al. 2006). Many transposable elements have been identified in *Vitis vinifera*. They belong to group I, e.g., the Ty1-copia-like *Vine1*, *Tvv1* transposon and the Ty3-gipsy-like *Gret1* transposon (Verriès et al. 2000; Pelsy and Merdinoglu 2002; Pereira et al. 2005) and group II, e.g., the *haT* transposon family (Benjak et al. 2008). Recently, 10 novel Ty1-copia-like retrotransposon families were characterized by an analysis of the whole-genome sequence of grapevine (Moisy et al. 2008). *Gret1* retroelement has been identified as playing a key role in generating berry color variation in *Vitis vinifera* (Kobayashi et al. 2004; Cadle-Davidson and Owens 2008). In addition, insertion of a *haT* transposable element in the promoter of the *TFL1A* gene was shown to cause an early phenotypic alteration affecting cluster ramification and development, delay in flower meristem specification as well as both flower and flower organ reiterations (Fernandez et al. 2010). The two grapevine genome sequencing programs recently allowed one to estimate the level of transposable elements from 21% (Velasco et al. 2007) to 41% (Jaillon et al. 2007), values similar to previous estimations (Tomkins et al. 2001). Thus it is likely that transposable elements have played a considerable role in the evolution of the *Vitaceae*.

1.5.4 Polyploid Origin of Grapevines

The high chromosome numbers ($2n$ = 38 and 40) of *V. vinifera* and *V. rotundifolia* suggest a polyploid origin of their genomes, as generally assumed in many plants (Lewis 1979). Cytological observations made during pollen meiosis of F_1 hybrids *V. vinifera* x *V. rotundifolia* led to the hypothesis of an allopolyploid origin of the grapevine with three basic genomes from which two are common to the subgenera *Vitis* and *Muscadinia* with haploid contents of $n = 6$ and $n = 7$ and the third different, with $n = 6$ for *Vitis* and $n = 7$ for *Muscadinia* (Patel and Olmo 1955). This leads to incomplete pairing during meiosis and high gamete sterility in the F_1 hybrids. It is yet to be explained whether only one of these triplicates has undergone large differentiation by chromosomal rearrangements during the speciation of *Vitis*. These observations on F_1 hybrids *Vitis* x *Muscadinia* should be compared with those of Shetty (1959) who frequently observed secondary associations of bivalents or multiple nucleolar bivalents during meïosis of *V. vinifera*. This hypothesis of paleoploidy was not confirmed by *in situ* hybridization on *V. vinifera* chromosomes, which revealed a single ribosomal locus (Haas et al. 1994; Haas and Alleweldt 2000). Moreover, the data acquired while developing genetic maps, intraspecific (Doligez et al. 2006) or interspecific (Lowe and Walker 2006) did not give any support to this hypothesis.

Meiotic analysis of haploid plants would be a way to support an allopolyploid origin of grapevine, but until now, all attempts to obtain such plants have been unsuccessful, whether they have used polyembryony (Bouquet 1980b), *in vitro* anther culture (Bouquet et al. 1982) or microspore culture (Sefc et al. 1997). Success in obtaining grapevine haploid plant by anther culture was reported by Zou and Li (1981) but not confirmed. In fact, anther-derived plantlets were clearly shown to be diploid and from somatic origin (Rajasekaran and Mullins 1983). From the 1990s, embryogenic callus obtained by *in vitro* anther culture proved to be a useful material for genetic engineering of grapevines (Bouquet et al. 2008). Apart from this special ability to somatic embryogenesis, it is worth noting that grapevines are particularly easy to propagate by *in vitro* techniques that can greatly ease genetic or sanitary improvement (Torregrosa et al. 2001).

The hypothesis of a polyploid origin of the grapevine finally found strong support with the publishing of the grapevine genome sequence (Jaillon et al 2007; Velasco et al. 2007) which showed that the grapevine genome is derived from a hexaploid ancestor with a probable ancestral haploid number of chromosomes of n = 7 (see Chapter 9).

1.5.5 Evolution and Diversification of the Genus Vitis

The presence of fossil pips of *Vitaceae* and *Vitis* in tertiary sediments is known since a long time (Kircheimer 1938; Tiffney and Barghoorn 1976). The most recent have been discovered in 55-million year-old northern-European tertiary sediments. Some (*V. rectisulcata*) are globular with a smooth chalaza, similar to the pips of the actual *Vitis* but the others (*V. longisulcata*) are oblong with a wrinkled chalaza, similar to the pips of the actual *Muscadinia* (Fairon-Demaret and Smith 2002). These recent findings confirm the previous distinction of the two species *V. teutonica* and *V. ludwigii* made by Kircheimer (1938). They suggest that separation of *Vitis* and *Muscadinia* may date from the early Tertiary and that *Muscadinia* likely became extinct in Europe during the Quaternary ice ages. It is yet to be explained why the *Muscadinia*, apparently widely distributed across the entire northern hemisphere during the Tertiary, did not undergo large diversification like the *Vitis* during the Quaternary and had a considerably reduced natural range limited to the North American continent. So far, there is no evidence that *Muscadinia* occurred in East Asia, though a wild vine found in India was previously identified as such (Syamal and Patel 1953), but was proved then to likely belong to the species *Ampelocissus latifolia*. The *Muscadinia* are actually represented by only three species, *V. rotundifolia*, *V. munsoniana* and *V. popenoei*, whose natural area is limited to the southeast of the USA and Mexico. However, *V. rotundifolia*, including a lot of ancient and modern cultivars (Olien 1990), and *V. munsoniana* are considered now by American botanists as varieties, separated only by the size of their berries (Comeaux et al. 1987). Furthermore, the "reality" of *V. popenoei* is debatable. *Muscadinia* could be thus be regarded as a relictual monospecific subgenus (or genus) that could make the transition between the genus *Vitis*, adapted to temperate climates, and the genus *Ampelocissus*, adapted to tropical climates. *Ampelocissus* and *Vitis* have morphological, anatomical and caryological similarities and are hypothesized to constitute sister clades (Soejima and Wen 2006). The strong resemblance of seeds of *Vitis rotundifolia* with those of *A. acapulcensis* from Mexico has been underlined (Chen and Manchester 2007). But the ability of *Ampelocissus* to hybridize with *Vitis* or *Muscadinia* is unknown.

The subgenus (or section) *Vitis* contains about 60 species. The determination of the number of "real" species of grapes, and their proper names, has considerable significance for genetic improvement. The systematics of *Vitis* has been a subject of controversy for more than a century. The first classification of *Vitis* was that of Planchon (1887) who placed the American and Asiatic species in separate series. Some of the later French ampelographers concentrated their interest on the relatively small number of species, which were useful germplasm in the struggle against Phylloxera, the predominant problem of French viticulture at the end of the 19th century. Consequently, they tended to give little attention to the species that were without practical interest and their view of the genus *Vitis* was somewhat restricted. By contrast, an extensive classification of North American species of *Vitis* was produced by Bailey (1934). The most recent and complete presentation of the genus is that of Galet (1988) who listed 59 species from which 56, belonging to the section *Vitis*, are distributed in 11 series (Table 1-2).

However, there is a tendency to reduce the number of "true"' species in the recent classifications of some American botanists, who for instance consider *V. lincecumii* to be a variety of *V. aestivalis*, and *V. berlandieri* a variety of *V. cinerea* (Rogers and Rogers 1978; Comeaux et al. 1987). Galet's classification includes 25 species from Asia, but the standing of some of them is still controversial. The revival of viticulture in China has stimulated interest in local genetic resources and accurate new information is now available (Luo and He 2001; Wan et al. 2008).

Vitis species are found mainly in the temperate zones of the northern hemisphere and are distributed almost equally between North America and eastern Asia, but some of them are typically subtropical (*V. caribeae, V. lanata*). Only one species, *V. vinifera* L., originated in Eurasia but has

Table 1-2 Systematics and geographical distribution of the genus *Vitis* (adapted from Galet 1988). Am. = America.

Series	Species	Origin	Series	Species	Origin
	Section *Vitis*			Section *Vitis*	
1. Candicansae	*V. candicans*	North Am. (East)	8. Flexuosae	*V. flexuosa*	Asia
	V. doaniana	North Am. (East)		*V. thunbergii*	Asia
	V. longii	North Am. (East)		*V. betulifolia*	Asia
	V. coriacea	North Am. (East)		*V. reticulata*	Asia
	V. simpsonii	North Am. (East)		*V. amurensis*	Asia
	V. champini	North Am. (East)		*V. piasekii*	Asia
2. Labruscae	*V. labrusca*	North Am. (East)		*V. embergeri*	Asia
	V. coignetiae	Asia		*V. pentagona*	Asia
3. Caribaeae	*V. caribaea*	North Am. (South)		*V. chunganensis*	Asia
	V. blancoii	North Am. (East)		*V. chungii*	Asia
	V. lanata	Asia		*V. piloso-nerva*	Asia
4. Arizonae	*V. arizonica*	North Am. (West)		*V. balsansaeana*	Asia
	V. californica	North Am. (West)		*V. hancockii*	Asia
	V. girdiana	North Am. (West)		*V. hexamera*	Asia
	V. treleasei	North Am. (West)		*V. pedicellata*	Asia
5. Cinereae	*V. cinerea*	North Am. (East)		*V. retordii*	Asia
	V. berlandieri	North Am. (East)		*V. seguinii*	Asia
	V. baileyana	North Am. (East)		*V. silvestrii*	Asia
	V. bourgeana	North Am. (South)		*V. tsoii*	Asia
6. Aestivalae	*V. aestivalis*	North Am. (East)		*V. bryoniifolia*	Asia
	V. lincecumii	North Am. (East)	9. Spinosae	*V. armata*	Asia
	V. bicolor	North Am. (East)		*V. davidii*	Asia
	V. gigas	North Am. (East)		*V. romanetii*	Asia
	V. rufotomentosa	North Am. (East)	10. Ripariae	*V. riparia*	North Am. (East)
	V. bourquina	North Am. (East)		*V. rupestris*	North Am. (East)
7. Cordifoliae	*V. cordifolia*	North Am. (East)	11. Viniferae	*V. vinifera*	Eurasia
	V. rubra	North Am. (East)		Section *Muscadinia*	
	V. monticola	North Am. (East)		*V. rotundifolia*	North Am. (East)
	V. illex	North Am. (East)		*V. munsoniana*	North Am. (East)
	V. helleri	North Am. (East)		*V. popenoei*	North Am. (East)

been spread throughout the world by humans. It is worth noting that the diversification of the species took place almost exclusively in the eastern parts of the continents. In the western parts, few species are found, *V. vinifera* in Eurasia, *V. californica* and *V. arizonica* in North America. This similarity is emphasized by the fact that some species located in similar climate regions have relatively close morphological characteristics. This is particularly the case with *V. labrusca* (north-east USA) and *V. coignetiae* (Japan) qualified as "vicarious" species (Levadoux et al. 1962). Resemblances between American and Asian species may be due to phylogenetic relationships but also to convergence resulting from independent adaptations to similar environments. In this regard, the affinities between *V. bourgeana* from Mexico and *V. reticulata* from western China are likely adaptation to desertic climates. Recent studies of the phylogenetic relationships in the *Vitis* genus including the subgenus *Muscadinia* suggest initial diversification of the genus in Asia and allow distinguishing American and Asiatic species based on cpDNA polymorphisms (J.P. Péros, pers. comm.).

Many species of *Vitis* are thought to have arisen during the Quaternary ice ages. According to De Lattin (1939), their distribution is consistent with the breaking up of large populations by the ice-fronts and the survival of small populations in "refuges", i.e., areas which were protected from the glaciations mainly by topography. Isolation and differing environmental conditions provided ideal circumstances for speciation. During the interglacial periods, there was selection for adaptation to new warmer environments. The formation and dissolution of refuges probably occurred several times during the Quaternary period, and at the end of the ice ages, the species of *Vitis* have acquired a remarkable diversity in morphological characters (Viala and Vermorel 1910). This diversity remained stable despite the absence of genetic barriers between the species of the subgenus *Vitis* that are all interfertile, and was maintained by barriers of geographical, ecological or phenological nature. *Vitis* species are therefore called *ecospecies* (Levadoux et al. 1962) and molecular analysis confirmed their low level of genetic differentiation (Di Gaspero et al. 2000; Pelsy 2007).

1.5.6 *Worth and Use of the Wild Germplasm of Vitis*

Despite the worldwide distribution of *V. vinifera,* its high sensitivity to many diseases led to the domestication of some more tolerant species in the United States during the 18th century, especially *V. labrusca* in the northeast (Booth 1911) and *V. rotundifolia* in the southeast, under the name of muscadine grapes (Olien 1990). In addition, it is worth noting some recent attempts of domestication of *V. coignetiae* in Japan (Okamoto et al. 2002). In fact, the *V. labrusca* varieties, and notably the cultivar "Concord", are obviously natural hybrids involving cultivars of *V. vinifera* imported by the first

European settlers (Tukey 1966). These natural hybrids were classified by Galet (1988) in the species *V. labruscana*. The cultivars, "Norton" and "Cynthiana", which are attributed to *V. aestivalis* have a similar origin (Reisch et al. 1993). By contrast, there is no doubt about the native origin of the muscadine cultivars (*V. rotundifolia*) because the genetic barriers between the subgenera *Vitis* and *Muscadinia* prevented natural hybridization.

The first rootstock cultivars used in France after the introduction of phylloxera, were vegetatively derived directly from *V. riparia* and *V. rupestris*, then from interspecific hybrids between those and *V. berlandieri* (Pouget 1990). Some other *Vitis* species were also used in rootstock breeding, such as *V. champini* (Lowe and Walker 2006), *V. longii* (Gray and Mortensen 1987), *V. cinerea* (Schmid et al. 2003), and *V. rotundifolia* (Bouquet et al. 2000).

But the main worth of wild *Vitis* species is to provide a large number of genes for resistance or tolerance to biotic and possibly abiotic stresses that can be introduced in the genome of *V. vinifera* by the conventional methods of hybridization, and possibly by genetic engineering in the near future. From this point of view, the most interesting species is by far *V. rotundifolia,* which shows high levels of resistance to many pests and diseases (Olmo 1986). Among them, we must cite the Phylloxera (Bouquet 1983), the root-knot nematodes (Bloodworth et al. 1980), the dagger nematode *Xiphinema index* that vectors the grape fanleaf virus (Bouquet et al. 2000), the anthracnose (Mortensen 1981), the powdery mildew (Bouquet 1986) and the downy mildew (Merdinoglu et al. 2003). The downy mildew genes *Rpv1* and *Rpv2* were mapped respectively on the chromosomes 12 and 18 (Wiedemann-Merdinoglu et al. 2006). The grapevine powdery mildew resistance gene *Run1* is closely linked to *Rpv1* and its genetic and physical mapping on chromosome 12 (Barker et al. 2005) enabled the positional cloning and possible introduction of the genes *Run1* and *Rpv1* in the genome of *V. vinifera* by genetic engineering (See Chapter 8).

But genes of resistance to downy and powdery mildews coming from other species of *Vitis* have been previously introduced in the old "French hybrids", and are actually used in breeding (Fisher et al. 2004). Many resistance gene analogs have been identified in *V. riparia, V. amurensis, V. cinerea* and *V. rupestris* and are candidates for disease resistance genes (Di Gaspero and Cipriani 2002; Mahanil et al. 2007). Special interest has been focussed on the downy mildew resistance gene *VRP1,* identified in *V. riparia* and mapped on the chromosome 10 (Kortekamp et al. 2008). The gene *Pdr1* of resistance to Pierce's disease has been identified and mapped on the chromosome 14 in *V. arizonica* (Riaz et al. 2008). This species was shown to be also resistant to *Xiphinema index* and the gene *XiR1* was mapped on the chromosome 19 (Xu et al. 2008). With regard to abiotic stresses, the interest of *V. amurensis* for introducing cold resistance genes into grape varieties grown under continental climatic conditions is debatable (Kovacs et al. 2003).

1.5.7 Origin and Genetic Diversity of *Vitis vinifera*

There is considerable diversity within *V. vinifera* (see Chapter 2). More than 9,600 varieties grown worldwide have been listed and described by Galet (2000) and 11,000 prime names appear in the *Vitis* International variety catalog (Maul et al. 2008). But an accurate knowledge of this diversity is hampered by a complex synonymy and the number of "real" varieties properly identified is probably closer to 5,000. A few cultivars that are included are widely grown in many grape growing countries, those that are of strictly national or regional importance and those which have practically disappeared from vineyards. The classical methods of ampelography are based on morphological descriptions coupled with computer technology (Boursiquot et al. 1987). They are now completed by the use of molecular markers, especially the microsatellite DNA markers (This et al. 2004).

Vitis vinifera took refuge in a number of "refuges" during the glaciations. The region stretching from the western Himalaya to the Caucasus is traditionally considered to be the center of origin and diversity according to Vavilov (Rives 1975). *Vitis vinifera* also took refuge in the circummediterranean zone, which explains the presence of wild vine populations in Europe and North Africa (Arnold et al. 1998; Lacombe et al. 2003). Their survival is endangered for many reasons, namely the possible competition with naturalized *Vitis* species or rootstocks, that have to be considered as invading species before representing a real threat (Arrigo and Arnold 2007). Native wild grapevines were sometimes referred as a particular subspecies, *V. vinifera subsp. silvestris*, distinct from the cultivated grapes *V. vinifera subsp. sativa*. But such a separation is not valid because the morphological traits that make the distinction (Levadoux 1956) are probably the result of domestication. However, the relationships between cultivated vines and wild vines have not yet been clearly established (This et al. 2006). Progress made in the use of nuclear and plastid DNA markers suggests the absence or a low level of gene flux between wild and cultivated compartments (Perret et al. 2000; This et al. 2001; Carreno et al. 2004; Snoussi et al. 2004; Di Vecchi-Staraz et al. 2009). Other reports showed some role of local domestication or genetic introgression from wild vines (Aradhya et al. 2003; Grassi et al. 2003; Sefc et al. 2003). These divergent results could be due to differences in size and composition of the samples. More extensive analyses of wild vines from a broad geographical area are crucial for understanding the role of *V. vinifera* ssp. *silvestris* in the domestication process. Uncertainty also remains about the place and period of this process and whether secondary independent domestication events also occurred (Arroyo-García et al. 2006; Imazio et al. 2006).

There is no doubt that hybridization played an important role in the diversification of grape varieties (This et al. 2006). Well-known varieties such as "Cabernet-Sauvignon", "Chardonnay", "Syrah" and "Merlot" are obviously the result of crossings between older varieties (Bowers and Meredith 1997; Bowers et al. 1999; Bowers et al. 2000; Vouillamoz and Grando 2006; Boursiquot et al. 2009). There is evidence that the variety "Gouais Blanc" (syn. "Heunisch weiss") originated from Dalmatia and was widely cultivated in the Middle Ages along with Pinot Noir, and played a key role in the diversification of French and European varieties (Boursiquot et al. 2004).

Negrul (1946) classified the *V. vinifera* varieties into three large eco-geographical groups called proles: proles occidentalis, identified by winegrape varieties with small berries, cultivated mainly in western Europe; proles orientalis, identified by tablegrape varieties with large berries, mainly cultivated in Asia Minor and the southern Mediterranean basin; and finally proles pontica, probably the most ancient group, identified by winegrape varieties cultivated around the Black Sea and in eastern Europe. Microsatellite markers supported Negrul's classification and brought to light the relationships between a number of French varieties of proles occidentalis with wild vines originating in the Southwest of France (Aradhya et al. 2003). According to their morphological similarities and geographical distribution, Levadoux (1956) and Bisson (1995) classified the main varieties of the *proles occidentalis* grown in France, in several groups called sorto-types, each of them divided in different sorto-groups.

In fact, the taxonomic classification of *V. vinifera* must be reviewed in the light of the molecular data (Boursiquot and This 1999). Within a sorto-type, e.g., the "Noiriens", the genetic differentiation between different sorto-groups, e.g., "Pinot", "Chardonnay" or "Gamay", is based on sexual reproduction. Within the sorto-group "Pinot", the differentiation between cultivars, e.g., "Pinot noir", "Pinot blanc", "Pinot gris", "Pinot meunier", only occurs as a result of the accumulation of spontaneous mutations occurring during countless cycles of vegetative propagation.

In grapevine, natural mutations at the gene level are relatively frequent. Some of these, which are linked to skin coloration, are well known and have played a significant role in the domestication and diversification of grape varieties (Walker et al. 2007). In perennial plants where artificial mutants are difficult to generate and screen, spontaneous mutants represent a unique resource to understand the genetic control of complex developmental patterns such as flower formation (Chatelet et al. 2007; see Chapter 3).

Mutations caused by chromosomal rearrangements (deletions, inversions, translocations) have not been identified as yet in *V. vinifera*, due to the small size of the chromosomes and the difficulties to perform classical banding of these . Mutations at the genome level are relatively few. As early as 1914, spontaneous tetraploid mutants have been reported in *V. vinifera* or in *V. labruscana* varieties (Olmo 1952). They were used intensively in tablegrape breeding, but without great success, except in East Asia where a tetraploid cultivar ("Kyoho") became the main variety grown in Japan and gained academic interest (Kobayashi et al. 2002). Aneuploids could be a useful tool for cytogenetic studies, but as yet, very few naturally-occurring aneuploids have been identified in grapevine. One chance seedling of *V. vinifera* with $2n + 1 = 40$ chromosomes was reported in a selfing of the variety "Mosel Riesling" (Husfeld 1932). In addition, one chance seedling of the cultivar "Kyoho" with $4n–1 = 75$ chromosomes was released in Japan as a commercial seedless cultivar called "Takao" (Ashikawa 1972). Attempts to obtain artificial aneuploid plants were done by self-pollinations and crosses of triploid grapes, using immature seed culture and *in vitro* embryo rescue (Park et al. 1999). Aneuploid forms of tetraploid grapes were induced *in vitro* by *para*-fluorophenylalanine treatment (Sarikhani and Wakana 2006). Attempts to obtain haploid plants ($n = 19$) have been previously discussed. A potential application of such haploid plants would be obtaining homozygote lines. Producing such lines needs many generations of selfing and is hampered in grapevine by strong inbreeding effect (Bronner and Oliveira 1990). However, near-homozygote lines could be obtained (Bronner and Oliveira 1990), one of them was used to develop the reference grapevine genome sequence (Jaillon et al. 2007; Chapter 9). Using rapid cycling mutants, Chaib et al. (2010) also developed several near-homozygote lines that are useful for forward and reverse genetics.

Mutations are harder to demonstrate using molecular markers, but the amplified fragment length polymorphism (AFLP) techniques (Scott et al. 2000), microsatellites (Riaz et al. 2002), and more recent techniques based on single nucleotide polymorphisms (SNPs) (Salmaso et al. 2004) or retrotransposons (Pelsy et al. 2003) now make it possible to detect genetic polymorphism derived from mutation among clones of the same variety. However, chimerism may hamper the identification and characterization of mutations in grapevines and has important implications for cultivar identity and genetic improvement (Franks et al. 2002; Riaz et al. 2002; Hocquigny et al. 2004).

1.6 Conclusion

Viticulture is characterized by the existence of many factors limiting to varietal innovation and grape breeding. Its ability to face the challenges of the future could thus be questioned. However, such a situation is being modified and there is great probability that progress in the knowledge of the genome, transcriptome, proteome and metabolome of the plant, based on the development of high performance analytical techniques, will be able to reconcile the opposite forces of tradition and innovation.

References

Anselm C, Chataigneaua M, Ndiaye M, Chataigneaua T, Schini-Kertha VB (2007) Grapenext term juice causes endothelium dependent relaxation via a redox-sensitive Src- and Akt-dependent activation of eNOS. Cardiovascul Res 73: 404–413.

Aradhya MK, Dangl GS, Prins BH, Boursiquot J-M, Walker MA Meredith CM (2003) Genetic structure and differentiation in cultivated grape, *Vitis vinifera* L. Genet Res Camb 81: 179–192.

Arnold C, Gillet F, Gobat J-M (1998) Occurrence of the wild vine *Vitis vinifera* ssp. *silvestris* in Europe. Vitis 37: 159–170. (In French with English summary).

Arrigo N, Arnold C (2007 Naturalised *Vitis* rootstocks in Europe and consequences to native wild grapevine. PLoS ONE 6: e521.

Arroyo-Garcia R, Ruiz-Garcia L, Boulling L, Ocete R, Lopez M-A, Arnold C et al. (2006) Multiple origins of cultivated grapevine (*Vitis vinifera* L. ssp. *sativa*) based on chloroplast DNA polymorphisms. Mol Ecol 15: 3707–3714.

Ashikawa K (1972) On the new grape cultivar 'Takao'. Jpn J Breed 22: 46–51.

Barker CL, Donald T, Pauquet J, Ratnaparkhe MB, Bouquet A, Adam-Blondon A-F, Thomas MR, Dry I (2005) Genetic and physical mapping of the grapevine powdery mildew resistance gene, *Run1*, using a bacterial artificial chromosome library. Theor Appl Genet 111: 370–377.

Bailey LH (1934) The species of grapes peculiar to North America. Gentes Herbarum 3: 151–244.

Benjak A, Forneck A, Casacuberta J-M (2008) Genome-wide analysis of the "cut-and-paste" transposons of grapevine. PLoS ONE 3: e3107.

Bennetzen JL (2000) Transposable element contributions to plant gene and genome evolution. Plant Mol Biol 42: 251–269.

Bisson J (1995) The principal ecogeographical groups in French grapevines assortment. J Int Sci Vigne Vin 29: 63–68 (In French with English summary).

Bisson LF, Waterhouse AL, Ebeler SE, Walker MA, Lapsley JT (2002) The present and the future of the international wine industry. Nature 418: 696–699.

Bloodworth PJ, Nesbitt WB, Barker KR (1980) Resistance to rootknot nematodes in *Euvitis* x *Muscadinia* hybrids. In: Univ Calif (eds) Proc 3rd Int Symp Grape Breeding June, Davis (California), USA 15–18 1980, pp 275–292.

Booth NO (1911) A suggestion in regard to the history of grape growth in America. Proc Am Soc Hort Sci 8: 105–112.

Boss PK, Buckeridge EJ, Poole A, Thomas MR (2003) New insights into grapevine flowering. Funct Plant Biol 30: 593–606.

Bouquet A (1980a) *Vitis* x *Muscadinia* hybridization: A new way in grape breeding for disease resistance in France. In: UCAL (eds) Proc 3rd Intern. Symp. Grape Breeding, Davis (California), USA, pp 42–61.

Bouquet A (1980b) Effect of some genetic and environmental factors on spontaneous polyembryony in grape (*Vitis vinifera* L.). Vitis 19: 134–150.

Bouquet A (1982) Origine et évolution de l'encépagement français à travers les siècles. Progr Agri Vitic 99: 110–121.

Bouquet A (1983) Phylloxera resistance of *Vitis vinifera* x *Muscadinia rotundifolia* hybrids. Vitis 22: 311–323 (In French with English summary).

Bouquet A (1986) Introduction dans l'espèce *Vitis vinifera* L. d'un caractère de résistance à l'oïdium (*Uncinula necator* Schw. Burr.) issu de l'espèce *Muscadinia rotundifolia* (Michx.) Small. C.R. 4ème Symp. Int. Genet. Vigne, Vérone (Italie) 13–18 April 1985. *VigneVini*, 13, supplemento al n° 12: 141–146.

Bouquet A (2008) The vine and its genetic diversity (a review). Proc 29th Int Congr Vine and Wine, 25–30 June 2006, Logroño, Spain. Progr Agri Vitic 125: 91–102 (in French with English summary).

Bouquet A, Piganeau B, Lamaison A-M (1982) Genotypic effect on *in vitro* production of callus, embryoids and plantlets from cultured anthers in *Vitis*. CR Acad Sci Paris 295: 569–574 (in French with English summary).

Bouquet A, Danglot Y, Torregrosa L, Bongiovani M, Castagnone-Sereno P, Esmenjaud D, Dalmasso A (2000) Breeding rootstocks resistant to grape fanleaf virus spread using *Vitis* x *Muscadinia* hybridization. Acta Hort 528: 517–526.

Bouquet A, Torregrosa L, Iocco P, Thomas MR (2008) Grapes. In: C Kole, TC Hall (eds) Compendium of Transgenic Crop Plants, vol 4. Transgenic Temperate Fruits and Nuts. Wiley-Blackwell, Oxford, UK, pp 189–232.

Boursiquot J-M, This P (1999) Essay to define the 'cépage'. Progr Agri Vitic 116: 359–361 (In French with English summary).

Boursiquot, J-M, Faber M-P, Blachier O, Truel P (1987) Computerization and statistical analysis of ampelographic data. Agronomie 7: 13–20 (In French with English summary).

Boursiquot J-M, Lacombe T, Bowers JE, Meredith CP (2004) Gouais, a key vine of the European wine heritage. Bull OIV 875: 5–19 (In French with English summary).

Boursiquot J-M, Lacombe T, Laucou V, Julliard S, Perrin F-X, Lanier N, Legrand D, Meredith CP, This P (2009) Parentage of Merlot and related winegrape cultivars of southwestern France: discovery of the missing link. Aust J Grape Wine Res 15: 144–155.

Bowers JE, Meredith CP (1997) The parentage of a classic wine grape, Cabernet-Sauvignon. Nat Genet 16: 84–87.

Bowers JE, Boursiquot J-M, This P, Chu K, Johansson H, Meredith CP (1999) Historical genetics: The parentage of Chardonnay, Gamay and other wine grapes of northeastern France. Science 285: 1562–1565.

Bowers JE, Siret R, Meredith CP, This P, Boursiquot J-M (2000) A single pair of parents proposed for a group of grapevine varieties in northeastern France. Acta Hort 528: 129–132.

Branas J (1932) Sur la caryologie des Ampélidées. CR Acad Sci Paris 194: 121–123.

Bronner A, Oliveira J (1990) Creation and study of the Pinot noir variety lineage. Proc 5th Int Symp Grape Breeding, St. Martin/Pflaz, Germany, Sept 12–16, 1989, Vitis (Spl issue) 69–80.

Cadle-Davidson MM, Owens CL (2008) Genomic amplification of the *Gret1* retroelement in white-fruited accessions of wild *Vitis* and interspecific hybrids. Theor Appl Genet 116: 1079–1094.

Carmona MJ, Chai J, Martinez-Zapater J-M, Thomas MR (2008) A molecular genetic perspective of reproductive development in grapevine. J Exp Bot 59: 2579–2596.

Carreno E, Lopez, M-A, Labra M, Rivera D, Sancha J, Ocete R, Martinez de Toda F (2004) Genetic relationship between some spanish *Vitis vinifera* L. ssp. *sativa* cultivars and wild grapevine populations (*Vitis vinifera* L. ssp. *sylvestris*): a preliminary study. Plant Genet Resour Newsl 137: 42–45.

Cebrii A (1950) Hybridation intergénérique *Ampelopsis* x *Vitis*. Vinod i Vin (Moscou) 7: 16–27.

Chaib J, Torregrosa L, Mackenzie D, Corena P, Bouquet A, Thomas MR (2010) The microvine —a model system for rapid forward and reverse genetics of grapevines. The Plant J 6: 1083–1092.

Chatelet P, Fernandez L, Laucou V, Sreekantan L, Lacombe T, Martinez-Zapater J-M, Thomas MR, Torregrosa L (2007) Characterization of *Vitis vinifera* L. somatic variants exhibiting abnormal flower development patterns. J Exp Bot 58: 4107–4118.

Chen I, Manchester SR (2007) Seed morphology of modern and fossil *Ampelocissus* (Vitaceae) and implications for phytogeography. Am J Bot 94: 1534–1553.

Chen WJ, Delmotte F, Richard-Cervera S, Douence L, Greif C, Corio-Costet M-F (2007) At least two origins of fungicide resistance in grapevine downy mildew populations. Appl Environm Microbiol 73: 5162–5172.

Chervin C, El-Kereamy A, Roustan J-P, Latche A, Lamon J, Bouzayen M (2004) Ethylene seems required for the berry development and ripening in grape, a non climacteric fruit. Plant Sci 167: 1301–1305.

Comeaux BL, Nesbitt WB, Fantz PR (1987) Taxonomy of the native grapes of North Carolina. Castanea 52: 197–215.

Coombe BG (1992) Research on the development and ripening of the grape berry. Am J Enol Vitic 43: 101–110.

Cramer GR, Ergül A, Grimplet J, Tillett RL, Tattersall EAR et al. (2007) Water and salinity stress in grapevines: early and late changes in transcript and metabolite profiles. Funct Integr Genom 7: 111–134.

Cronquist A (1981) An integrated system of classification of flowering plants. Columbia Univ Press, New York, USA.

da Silva FG, Iandolino A, Al-Kayak F, Bohlmann MC, Cushman MA et al. (2005) Characterizing the grape transcriptome. Analysis of expressed sequence tags from multiple *Vitis* species and development of a compendium of gene expression during berry development. Plant Physiol 139: 574–597.

De Lattin G (1939) Uber den Ursprung und die Verbeitung der Reben. Zuchter 11: 217–225.

Deluc LG, Grimplet J, Wheatley MD, Tillett RL, Quilici DR, Osborne C, Schooley DA, Schlauch KA, Cushman JC, Cramer GR (2007) Transcriptomic and metabolite analyses of Cabernet Sauvignon grape berry development. BMC Genom 8: 429.

Délye C, Laigret F, Corio-Costet M-F (1997) A mutation in the 14α-demethylase gene of *Uncinula necator* that correlates with resistance to a sterol biosynthesis inhibitor. Appl Environm Microbiol 63: 2966–2970.

Detjen LR (1919) The limits in hybridization of *Vitis rotundifolia* with related species and genera. NC Agri Expt Sta Bull n°12.

Deytieux C, Geny L, Lapaillerie D, Claverol S, Bonneu M, Donèche B (2007) Proteome analysis of grape skins during ripening. J Exp Bot 58: 1851–1862.

Di Gaspero G, Cipriani G (2002) Resistance gene analogs are candidate markers for disease resistance genes in grape (*Vitis* sp.). Theor Appl Genet 106: 163–172.

Di Gaspero G, Peterlunger E, Testolin R, Edwards KJ, Cipriani G (2000) Conservation of microsatellite loci within the genus *Vitis*. Theor Appl Genet 101: 301–308.

Di Gaspero G, Cipriani G, Adam-Blondon A-F, Testolin R (2007) Linkage maps of grapevine displaying the chromosomal locations of 420 microsatellite markers and 82 markers for R gene candidates. Theor Appl Genet 114: 1249–1263.

Di Vecchi-Staraz M, Laucou V, Bruno G, Lacombe T, Gerber S, Bourse T, Boselli M, This P (2009) Low level of pollen-mediated gene flow from cultivated to wild grapevine: consequences for the evolution of the endangered subspecies *Vitis vinifera* L. ssp. *silvestris*. J Hered 100: 66–75.

Doazan J-P (2000) Evolution of grapevine variety range in France for quality wine. Acta Hort 528: 655–658.

Doligez A, Adam-Blondon A-F, Cipriani G, Di Gaspero G, Laucou V, Merdinoglu D, Meredith CP, Riaz S, Roux C, This P (2006) An integrated SSR map of grapevine based on five mapping populations. Theor Appl Genet 113: 369–382.

Eibach R, Töpfer R (2003) Success in resistance breeding: "Regent" and its steps into the market. Acta Hort 603: 687–691.

Fairon-Demaret M, Smith T (2002) Fruit and seeds from the Tienen formation at Doormal Palaeocene-Eocene transition in Eastern Belgium. Rev Paleobot Palynol 122: 47–62.

Fernandez L, Torregrosa L, Terrier N, Sreekantan L, Grimplet J, Davies C, Thomas MR, Romieu C, Ageorges A (2007) Identification of genes associated with flesh morphogenesis during grapevine fruit development. Plant Mol Biol 63: 307–323.

Fernandez L, Torregrosa L, Segura V, Bouquet A, Martinez-Zapater JM (2010) Transposon-induced gene activation as a mechanism generating cluster shape somatic variation in grapevine. Plant J 61: 545–557.

Fisher B-M, Salakhutdinov I, Akkurt M, Eibach R, Edwards KJ, Topfer R, Zyprian E (2004) Quantitative trait locus analysis of fungal disease resistance factors on a molecular map of grapevine. Theor Appl Genet 108: 501–515.

Franks T, Botta R, Thomas MR (2002) Chimerism in grapevines: implications for cultivar identity, ancestry and genetic improvement. Theor Appl Genet 104: 192–199.
Galet P (1988) Cépages et Vignobles de France. Tome 1. Les Vignes américaines. Imp. Dehan, Montpellier, France.
Galet P (2000) Dictionnaire encyclopédique des cépages. Hachette Ed, Paris, France.
German JB, Walzem RL (2000) The health benefits of wine. Annu Rev Nutr 20: 561–593.
Grassi F, Labra M, Imazio S, Spada A, Sgorbati S, Scienza A, Sala F (2003) Evidence of a secondary grapevine domestication center detected by SSR analysis. Theor Appl Genet 107: 1315–1320.
Gray DJ, Mortensen JA (1987) Initiation and maintenance of long term somatic embryogenesis from anthers and ovaries of *Vitis longii* 'microsperma'. Plant Cell Tiss Org Cult 9: 73–80.
Haas H-U, Alleweldt G (2000) The karyotype of the grapevine (*Vitis vinifera* L.). Acta Hort 528: 247–255.
Haas H-U, Budahn H, Alleweldt G (1994) *In situ* hybridization in *V. vinifera* L. *Vitis* 33: 251–252.
Hocquigny S, Pelzy F, Dumas V, Kindt S, Heloir M-C, Merdinoglu D (2004) Diversification within grapevine cultivars goes through chimeric states. Genome 47: 579–589.
Husfeld B (1932) Uber die züchtung plasmoparawiderstandfähiger reben. Gartenbauwiss 7: 1.
Imazio S, Labra M, Grassi F, Scienza A, Failla O (2006) Chloroplast microsatellites to investigate the origin of grapevine. Genet Resour Crop Evo. 53: 1003–1011.
Ingrouille MJ, Chase MW, Fay MF, Bowmann D, Van der Bank M, Bruijn DE (2002) Systematics of Vitaceae from the point of view of plastid *rbcL* DNA sequence data. Bot J Linn So. 138: 421–432.
Iriti M, Rossoni M, Faoro F (2006) Melatonin content in grape: myth or panacea? J. Food Sci Agri 86: 1432–1438.
Jaillon O, Aury JM, Noel B, Policriti A, Clepet C, Casagrande A, Choisne N, Aubourg S, Vitulo N, Jubin C, Vezzi A, Legeai F, Hugueney P, Dasilva C, Horner D, Mica E, Jublot D, Poulain J, Bruyère C, Billault A, Segurens B, Gouyvenoux M, Ugarte E, Cattonaro F, Anthouard V, Vico V, Del Fabbro C, Alaux M, Di Gaspero G, Dumas V, Felice N, Paillard S, Juman I, Moroldo M, Scalabrin S, Canaguier A, Le Clainche I, Malacrida G, Durand E, Pesole G, Laucou V, Chatelet P, Merdinoglu D, Delledonne M, Pezzotti M, Lecharny A, Scarpelli C, Artiguenave F, Pè ME, Valle G, Morgante M, Caboche M, Adam-Blondon AF, Weissenbach J, Quétier F, Wincker P (2007). The grapevine genome sequence suggests ancestral hexaploidization in major angiosperm phyla. Nature 449: 463–467.
Jansen RK, Kaittanis C, Saski C, Lee SB, Tomkins J, Alverson AJ, Daniell H (2006) Phylogenetic analyses of *Vitis* (Vitaceae) based on complete chloroplast genome sequences: effects of taxon sampling and phylogenetic methods on resolving relationships among rosids. BMC Evol Biol 6: 32.
Johnson H (1989) The story of wine. Mitchell Beazley Publishers, New York, USA.
Jones GV, White MA, Cooper OR, Storchmann K (2005) Climate change and global wine quality. Climat Change 73: 319–343.
Kirchheimer F (1938) Beiträge zur näheren Keentnis von VitaceenSamenformen tertiären Alters. Planta 28: 582–598.
Kobayashi S, Ishimaru M, Hiraoka K, Honda C (2002) *Myb*-related genes of the Kyoho grape (*Vitis labruscana*) regulate anthocyanin biosynthesis. Planta 215: 924–933.
Kobayashi S, Goto-Yamamoto N, Hirochika H (2004) Retrotransposon-induced mutations in grape skin color. Science 304: 982.
Kortekamp A, Welter L, Vogt S, Knoll A, Schwander F, Töpfer R, Zyprian E (2008) Identification, isolation and characterization of a CC-NBS-LRR candidate disease resistance gene family in grapevine. Mol Breed 22: 421–432.

Kovacs LG, Byers PL, Kaps ML, Saenz J (2003) Dormancy, cold hardiness and spring frost hazard in *Vitis amurensis* hybrids, under continental climatic conditions. Am J Enol Vitic 54: 8–14.
Lacombe T, Laucou V, Di Vecchi M, Bordenave L, Bourse T, Siret R, David J, Boursiquot J-M et al. (2003) Inventory and characterization of *Vitis vinifera* L. ssp. *silvestris* in France. Acta Hort 603: 553–557.
Lavie P (1970) Contribution à l'étude systématique des Vitacées. Thèse Université de Montpellier, France.
Ledbetter CA, Ramming DW (1989) Seedlessness in grapes. Hort Rev 11: 159–184.
Lewis WH (1979) Polyploidy in Angiosperms: dicotyledons. Basic Life Sci 13: 241–268.
Levadoux L (1956) Les populations sauvages et cultivées de *Vitis vinifera* L. Ann Amélior Plantes 1: 59–118.
Levadoux L, Boubals D, Rives M (1962) Le genre *Vitis* et ses espèces. Ann Amélior Plantes 12: 19–44
Lodhi MA, Reisch BI (1995) Nuclear content of *Vitis* species, cultivars and other genera of the Vitaceae. Theor Appl Genet 90: 11–16.
Lowe KM, Walker MA (2006) Genetic linkage map of the interspecific grape rootstock cross Ramsey (*Vitis champinii*) x Riparia Gloire (*Vitis riparia*). Theor Appl Genet 112: 1582–1592.
Luo S, He P (2001) Discrimination of wild grapes native to China by RAPD markers. Vitis 40: 163–168.
Mahanil S, Reisch BL, Owens CL, Thipyajong P, Laosuwan P (2007) Resistance gene analogs from *Vitis cinerea*, *V. rupestris* and *Vitis* hybrid Horizon. Am J Enol Vitic 58: 484–493.
Maul E, Topfer R, Eibach R (2008) *Vitis* International Variety Catalogue: *http://www.vivc.bafz.de*
McGovern PE (2003) Ancient Wine: The Search for the Origins of Viniculture. Princeton Univ Press, Princeton, NJ, USA.
Merdinoglu D, Wiedemann-Merdinoglu S, Coste P, Dumas V, Haetty S, Butterlin G, Greif C, Adam-Blondon A-F, Bouquet A, Pauquet J (2003) Genetic analysis of downy mildew resistance derived from *Muscadinia rotundifolia*. Acta Hort 603: 451–456.
Millardet A (1901) Note sur la fausse hybridation chez les Ampelidées. Rev Vitic 11: 677–680.
Moisy C, Garrison KI, Meredith CP, Pelsy F (2008) Characterization of ten novel Ty1/*copia*-like retrotransposon families of the grapevine genome. BMC Genom 9: 469.
Morgan AF, Kimmel L, Field A, Nichols PF (1934) The vitamin content of Sultanina (Thompson seedless) grapes and raisins. J Nutr 9: 369–382.
Mortensen JA (1981) Sources and inheritance of resistance to anthracnose in *Vitis*. J Hered 72: 423–426.
Moroldo M, Paillard S, Marconi R, Legeai F, Canaguier A et al. (2008) A physical map of the heterozygous grapevine 'Cabernet-Sauvignon' allows mapping candidate genes for disease resistance. BMC Plant Biol 8: 66.
Mullins MG, Bouquet A, Williams LE (1992) Biology of the Grapevine. Cambridge Univ Press, Cambridge, UK.
Naito K, Cho E, Yang GJ, Campbell MA, Yano K, Okumuto Y, Tanisaka T, Wessler SR (2006) Dramatic amplification of a rice transposable element during recent domestication. Proc Natl Acad Sci USA 103: 17620–17625.
Negrul AM (1946) Origin of the Cultivated Grapevine and its Classification. Ampelographia SSSR, Moscow 1: 159–216 (In Russian).
Nuñez DR, Walker MJ (1989) A review of paleobotanical findings of early *Vitis* in the Mediterranean and of the origins of cultivated grapevines, with special reference to prehistoric exploitations in the western Mediterranean. Rev Paleobot Palynol 61: 205–237.
Okamoto G, Ueki K, Ichi T, Aoki H, Fujiwara M, Hirano K (2002) Juice constituents and skin pigments in *Vitis coignetiae* Pulliat grapevines. Vitis 41: 161–162.
Olien WC (1990) The Muscadine grape: botany, viticulture, history and current industry. HortScience 25: 732–739.

Olmo HP (1952) Breeding tetraploid grapes varieties. Proc Am Soc Hort Sci 59: 285–290.

Olmo HP (1986) The potential role of (*V. vinifera* x *rotundifolia*) hybrids in grape variety improvement. Experientia 42: 921–926.

Papadakis A, Reustle G, Roubelakis-Angelakis KA 2001. Protoplast technology in grapevine. In: KA Roubelakis-Angelakis (ed) Molecular Biology and Biotechnology of the Grapevine, Kluwer Academic Publ, Dordrecht, The Netherlands, pp 353–392.

Park SM, Hiramatsu M, Wakana A (1999) Aneuploid plants derived from crosses with triploid grapes through immature seed culture and subsequent embryo culture. Plant Cell Tiss Org Cult 59: 125–133.

Patel GI, Olmo HP (1955) Cytogenetics of *Vitis*: I. The hybrid *V. vinifera* x *V. rotundifolia*. Am J Bot 42: 141–159.

Pelsy F (2007) Untranslated leader region polymorphism of *Tvv1*, a retrotransposon family, is a novel marker useful for analyzing genetic diversity and relatedness in the genus *Vitis*. Theor Appl Genet 116: 15–27.

Pelsy F, Merdinoglu D (2002) Complete sequence of *Tvv1*, a family of Ty1 *copia*-like retrotransposons of *Vitis vinifera* L., reconstituted by chromosome walking. Theor Appl Genet 105: 614–621.

Pelsy F, Schehrer L, Merdinoglu D (2003) Development of retrotransposon-based molecular markers (S-SAP). Acta Hort 603: 83–87.

Pereira H-S, Barao A, Delgado M, Morais-Cecilio L, Viegas W (2005) Genomic analysis of grapevine retrotransposon 1 (Gret1) in *Vitis vinifera*. Theor Appl Genet 111: 871–878.

Perret M, Arnold C, Gobat J-M, Küpfer P (2000) Relationships and genetic diversity of wild and cultivated grapevines (*Vitis vinifera* L.) in central Europa based on microsatellite markers. Acta Hort 528: 155–159.

Planchon J-E (1887) Monographie des Ampelidées vraies. In: A de Candol (ed) Monographiae phanerogamarum, vol 5, Editions Masson, Paris, France, pp 306–356.

Pollefeys P, Bousquet J 2003. Molecular genetic diversity of the French-American grapevine hybrids cultivated in North America. Genome 46: 1037–1048.

Pouget R (1990) Histoire de la lutte contre le Phylloxéra de la vigne en France. INRA, OIV (eds), Editions Quae, Paris, France, pp 1–157.

Raj AS, Seethaiah L (1969) Karyotype analysis and meiotic studies in three varieties of grape (*Vitis vinifera* L.). Cytologia 34: 475–483.

Rajasekaran K, Mullins MG (1983) The origin of embryos and plantlets from cultured anthers of hybrid grapevines. Am J Enol Vitic 34: 108–113.

Reisch BI, Goodman RN, Martens MH, Weeden NF (1993) The relationship between Norton and Cynthiana red wine cultivars derived from *Vitis aestivalis*. Am J Enol Vitic 44: 441–444.

Renaud S, De Lorgeril M (1992) Wine, alcohol, platelets and the French paradox for coronary heart disease. Lancet 339: 1523–1526.

Reustle G, Harst M, Alleweldt G (1995) Plant regeneration of grapevine (*Vitis* sp.) protoplasts isolated from embryogenic tissue. Plant Cell Rep 15: 238–241.

Riaz S, Garrison KE, Dangl GS, Boursiquot J-M, Meredith CP (2002) Genetic divergence and chimerism within ancient asexually propagated winegrape cultivars. J Am Soc Hort Sci 127: 508–514.

Riaz S, Tenscher AC, Rubin J, Graziani R, Pao SS, Walker MA (2008) Fine-scale genetic mapping of two Pierce's disease resistance loci and a major segregation distortion region on chromosome 14 of grape. Theor Appl Genet 117: 671–681.

Rives M (1975) Les origines de la vigne. La Recherche 53: 120–129.

Rogers DJ, Rogers CF (1978) Systematics of North American grape species. Am J Enol Vitic 29: 73–78.

Rossetto M, Jackes BR, Scott KD, Henry RJ (2002) Is the genus *Cissus* (Vitaceae) monophyletic? Evidence from plastid and nuclear ribosomal DNA. Syst Bot 27: 522–533.

Salinari F, Giosue S, Tubiello F-N, Rettori A, Rossi V, Spanna F, Rozenzweig C, Gullino M-L (2006) Downy mildew (*Plasmopara viticola*) epidemics on grapevine under climate change. Glob Change Biol 12: 1299–1307.

Salmaso M, Faes G, Segala C, Stefanini M, Salakhutdinov I, Zyprian E, Töpfer R, Grando M-S, Velasco R (2004) Genome diversity and gene haplotypes in the grapevine (*Vitis vinifera* L.), as revealed by single nucleotide polymorphisms. Mol Breed 14: 385–395.

Sarikhani H, Wakana A (2006) *In vitro* induction of aneuploid forms of tetraploid grapes by *para*-fluorophenylalanine. J Fac Agri Kyushu Univ 51: 257–260.

Savolainen V, Fay MF, Albach DC, Backlund A, van der Bank M, Cameron KM et al. (2000) Phylogeny of the eudicots: a nearly complete familial analysis based on *rbcL* gene sequences. Kew Bull 55: 257–309.

Schmid J, Manty F, Rühl EH (2003) Utilizing the complete phylloxera resistance of *Vitis cinerea* Arnold in rootstock breeding. Acta Hort 603: 393–400.

Scott KD, Ablett EM, Lee LS, Henry RJ (2000) AFLP markers distinguishing an early mutant of Flame seedless grape. Euphytica 113: 245–249.

Sefc KM, Ruckenbauer P, Regner F (1997) Embryogenesis in microspore culture of *Vitis* subspecies. Vitis 36: 15–20.

Sefc KM, Steinkellner H, Lefort F, Botta R, Da Câmara Machado A, Borrego J, Maletić E, Glössl J (2003) Evaluation of genetic contribution of local wild vines to European grapevine cultivars. Am J Enol Vitic 54: 15–21.

Shetty BV (1959) Cytotaxonomical studies in Vitaceae. Bibl Genet 18: 167–272.

Small JK (1903) *Vitis* and *Muscadinia*. In: JK Small (ed) Flora of the Southeastern United States. New-York, USA, pp 752–757.

Snoussi H, Harbi ben Slimane M, Ruiz-Garcia L, Martinez-Zapater J-M, Arroyo-Garcia R (2004) Genetic relationship among cultivated and wild grapevine accessions from Tunisia. Genome 47: 1211–1219.

Soejima A, Wen J (2006) Phylogenetic analysis of the grape family (Vitaceae) based on three chloroplast markers. Am. J Bot 93: 278–287.

Soltis DE, Soltis PS, Chase MW, Mort ME, Albach DC, Zanis M, Savolainen V et al. (2000) Angiosperm phylogeny inferred from 18S rDNA, rbcL and atpB sequences. Bot J Linn Soc 133: 381–461.

Syamal NB, Patel GI (1953). A wild species of grape in India. Proc Am Soc Hort Sci 62: 228–230.

Terrier N, Glissant D, Grimplet J, Barrieu F, Abbal P, Couture C, Ageorges A et al. (2005) Isogene specific oligo arrays reveal multifaceted changes in gene expression during grape berry (*Vitis vinifera* L.) development. Planta 222: 832–847.

Tesniere C, Pradal M, El-Kereamy A, Torregrosa L, Chatelet P, Roustan J-C, Chervin C (2004) Involvement of ethylene signalling in a non-climacteric fruit: new elements regarding the regulation of ADH expression in grapevine. J Exp Bot 55: 2235–2240.

This P, Roux C, Parra P, Siret R, Bourse T, Adam A-F, Yvon M, Lacombe T, David J, Boursiquot J-M (2001) Characterization of genetic diversity in a population of wild grapes from Pic Saint Loup area and its relationship with the cultivated grapes. Genet Select Evol 33: 289–304 (In French with English summary).

This P, Jung A, Boccaci P, Borrego J, Botta R, Costantini L, Crespan M, et al. (2004) Development of a standard set of microsatellite reference alleles for identification of grape cultivars. Theor Appl Genet 109: 1448–1458.

This P, Lacombe T, Thomas MR (2006) Historical origins and genetic diversity of wine grapes. Trends Genet 22: 511–519.

Thomas H, Beveridge J, Girard B, Kopp T, Dover JCG (2005) Yield and composition of grape seed oils extracted by supercritical carbon dioxide and petroleum ether: varietal effects. J Agri. Food Chem 53: 1799–1804.

Tiffney BH, Barghoorn ES (1976) Fruits and seeds of Brandon lignite. I. Vitaceae. Rev Palaeobot Palynol 22: 169–191.

Tomkins MR, Peterson DG, Yang TJ, Ablett ER, Henry RJ, Lee LS, Holton TA, Waters D, Wing RA (2001) Grape (*Vitis vinifera* L.) BAC library construction, preliminary STC analysis and identification of clones associated with flavonoid and stilbene biosynthesis. Am J Enol Vitic 52: 287–291.

Torregrosa L, Bouquet A, Goussard P-G (2001) *In vitro* culture and propagation of grapevine. In: KA Roubelakis-Angelakis (ed) Molecular Biology and Biotechnology of the Grapevine, Kluwer Academic Publ, Dordrecht, The Netherlands, pp 281–326.

Tukey HB (1966) The story of the Concord grape. Fruit Var Hort Digest 20: 54–55.

Velasco R, Zharkikh A, Troggio M, Cartwright DA, Cestaro A, Pruss D, Pindo M, Fitzgerald LM, Vezzulli S, Reid J, Malacarne G, Iliev D, Coppola G, Wardell B, Micheletti D, Macalma T, Facci M, Mitchell JT, Perazzolli M, Eldredge G, Gatto P, Oyzerski R, Moretto M, Gutin N, Stefanini M, Chen Y, Segala C, Davenport C, Demattè L, Mraz A, Battilana J, Stormo K, Costa F, Tao Q, Si-Ammour A, Harkins T, Lackey A, Perbost C, Taillon B, Stella A, Solovyev V, Fawcett JA, Sterck L, Vandepoele K, Grando SM, Toppo S, Moser C, Lanchbury J, Bogden R, Skolnick M, Sgaramella V, Bhatnagar SK, Fontana P, Gutin A, Van de Peer Y, Salamini F, Viola R (2007) A high quality draft consensus sequence of the genome of a heterozygous grapevine variety. PLoS ONE 2: e1326.

Verriès C, Bès C, This P, Tesnière C (2000) Cloning and characterization of *Vine-1*, a LTR-retrotransposon-like element in *Vitis vinifera* L. and other *Vitis* species. Genome 43: 366–376.

Viala P, Vermorel V (1910) Ampélographie. Tome 1. Masson et Cie, Paris, France.

Vouillamoz J-F, M-S Grando (2006) Genealogy of wine grape cultivars: 'Pinot' is related to 'Syrah'. Heredity 97: 102–110.

Vouillamoz J-F, Schneider A, Grando M-S (2007) Microsatellite analysis of Alpine grape cultivars (*Vitis vinifera* L.) alleged descendants of Pliny the Elder's *Raetica* are genetically related. Genet Resour Crop Evol 54: 1095–1104.

Walker AR, Lee E, Bogs J, McDavid DA, Thomas MR, Robinson SP (2007) White grapes arose through the mutation of two similar and adjacent regulatory genes. Plant J 49: 772–785.

Wan Y, Schwaninger H, Simon CJ, Wang Y, He P (2008) The eco-geographic distribution of wild germplasm in China. *Vitis* 47: 77–80.

Weaver RJ (1958) Effect of gibberellic acid on fruit set and berry enlargement in seedless grapes of *Vitis vinifera*. Nature 181: 851–852.

Wen J, Nie ZL, Soejima A, Meng Y (2007) Phylogeny of *Vitaceae* based on nuclear *GAI1* gene sequences. Can J Bot 85: 731–745.

Weidemann-Merdinoglu S, Prado E, Coste P, Dumas V, Butterlin G, Bouquet A, Merdinoglu D (2006) Genetic analysis of resistance to downy mildew from *Muscadinia rotundifolia*. Proc 9th Int Conf Grape Genetics and Breeding, 2006, July 2–6, Udine, Italy (abstract poster).

Wylie AP (1871) Hybridization of *rotundifolia* grapes. Am Pomol Soc Pro. 13: 113–116.

Xu K, Riaz S, Roncoroni N-C, Jin Y, Hu R, Zhou R, Walker MA (2008) Genetic and QTL analysis of resistance to *Xiphinema index* in a grapevine cross. Theor Appl Genet 116: 305–311.

Zou CJ, Li PF (1981) Induction of pollen plant of grape (*Vitis vinifera* L.). Acta Bot Sin 23: 79–81 (In Chinese).

2

Natural Variation in *Vitis*

Patrice This,[1,*] *José M. Martínez Zapater,*[2] *Jean-Pierre Péros,*[1] and *Thierry Lacombe*[1]

ABSTRACT

The natural diversity in the genus *Vitis* and more specifically for *V. vinifera* is important both at the morphological and molecular level. Much work has been devoted in the past years to the development of evaluation methods for many traits from composition of berries to disease resistance and biotic and abiotic stress tolerance. *Vitis vinifera* shows considerable variation in morphology and represents a huge reservoir of diversity for several traits in particular quality related traits and abiotic stresses. Similarly, the *Vitis* species show considerable variation in their habitat and wide morphological diversity. They represent a big reservoir for biotic stress tolerances to many diseases and to abiotic stresses.

The genetic diversity at the molecular level, mostly recorded by SSR markers and more recently by SNP, is also important, and similar to that found in maize or rice. Heterozygosity of grape is very high, thus rendering the analysis and use of this diversity more difficult. Diversity in important genes linked to agronomical traits is still poorly documented and may be of interest in helping the grape community to understand the functioning of the grape genome and to speed up the development of new cultivars tolerant or resistant to diseases, adapted to a changing environment and of optimal quality.

Keywords: Phenotypic diversity, SSR, SNPs, Chloroplastic DNA, *Vitis vinifera*, *V.v. sylvestris*

[1]UMR INRA 1334 CIRAD Montpellier Supagro AGAP, 2 Place Pierre Viala, 34060 Montpellier cedex 1, France.
[2]Instituto de Ciencias de la Vid y del Vino (CSIC, Universidad de La Rioja, Gobierno de La Rioja), C/Madre de Dios 51, 26006 Logroño, Spain.
*Corresponding author: *patrice.this@supagro.inra.fr*

2.1 Introduction

The genomic revolution has been the starting point for renewing attention on the extent of naturally occurring variation in many plant species and in particular in the model species *Arabidopsis thaliana* (Alonso-Blanco and Koornneef 2000). Because grape is experiencing similar genomic breakthroughs, the use of genetic diversity in this species may greatly improve our understanding of economically important quantitative traits. Although naturally induced variants have been analyzed (Boss and Thomas 2002; Chatelet et al. 2007; Fernandez et al. 2010) and used for genetic and physiological studies (Kobayashi et al. 2004; Fernandez et al. 2006, 2010, see Chapter 3), the main focus in this chapter will be on natural diversity. Despite the number of genera and species in the Vitaceae family (see Chapter 1), most of the useful resources for viticulture belong to the genus *Vitis*. We will, therefore, emphasize on this genus and most specifically on *V. vinifera* L. in this chapter.

Due to the long history of grapevine cultivation (Olmo 1995; Zohary 1995; McGovern 2003) and the vegetative propagation enabling conservation of cultivars over a long period of time, *V. vinifera* L. diversity is important: the number of cultivars available today is estimated from 6,000 to 11,000 (Galet 2000; Maul et al. 2008). As grape species are predominantly outcrossing they can be maintained by vegetative propagation, most of the accessions are heterozygous plants maintained in the field and conserved in germplasm collections even if alternative ways of conservations such as *in vitro* collections (Bouquet and Boursiquot 1999), or cryoconservation (Plessis et al. 1993; Matsumoto and Sakai 2003; Wang et al. 2004; Chatelet et al. 2009) have been proposed. Very few published papers were based on the analysis of plants collected in wild conditions and they usually concern either *V. vinifera* subspecies *sylvestris*, the wild compartment of *V. vinifera* (Soylemezoglu et al. 2001; This et al. 2001; Grassi et al. 2003a, b, 2006, 2008; Lacombe et al. 2003; Pavek et al. 2003; Snoussi et al. 2004; Arroyo-Garcia et al. 2006; Cunha et al. 2007a, b; Arrigo and Arnold 2007) or American and Asian *Vitis* species (Pavek et al. 2003; Wan et al. 2007, 2008).

In the present chapter, we will present the extent of naturally occurring variation both at the phenotypic and molecular level among cultivars or species, but will not consider somatic variation since it has been presented elsewhere in this book (see Chapter 3). The pattern of genetic structure also greatly affects the usefulness of naturally occurring allelic variation for the analysis of complex traits, this aspect will be covered too. Finally, we will also discuss a few uses of naturally occurring variation in grape.

2.2 Phenotype-based Diversity

Despite the economical importance of grape, and the fact that natural variation has already been successfully used to decipher traits or metabolic pathways in grape (Boss et al. 1996; Debolt et al. 2006; Lijavetzky et al. 2006; This et al. 2007), the analysis of large sets of genetic resources at the morphological level has not been intensive. One of the reasons might be the complexity of the methods available so far or the fact that phenotyping grape is expensive, time consuming and requires a lot of space. The interactions between rootstocks and scions (Zhang et al. 2002) or between traits (Lopez-Miranda and Yuste 2004) may also be limiting factors. Finally, while much work has been performed from quantitative trait loci (QTL) analysis to genes (Costantini et al. 2008; also see Chapters 4 and 6 in this book) and as resources are limited, only a few extensive works on genetic diversity have been possible.

Most of the work in the past years has been devoted to the development of methods for many traits from composition of berries to disease resistance and abiotic stresses tolerance (Table 2-1) but development of rapid methods (such as for anthocyanins, Grant and Helleur 2008) and non-destructives ones (Agati et al. 2007) should still be a priority in order to speed up the analysis of genetic resources. The nutritional interest of natural compounds such as the antioxidant activity of polyphenols (Cooke et al. 2005; Heinonen 2007; Mcghie and Walton 2007) will also increase the potential interest of genetic resources screening. Nevertheless, available genetic resources within *Vitis* genus display important phenotypic diversity useful for genomic works. A very interesting resource is the International Vitis database (*http://www.genres.de/eccdb/vitis/*), where one can find data on 12436 accessions of *Vitis vinifera* and 4462 accessions of *Vitis* species.

Vitis vinifera L.—the most important species in terms of grape production —shows considerable variation in morphology (Fig. 2-1) and represents a huge reservoir of diversity for several traits.

Most of the recent published analyses on *V. vinifera*—the standard of grape quality today—have been conducted for plant morphology, quality related traits, i.e., berry composition, phenology and abiotic stresses (Table 2-2).

The study of plant morphology, mainly leaf, shoots, bunch morphology (also called ampelography) have been the sole means of identification of grape cultivars until recently (Boursiquot and This 1996) and have thus received significant consideration, and are still used today for identification (Pavek et al. 2003; Ortiz et al. 2004; Arrigo and Arnold 2007) especially during *in situ* plant collection. Extensive work of harmonization have been performed within GenRes81 project (*http://www.genres.de/eccdb/vitis*) that led to the publication by OIV of the 2nd edition of the OIV Descriptor List

Table 2-1 Example of works devoted to the development of methods for screening genetic resources.

Trait of interest	Reference
Anthocyanin content in grape leaves as physiological parametre	Steele et al. 2008
Canopy reflectance	Stamatiadis et al. 2006
Esca infection	Christen et al. 2007
Fruit shape	Wycislo et al. 2008
Limestone tolerance	Ollat et al. 2003
Mildew resistance	Liu et al. 2003
Morphology: Leaf graphic reconstruction	Martinez and Grenan 1999
EFA/ANN	Mancuso 1999
Phenology and starch accumulation	Lebon et al. 2005
Pierce disease resistance	Fritschi et al. 2008
Plasmopara infection	Diez-Navajas et al. 2007
Resistance to anthracnose	Yun et al. 2007
Resistance to downy mildew	Kortekamp and Zyprian 2003 Valsesia et al. 2005
Resistance to powdery mildew	Peros et al 2006
Resistance to Phaeoacremonium and Phaeomoniella	Santos et al. 2005
Resistance to UV-C	Bonomelli et al. 2004
Salt resistance	Sinclair and Hoffmann 2003
Sap flow	Braun and Schmid 1999
Shoot architecture	Louarn et al. 2007
Stilbene content	Poutaraud et al. 2007
Stomatal opperturre	Jones et al. 2002
Tanins composition	Perret et al. 2003a Perret et al. 2003b
Water stress deficit	Pellegrino et al. 2005 Yamane et al. 2009
Water Transport	Silva-Contreras et al. 2008

for Grape Varieties and *Vitis* Species (OIV 2008) and is still underway (in particular within the European project Grapegen06; *http://www1.montpellier.inra.fr/grapegen06/*) in order to optimize the number of morphological descriptors. Other traits such as pollen or seed morphology, phenology have also been analyzed. The quality related traits, mostly berry composition, concerned flavonoids and also cell wall composition, melatonin and terpenes. Interestingly, because of climate evolution scenarios, many works have been performed on abiotic stresses in *V. vinifera,* mainly drought tolerance. Data collected from the INRA grape collection at Vassal (Boursiquot et al. 1995) describe the important diversity for many traits in *V. vinifera,* with high ranges of variation for traits of interest such as berry weight, bunch weight, fertility or phenology (Table 2-3).

Figure 2-1 Illustration of morphological variation (A) Variation in berry shape and size in several *V. vinifera* cultivars (B) Illustration of morphological variation in *Cissus*, *Tetrastigma* and *Vitis* species.

Color image of this figure appears in the color plate section at the end of the book.

Table 2-2 Traits showing variation among accessions of *Vitis* species, *V. vinifera* cultivars and rootstock varieties.

Trait type	Species	Trait of interest	# accessions/ cultivars	Reference
Agronomy	Both	Phenology	6 cvs	Da Silva et al. 2008
	V. vinifera	Phenology	4 cvs	Barbeau et al. 2004
	V. vinifera	Yield, fruit composition, and vegetative growth	23 cvs	Shellie 2007
	Vitis sp.	Phenology, berry and bunch weight, berry composition, wine production	56 accessions of 12 *Vitis* species	Wan et al. 2008
Morphology	*V. v. sylvestris*	Ampelography	3 wild pop. 114 accessions	Cunha et al. 2007a Soylemezoglu et al. 2001
	V. vinifera	Ampelography	9 cvs 15 cvs 10 cvs (35 descr.) 16 cvs (18 descr.) 20 cvs 621 accessions	Asensio et al. 2002 Martinez et al. 2003 Muganu et al. 2009 Zdunic et al. 2008 Santiago et al. 2005 Ortiz et al. 2004
		Phyllometry	13 cvs	Tomazic and Korosec-Koruza 2003
		Pollen morphology	8 cvs 27 cvs	Maria et al. 1994 Roytchev et al. 1994
	Both	Seed morphology	5 *Vitis*, 92 cvs and wild *sylvestris*	Rivera et al. 2007 Nakagawa et al. 1991
	Vitis rupestris	Ampelography	114 acc. *in situ*	Pavek et al. 2003
	Vitaceae	Shoot Architecture	13 + 21 genera	Gerrath and Posluszny 2007

Table 2-2 contd....

Table 2-2 contd....

Trait type	Species	Trait of interest	# accessions/ cultivars	Reference
Berry composition	Both	Monoterpenes	2 cvs	Duchêne et al. 2009
		Anthocyanin	110 cvs	Liang et al. 2008
	V. vinifera	Anthocyanin	4 cvs 12 cvs 17 cvs 50 cvs 5 cvs	Romero-Cascales et al. 2005 Gomez-Alonso et al. 2007 Kallithraka et al. 2005 Pomar et al. 2005 Guerrero et al. 2009
		extractable skin and seeds tanins	7 cvs	Mattivi et al. 2009
		Cell wall composition	4 cvs	Ortega-Regules et al. 2008
		Flavonols	91 cvs 7 cvs	Mattivi et al. 2006 Castillo-Munoz et al. 2007
		Melatonin content	8 cvs	Iriti et al. 2006
	Vitis ssp.	Non-anthocyanin phenols	10 cvs	Montealegre et al. 2006
		Skin cracking	13 cvs	Yamamoto and Satoh 1994
		Tanin composition	1300 wines 5 cvs	Harbertson et al. 2008
		Flavonoids	22 species	Moore and Giannasi 1994
	Muscadinia	Berrry protein content	28 cvs	Mazhar et al. 2002
		Polyphenols	10 cvs	Pastrana-Bonilla et al. 2003
Biotic resistance	Both	Botrytis resistance	8 cvs	Jung et al. 2008
		Crown Gall resistance	17 cvs	Mahmoodzadeh et al. 2004
		Phylloxera resistance	6 cvs	Kellow et al. 2002
		Powdery mildew	12 cvs	Boso and Kassemeyer 2008
		Powdery mildew	79 cvs	Gee et al. 2008
	Rootstocks	Armillaria root disease resistance	8 cvs	Baumgartner and Rizzo 2006
		PD resistance/ adaptation to env.	10 cvs	Lu et al. 2008
	V. sylvestris	Leaf morphology and Phyloxera resistance	15 accs	This et al. 2001
	Vitis sp.	Anthracnose	56 acc. of 13 wild Asian *Vitis* Sp.	Li et al. 2008
		Botrytis resistance	32 accessions *Vitis* species	Cadle-Davidson 2008
		Pierce disease	*Muscadine* and 10 species (17 accs)	Fritschi et al. 2007
		Powdery and downy mildew	66 accs of 13 *Vitis* species	Wan et al. 2007

Trait type	Species	Trait of interest	# accessions/ cultivars	Reference
Abiotic resistance	*V.v. sylvestris*	Mildew tolerance, salt and lime tolerance	26 wild populations	Ocete et al. 2008
	V. vinifera	dynamics of soluble solids, water in berries	12 cvs	Sadras et al. 2008
		Fe deficiency	4 cvs	Ksouri et al. 2007
		G X E interaction	2 cvs /2 rootstocks 5 cvs	Stefanini et al. 2007 Storchi et al. 2007
		drought tolerance	2 cvs 2 cvs	Schultz 2003 Medrano et al. 2003
	Both	Freezing tolerance	102 cvs	Korbuly et al. 2004
	Rootstocks	Salt tolerance	7 rootstock cvs	Walker et al. 2002, 2004, 2007
			4 cvs, 7 rootstocks	Agaoglu et al. 2004
	Muscadinia	vigor, resistance to diseases, yield, and fruit quality	20 cvs	Stringer et al. 2008

Similarly, the *Vitis* species show considerable diversity in habitat and in morphology (Fig. 2-1). These species have extensively been used for breeding (This et al. 2006a) since the phylloxera crisis (Pouget 1990). They represent a big reservoir for biotic stress tolerances to many diseases and pests (mildew, Botrytis, Phylloxera, etc.) and to abiotic stresses (Table 2-2).

One interesting question, however, arises because of the heterozygous nature of grape: whether natural variation conserved for grape is more important than what could be obtained from artificial simple crosses or complex genealogies (Perterlunger et al. 2003; Ebadi et al. 2009). Even if QTL analysis have been performed for a long period of time, it is still difficult to answer this question since no report has compared diversity in natural and artificial populations. Comparison of published works is not easy because of the differences in methods and/or traits analyzed and because of the very large effect of environment on these traits (Bureau et al. 2000; Jones and Davis 2000; Tesic et al. 2002; Duchene et al. 2003; Cortell et al. 2005; Ortega-Regules et al. 2006). Nevertheless, since the collections are mostly based on cultivars selected for cultivation, the human selection may have captured only part of the diversity that can be created by crosses. Use of the knowledge of grape diversity and in particular the genetic distance between cultivars may help to select the best parents in order to increase the morphological diversity available today for genomic works. Nevertheless, due to the long period of juvenility in grape, collections will be useful more rapidly for genomic works.

Table 2-3 Quantitative variation of *V. vinifera* morphological traits in INRA's vassal grapevine collection: Data on 6,401 accessions (according to Boursiquot et al. 1995).

Traits[1]	Units	Level of variation		
Utilization of grape	%	Wine: 56.7	Table: 33.9	Wine-Table: 9.5
OIV 151 Flower: sexual organs	%	Hermaphrodite: 92.1	Female: 7.8	Male: 0.2
OIV 225 Berry: color of skin	%	Blue black: 36.9 Dark red violet: 1.5	Gray: 1.2 Red: 4.3 Rose: 6.3	Green-yellow: 49.7
OIV 231 Berry: coloration of flesh	Number of pulp colored genotypes	49		
OIV 241 Berry: formation of seeds	Number of seedless genotypes	50		
OIV 301 Time of bud burst	Day/Chasselas = 21 March	Min: - 12	Max: + 26	Mean: + 4.3
OIV 304 Time of physiological maturity	Week/Chasselas = 14 August	Min: - 2.7	Max: + 7.7	Mean: + 3.1
OIV 006 Shoot attitude (before tying)	%	Erect: 22.3 Semi-erect: 12.3	Horizontal: 53.5	Drooping: 4.6 Semi-drooping: 7.3
OIV 155 Shoot: fertility of basal buds	Nb of inflorescence for 10 buds after pruning	Min: 1	Max: 35	Mean: 9.08
OIV 503 Berry: weight	g	Min: 0.5	Max: 11.5	Mean: 3.1
OIV 502 Bunch: weight	g	Min: 50	Max: 1750	Mean: 315.5

[1]According to the 2nd Edition of the OIV Descriptor List for Grape Varieties and *Vitis* Species (OIV 2008).

2.3 Genotype-based Diversity Analysis

Correlated to the phenotypic variation, a high degree of molecular diversity has been revealed in grape using several molecular markers and by the analysis of both chloroplast and nuclear DNA. Since these two types of DNA have very different rates and mechanisms of evolution, we will present them separately.

2.3.1 Chlorotype Genetic Diversity in Grapevine

Chloroplast genetic diversity has been used in phylogenetic and phylogeographic studies (Olmstead and Palmer 1994) as well as to elucidate the relative contribution of seed and pollen flow to the genetic structure of natural populations (Provan et al. 2001). Their use is based on specific features of the chloroplast genome as well as its conserved gene order and coding sequences in different species and its general lack of heteroplasmy and recombination. Furthermore, chloroplasts are uniparentally transmitted in most species (usually maternal in angiosperms and paternal in gymnosperms).

The low mutations rates observed in the chloroplast genome represent a drawback to their wide application in the study of population history and dynamics within a given species. However, this problem has been overcome by the identification of variable intergenic regions and introns flanked by conserved sequences in many species as well as by the identification of chloroplast microsatellites which consist of mononucleotide repeats. Chloroplast microsatellites have been found in all plant species analyzed and they frequently are highly polymorphic (Provan et al. 2001). One problem associated with chloroplast microsatellites is their high homoplasy due to the recurrent generation of alleles of the same length, that creates alleles which being identical by state are not identical by descent. High levels of homoplasy can confound estimates of population differentiation and the recurrent generation of alleles could mimic gene flow (Goldstein and Pollock 1997). The risk is however reduced in intraspecific analysis (Arnold et al. 2002).

As in other angiosperms, grapevine chloroplasts are maternally inherited (Strefeler et al. 1992; Arroyo García et al. 2002) and therefore transmitted through seeds and cuttings. The chloroplast genome of grape is 160,928 bp in length and its gene content and gene order are identical to many other unarranged angiosperm chloroplast genomes (Jansen et al. 2006). Genetic diversity at the grape chloroplast has so far only been analyzed at the level of chloroplast microsatellite loci. Polymorphisms were searched for with 54 chloroplast microsatellite markers corresponding to 34 different loci in sample sets of four *Vitis* species (*Vitis berlandieri* Planchon, *V. riparia* Mich.,

V. rupestris Scheele and *V. vinifera* L.), using primer pairs developed for tobacco (Bryan et al. 1999; Weising and Gardner 1999; Chung and Staub 2003) and *Arabidopsis* (Provan 2000) by Arroyo-García et al. (2006). Nine loci were initially found polymorphic due to differences in the number of mononucleotide repeats in poly T/A stretches (Arroyo-García et al. 2006), which after comparison with the complete chloroplast genome sequence (Jansen et al. 2006) corresponded to five different loci: cpSSR3 (equivalent to NTCP-8), cpSSR5 (equivalent to NTCP-12 and ccSSR5), cpSSR10 (equivalent to ccSSR14), ccSSR9 and ccSSR23. These loci were genotyped in a sample of more than 1,200 genotypes of *V. vinifera* which uncovered the presence of two to three alleles per polymorphic locus and a total of eight chlorotypes. Among them, only four (A, B, C and D) had global frequencies greater than 5% (Table 2-4). Chlorotype diversity is moderate in grapevine with *h* values reaching 0.44 in the most diverse populations or cultivars groups that contrast with average *h* values of 0.55 reported in *Arabidopsis* (Picó et al. 2008) or *h* values higher than 0.95 observed in *Pinus sylvestris* (Provan et al. 1998).

Table 2-4 Grapevine chlorotypes.

Chlorotype	Chloroplast microsatellite loci				
	cpSSR3*	cpSSR5*	cpSSR10*	ccSSR9*	ccSSR23
A	106	105	114	166	280
B	106	105	115	165	281
C	106	105	116	165	282
D	107	104	115	165	281
E	107	104	116	165	281
F	107	105	115	165	281
G	106	105	114	165	280
H	106	105	115	166	280

The eight chlorotypes can be identified using only four loci (for example, those marked with*).

Although the small number of effective chlorotypes identified limits their use in population gene flow studies, their distributions in populations of *V. vinifera* collected in the wild and in cultivated plants from different regions and uses display some characteristics that can help in following the maternal origin of cultivars. In this way, chlorotype A which has been mainly found in western and central European wild populations is absent in Asian populations; whereas chlorotype C is more frequent in eastern Eurasian populations (Arroyo-Garcia et al. 2006). Chlorotype D is detected in eastern and central Mediterranean populations and chlorotype B seems to be present at low frequency in all regions. Similarly, the distribution of chlorotypes in cultivated varieties also follows patterns suggesting the participation of both western and eastern native germplasm in the origin of current cultivars as well as a relevant dissemination of cultivars following a predominant east to west direction (Arroyo-García et al. 2006). In this way chlorotype A which is very frequent in western European wine cultivars, whereas chlorotype C is more characteristic of eastern cultivars and more frequent among table grapes (Arroyo-García et al. 2006). This pattern of distribution has also been confirmed in northern African cultivated and wild samples where table grape cultivars frequently bear chlorotype C whereas wild populations bear chlorotype A (Snoussi et al. 2004; Laiadi et al. 2009). Parallel analyses of chlorotype distribution have also been performed in additional cultivated and wild samples along the Mediterranean basin although only based on two polymorphic chloroplast microsatellites. Although chlorotype names and alleles are different due to different analytical methods, the results follow the same trend. Only six chlorotypes were detected in samples collected all along the Mediterranean, with chlorotype VI (equivalent to Chlorotype A) being very frequent in the Iberian Peninsula (Imazio et al. 2006). Furthermore, as described for chlorotype A, chlorotype VI is neither detected in cultivated nor in wild samples from Iran where chlorotypes I and III (equivalent to D and C) are the most represented (Baneh et al. 2007).

Therefore, in spite of the low genetic diversity observed in grape chlorotypes, they can be useful tools to help unravel the origin of specific cultivars and their dissemination patterns. Future analyses will help understanding chloroplast nucleotide diversity at loci with lower homoplasy which can be very important in both confirming these data and providing information on phylogenetic relationships among different species in the genus *Vitis*.

2.3.2 Nuclear Diversity in Grapevine

The grape genome has been sequenced recently (Jaillon et al. 2007; Velasco et al. 2007). It is 467.5 Mbp long. The 30,434 genes represent about 43% of DNA but their coding sequences only 6.3% (33.6 Mbp) and the repeated sequences cover 185 Mbp (Jaillon et al. 2007). In the past decade, due to an important international effort, our understanding of grape nuclear diversity really improved, even if this vision is still limited because most of the works were performed on small samples and with different markers. In terms of marker evolution, it is however important to differentiate the analyses done with neutral markers and the analyses of gene allelic diversity. In this regard, random amplified polymorphic DNA (RAPD), amplified fragment length polymorphism (AFLP), cleaved amplified polymorphic sequence (CAPS) and simple sequence repeat (SSR) markers, which have been applied to grape diversity (for a recent review, see This et al. 2006b), should be considered as non-coding and thus mostly unselected. On the other hand, restriction fragment length polymorphism (RFLP) and single nucleotide polymorphisms (SNPs) have been developed for both coding or non coding gene regions and can be more useful as functional markers. This selected distinction versus neutral DNA is however questionable since the data on linkage disequilibrium in plants tends to show that big regions of the genomes (including SSR markers) may be under selection because of "hitchhiking" with selected genes (Palaisa et al. 2003, 2004).

With few exceptions such as the use of retrotransposon based markers (Labra et al. 2004; Pelsy 2007; Moisy et al. 2008; Cadle-Davidson and Owens 2008), AFLP (Martinez et al. 2003; Ergül et al. 2006; Baneh et al. 2007; Upadhyay et al. 2007), RAPD (Ulanovsky et al. 2002; Ercisli et al. 2008; Karatas et Agaoglu 2008) or isozymes (Ortiz et al. 2004), most of the recent works were either performed using SSR markers or SNPs.

2.3.2.1. SSR Diversity in Grapevine

The main arguments for the use of SSR markers were their availability, their ease of technical automation and the possibilities to compare data between laboratories.While the first two characteristics of SSR markers were promising, the comparison of data between laboratories was not as simple as initially presumed requiring coding of data according to reference cultivars prior to comparisons (This et al. 2004).

SSR markers are numerous; besides the first set of markers developed in the 1990s corresponding mainly to dinucleotide (Thomas et al. 1994; Bowers et al. 1996, 1999a; Sefc et al. 1999), additional dinucleotide markers have been developed (Lefort et al. 2002; Merdinoglu et al. 2005; Di Gaspero et al. 2005; Goto-Yamamoto et al. 2006) as well as highly polymorphic markers (Lefort et al. 2003), many of them through the *Vitis* Microsatellite Consortium launched by C.P Meredith, under the coordination of Agrogene (Moissy, France). New markers developed from expressed sequence tags (ESTs) (Scott et al. 2000; Decroocq et al. 2003) showed great promise and seemed more transferable across species than non-coding ones. Finally, markers based on long (tri- to penta-) nucleotides were also recently developed in order to simplify the identification of cultivars. They displayed a similar level of diversity but are easier to score (Cipriani et al. 2008). The total number of markers developed so far is high, but it still represents a small proportion of the microsatellites in the grape genome. A total of 239,634 microsatellite elements were identified in PN40024 (Jaillon et al. 2007), including 26,962 perfect microsatellite SSR markers containing tri-, tetra-, and penta-nucleotide repeats (Cipriani et al. 2008) and 88,909 SSR markers were identified on the ENTAV 115 Pinot sequence (Velasco et al. 2007).

Once developed, the markers can indeed be automatized and most of the works are now done on automatic sequencers (Table 2-5). Another step was recently proposed—high resolution amplicon melting analysis (MacKay et al. 2008)—in order to reduce the analysis time. Besides this automated final step, all the steps in the analysis process can also be optimized (Laucou et al. 2010) to enable the analysis of thousands of accessions (LeCunff et al. 2008).

Analysis of recent grape literature including a few older analyses revealed that cultivars from most of the regions of grape cultivation (western and eastern Europe, Middle and Far East, America) have been analyzed (Table 2-5). The analyses are usually performed with up to 20 SSR markers for diversity studies (Mean number = 9.2 for the works listed in Table 2-5).

Even if up to 5,550 accessions were analyzed, the data are very difficult to compare because identity and number of accessions and marker loci greatly differ between the studies. Nevertheless, even if no international effort on marker homogenization has been developed—the race for new methodologies will probably not solve this problem—many works have been performed with a common set of markers (Table 2-5) composed of VVMD 5, 7, 27 (Bowers et al. 1996), VVS 2 (Thomas et al. 1994), VrZag 62, 79 (Sefc et al. 1999), for which a partial effort of homogenization was conducted (This et al. 2004). Many works were based on the VVMD (Bowers et al. 1996, 1999a) and the VrZag series (Sefc et al. 1999) along with VVMC and VVI (Merdinoglu et al. 2005) series.

The nuclear SSR diversity revealed for *V. vinifera* is high: in a different analysis, the mean number of alleles varies from 8 to 11 (Table 2-5) for sample sizes from 50 to 316 individuals and gene diversity is comprised between 0.6 and 0.85, averaging 0.767. A work on Turkish cultivars revealed up to 14.66 alleles per primers on a small number of cultivars (Karatas and Agaoglu 2008) but the origin of the cultivars may explain the difference in this work. Finally, a mean number of 16.3 alleles per locus and a gene diversity of 0.76 was revealed using 20 SSRs on 2,262 *V. vinifera* accessions (Le Cunff et al. 2008), the largest sample of cultivars analyzed so far in a single work, with a variation ranging from 6 to 36 alleles per locus (Laucou et al. 2010).

This diversity is comparable or slightly lower to the one observed for Arabidopsis (Picó et al. 2008) for natural populations in the Iberian Peninsula (18.7 alleles/locus, gene diversity = 0.82) or a set of selected accession of wordlwide origin (14.4 alleles/locus, gene diversity = 0.83), for wild populations of wild rice in China (gene diversity = 0.860, Gao et al. 2006) and for maize (14.8 alleles/locus, gene diversity = 0.79, Vigouroux et al. 2005). It is nevertheless higher than that for wheat (gene diversity = 0.609, Peng et al. 2009; gene diversity = 0.650, Roussel et al. 2005) or for tomato (gene diversity = 0.496, Ranc et al. 2008).

Table 2-5 SSR analysis of *Vitis* accessions.

Region Taxa	Obj.[1]	# locus	Common set [2]	Other locus	Cultivars (acc.[3])	Method[4]	Na[5]	Mean[6] (Min Max)	He[7]	Ho[8]	Article
Algeria	I	12	Y	VVMD 21, 24, 25, 28, 31, 32	31 (27)	A	95	7.92	0.80	0.80	Laiadi et al. 2009
Carpatians	P, I	12	Y	Scu08vv VVMD 21, 25, 28, 31, 36 VrZag 47, 62, 79, 89, 112	115	A	104	8.60 (4–17)	0.79	0.73	Galbacs et al. 2009
Diverse	S	8	Except VrZag 62, 79	VVMD 6, 28, 31, 32	266 (366)	M	94	11.75 (5–19)	0.77	0.76	Aradhya et al. 2003
		4	VVMD 5, 7	VVMD 6, 8	77	M	30	7.50 (5–11)	-	-	Bowers et al. 1996
	I	10	Y	VVMD 28 VVS 29 , VrZAg 67, 83, 112	22 cvs + 4 species	A		9.22 (5–12)	0.82	0.79	Fernandez et al. 2008
	P, I, S	20	Except VrZag 62 79	VVMD 21, 24, 25, 27, 28, 32 VMC 1b11, 4f3, VVI B01, H54, N16, N73, P31, P60, Q52, V37, V67	2323 (4425)	A	338	16.9 (6–36)	0.77	0.76	Laucou et al. 2010 LeCunff et al. 2008
	I	14	Except VVMD 5, 7, 27	VVMD 30 , VVS 29 VrZAG 25, 67, 79, 93, Uch 12, 29 VMC 1e11, 2a12, 3a9, 5g7, 8a7	58	A	129	8.60 (1–13)	0.70	0.66	Pelsy 2007
	S	9	Y	VrZag 21, 47, 64	160	-	88	9.80 (4–13)		0.82	Sefc et al. 2000
Greece	I	11	Y	VVMD 6 VrZag 21, 47, 64, 67, 83	50	-	87	7.90			Lefort and Roubelaskis 2002
Israel	I	6	Y		25	A	47	7.80 (6–11)	-	-	Klein et al. 2008

Region Taxa	Obj.[1]	# locus	Common set [2]	Other locus	Cultivars (acc.[3])	Method[4]	Na[5]	Mean[6] (Min Max)	He[7]	Ho[8]	Article
Italy	P, I, S	12	Y	VVMD 25, 28, 36 VVS 5, VrZag 67, 64	30 (32)	A	104	8.67 (7–12)	0.79	0.83	Boccacci et al. 2005
Italy (Tuscany)	P, S	20	Except VrZag 62, 79	VVMD21, 24, 25, 27, 28, 32 VMC 1b11, 4f3, VVI B01, H54, N16, N73, P31, P60, Q52, V37, V67	78	A	-	-	0.77	0.76	Di Vecchi Staraz et al. 2007
Italy	P, I	30	Y	VVMD 8, 17,25, 36, UDV 27 109 VrZag 21, 47, 67, 83, 93, 112 VMC 2A3, 2C7, 5C5, 2G2 VVI P10, Q61, P77, S70, U37, V16, V37	28	A	137	4.60 (2–10)	0.66	0.63	Salmaso et al. 2008
Portugal	I	6	Y		51	A	49	8.17 (7–11)	-	-	Almadanim et al. 2007
Portugal	I	11	Y	VVMD 6 VrZag 21, 47, 64, 67, 83	46	A	94	8.55 (5–11)	-	-	Lopes et al. 2006
Portugal	I	11	Y	VVMD 6 VrZag 21, 47, 64, 67, 83	49	A	77	7.00 (5–9)		0.76	Lopes et al. 1999
Slovenia		21	Y	VVMD 8, 17, 24, 25, 31, 32, 36 VrZag 21, 47, 67, 83,112 Uch 11, 29	33	A	169	8.00	0.78	0.8	Stajner et al. 2008
South America	I, P	9	Y	VrZAG 21, 47, 64, 83	79	A	39	4.30 (3–5)		0.69	Tapia et al. 2007
Spain	I	6	Y		176 (318)	A	66	11.00 (9–13)	0.81	0.84	Martin et al. 2003
Spain	I	6	Y		56 (272)	A	53	8.33 (7–10)	-	0.83	Martin et al. 2006

Table 2-5 contd....

Table 2-5 contd....

Region Taxa	Obj.[1]	# locus	Common set [2]	Other locus	Cultivars (acc.[3])	Method[4]	Na[5]	Mean[6] (Min Max)	He[7]	Ho[8]	Article
Spain	I, P, S	6	Y		25	M	58	9.67 (6–13)	0.81	0.82	Martınez et al. 2006
Spain	I	12	Y	VVMD 21, 28, 36 VrZag 64, 67, 83	12	A	51	4.25 (3–5)	-	-	Ortiz et al. 2004
Spain	I, P, S	5	Except VVMD 27	VrZag 47	65	A	52	8.67 (7–10)	0.79	0.84	Santana et al. 2008
Spain (La mancha)	I	12	Y	VVMD 21 ,28, 36 VrZag 64, 67, 83	39 (73)	A	93	8.00 (4–10)	0.78	0.84	Fernadez-Gonzales et al. 2007
Table grape	I	8	Except VrZag 62 79	VVMD 6 VVS 1, 3, 4, 5	43	A	48	6.00 (2–8)		0.65	Sanchez-Escribano et al. 1999
Tunisia	I, P, S	10	Y VrZag 62 79	VVMD 21, 24, 27, 28, 32 VVI P31, P60	61	A	84	8.4 (6–11)	0.77	0.85	Zoghlami et al. 2009
Turkey	I	6	Y		16 (39)	A	88	14.66 (10–21)	0.85	0.93	Karatas et al. 2007
V.v sylvestris (France)	S		Y	VVMD 6, 21, 25	101	A	65	6.50		0.58	Arrigo and Arnold 2007
V.v sylvestris	S	6	Y		310	A	-	(11–19)	0.62	0.67	Grassi et al. 2008
V.v sylvestris (Italy)	S	6	Y		418	A	-		-	-	Grassi et al. 2006
V.v sylvestris (Tunisia)	S	9	Y	VrZag 21, 47, 64, 83	109	M	66	7.33 (9–11)		0.72	Snoussi et al. 2004
Ukrenia (both)	P, I, S,	6	Y		47 cvs + 37 hyb	A	110	8.09	0.85		Heuertz et al. 2008
Rootstocks	I		Y	VVS 1, 5, 29 VrZag 47	92 (208)	A	154	17.10 (9–23)	0.84	0.78	de Andrès et al. 2007

Region Taxa	Obj.[1]	# locus	Common set [2]	Other locus	Cultivars (acc.[3])	Method[4]	Na[5]	Mean[6] (Min Max)	He[7]	Ho[8]	Article
Rootstocks	I	7	Except VrZag 62 79 VVMD27	VVMD 6, 28, 31 VVS 29	58		123	17.6 (6–23)	-	-	Lin and Walker 1998
Rootstocks	I	10	Except VrZag 62 79, VVMD27	VVMD 28, 32, 36 VVS 1, 3, 4, 29	19	A	-	-		0.82	Sefc et al. 1998a
Rootstocks	I	7 7	VVMD 5, 7,	VVMD 3, 4, 7, 17, 24, 31	21	-	54	8.00 (5–12)		0.77	Upadhyay et al. 2007
V. rupestris populations	S	4	VVMD 7, VVS 2	VVMD 6, VVS 4	113	A	47	11.75 (6–16)		0.5*	Pavek et al. 2003
Vitis sp.	I	5	VVMD 7, VVS 2	VVMD 6, 8 VVS 4	110	A	122	21.00 (16–38)		0.63	Lamboy and Alpha 1998
Vitis sp.	I	14	Except VVMD 5, 7, 27	VVMD 30, VVS 29 , Uch 12, 29 VrZAG 25, 67, 79, 93 VMC 1e11, 2a12, 3a9, 5g7, 8a7	94	A	322	21.50 (3–34)	0.81	0.62	Pelsy 2007

1: Objectives are: I: Identification, P: parentage relationships, S: analysis of grape structure.
2: Common set is composed of VVMD 5, 7, 27 (Bowers et al. 1996), VVS 2 (Thomas et al. 1994), VrZag 62, 79 (Sefc et al. 1999) Y = Yes.
3: Number of accessions when indicated.
4: A = Automatic sequencer, M = Manual sequencing.
5: Number of alleles.
6: Mean number of alleles per locus.
7: Expeted heterozygosity.
8: Observed heterozygosity.

This diversity may be an underestimation since more than two alleles were observed for several SSRs and were confirmed by sequencing (Martinez et al. 2006; Salmaso et al. 2008) and SNP variation in flanking regions was reported (Arnold et al. 2002; Fernandez et al. 2008).

Several published data indicated good transferability of SSR markers within the genus *Vitis* (Thomas et al. 1994; Lamboy and Alpha 1998; Lin and Walker 1998; Sefc et al. 1998a, 1999) and to a lesser extend to Vitaceaes (Rossetto et al. 2001, 2002a). However, the sequencing of some microsatellite loci suggested that the polymorphism did not correspond only to a variation in the number of repeats (Sefc et al. 1999; Di Gaspero et al. 2000). The repeat motifs showed major changes in their architecture and the flanking regions exhibited long indels and base substitutions (Arnold et al. 2002; Rosetto et al. 2002b; Fernandez et al. 2008, Peros et al. 2010). As a consequence, nuclear polymorphism was underestimated using only the size of electromorphs and size homoplasy—identity of size but not of descend—which might complicate species delimitation.

Nevertheless, the diversity revealed by SSRs is even higher for *Vitis* species and/or hybrids (mean number of alleles > 17, gene diversity = 0.83, Table 2-5), and is probably even higher considering homoplasy of microsatellites but also the diversity in sequences of microsatellite flanking regions already mentioned.

The grape genome is highly heterozygous: in *V. vinifera*, observed heterozygosity (*Ho*) ranged from 70 to 93% according to the studies while it is lower for *V.v sylvestris* (Ho = 0.65) and for the other *Vitis* species (Ho = 0. 68, Table 2-5). This value nevertheless varies according to the cultivars (from 20 to 100%, Laucou et al. 2010) the nearly homozygous grape accessions PN40024 with a residual heterozygosity of 7% being probably among the most homozygous accessions available for functional genomics, besides the other lines obtained from Bronner and Oliviera (1991). The discovery of spontaneous short-life variants (Boss and Thomas 2002) will certainly help create near homozygous lines useful for genomic works (Chaïb et al. 2010).

2.3.2.2 *SNP Diversity in Grapevine*

Diversity in genes because this polymorphism can be directly associated with variation at the phenotype level (Lijavetzky et al. 2006; This et al. 2007), has also received much attention. The number of SNPs revealed on ENTAV 115 Pinot genome is very high (1,751,176 SNP, Velasco et al. 2007) and since the identification and analysis of SNPs can easily be automated (Pindo et al. 2008; Troggio et al. 2008) the analysis of SNP diversity should become commonplace for grape. Before the publication of the grape sequence, diversity by resequencing DNA fragments in grape cultivars has been performed on nine accessions for 25 genes (Salmaso et al. 2004), on 11 accessions for 230 genes (Lijavetzky et al. 2007) and on 91 accessions for four genes (Le Cunff et al. 2008). The SNP diversity in *V. vinifera* is high since diversity of one SNP per 129 bases (Salmaso et al. 2004), one SNP every 64 bp (Lijavetzky et al. 2007) up to one SNP per 49 bases (Le Cunff et al. 2008) were revealed. This diversity is comparable to that of maize presenting SNP frequencies corresponding to one SNP every 60 (Ching et al. 2002) to 104 bp (Tenaillon et al. 2001) but more important than for tomato (one SNP every 139 bp, Labate et al. 2009) and much more important that for Arabidopsis (one SNP every 336 bp, Schmid et al. 2003) or wheat (one SNP every 540 bp, Somers et al. 2003).

Interestingly a high number of SNPs were observed in exons even if more diversity was revealed in non coding regions with 16% of non silent changes (Lijavetzky et al. 2007). Nucleotide diversity in grape ($\pi = 0.0051$) was found to be similar to values observed in highly polymorphic plant species such as maize. Diversity values (expected heterozygosity) for SNP are generally low due to their bi-allelic nature. In grapevine, SNP diversity values ranged from 0 to 0.66 with a mean value of 0.30 (Lijavetzky et al. 2007), which is slightly higher than the mean value reported for maize (0.26, Ching et al. 2002). Consistently, nucleotide diversity values observed in grapevine ($\theta = 0.0046$, $\pi = 0.0051$) were similar to those observed in maize ($\theta = 0.0096$, Tenaillon et al. 2001; $\pi = 0.0063$, Ching et al. 2002) and ~ 5-fold higher than those reported for soybean ($\theta = 0.00097$, $\pi = 0.0012$, Zhu et al. 2003), tomato ($\theta = 0.0015$, $\pi = 0.001$, Labate et al. 2009) or humans ($\theta = 0.0008$, Halushka et al. 1999).

The diversity present in grape is even higher since Indels have also been observed with a relatively high frequency: 52 Indels identified by Lijavetzky et al. (2007) while only two Indels were identified by Salmaso et al. (2004).

When considering haplotypes instead of single SNPs, an average of approximately 6.04 different haplotypes per gene were recorded. The most common situation was the presence of a major haplotype (average frequency = 0.49), with the average cumulative frequency of the first three haplotypes being 0.83, followed by a series of minor haplotypes (Lijavetzky et al. 2007).

As for SSR markers, SNP diversity is higher when considering interspecific variation: the number of SNPs varies from one SNP every 78 bp when considering *Vitis* accessions to one SNP every 129 bp when only considering *V. vinifera* (Salmaso et al. 2004).

The data on SNPs in grape increased tremendously with the grape genome sequencing: a total of 1,751,176 SNPs were disovered between both haplotypes of Pinot Noir, clone ENTAV 115 and over a millions in/dels were annotated (Velasco et al. 2007). Additional SNPs were also revealed by resequencing 17 accessions of *Vitis* ssp. and *V. vinifera* accessions (Myles et al. 2010). Portions of the genome not represented in the current sequence (Jaillon et al. 2007) or long polymorphic fragments (inserted or absent) between accessions were also observed. The number of SNPs transferable from the Pinot sequence decreases when genetic distance between the accessions and Pinot N. increases (Vezzulli et al. 2008).

Grapevine SNP show lower diversity values than SSR and therefore are less informative markers (average polymorphism information content —PIC—for SNPs is 0.25 (Lijavetzky et al. 2007) as compared to 0.70 for microsatellite. This potential drawback of SNP can be overcome either by using larger sets of markers or by considering haplotypes structure for each locus in place of single SNPs. When haplotypes are considered for each locus, the genetic diversity value rises more than 2-fold (0.64, Lijavetzky et al. 2007) reaching similar values as those reported for grapevine microsatellites. Furthermore, the problem of homoplasy may be overcome with SNP markers, while multicopy family analysis may become a problem.

2.4 Pattern of Genetic Structure

Molecular markers have been of great value for identification purposes since they show very high identification power (PI) up to $7.4\ 10^{-26}$ (Table 2-5). They have also been very useful for the analysis of phylogeny in *Vitis* genus and of the pattern of genetic structure.

2.4.1. Phylogenic Analysis of Vitis Species

Despite the wide variation in morphology, the subgenus *Vitis* was suspected to include genetically close ecospecies (Levadoux et al. 1962; Mullins et al. 1992) and studies with nuclear markers indeed suggested the absence of a clear cut separation between American and Asian species (Di Gaspero et al. 2000; Pelsy 2007). Low genetic differentiation within *Vitis* was also apparent using sequences of chloroplastic (Rossetto et al. 2001; Ingrouille et al. 2002; Rossetto et al. 2002b; Soejima and Wen 2006) and nuclear DNA (Rossetto et al. 2002a; Wen et al. 2007) but these studies compared a limited number of species. In a recent study (Péros et al. 2010), we obtained a large sample of species for a first chloroplast phylogeny of the genus *Vitis*. These results suggest that the genus first diverged in Asia and then dispersed in both Europe and North America. Two dispersal events probably occurred in North America since *V. californica* had a derived chlorotype different from those observed in all other American species. In the same work, the nuclear microsatellite data partly confirmed the cpDNA data but did not enable us to differentiate certain American species from the Asian pool. Based on nuclear SSR, several differentiated groups of species were found in North America, whereas Asian species formed a more consistent group. The lack of reproductive barriers in the *Vitis* subgenus, the low level of polymorphism observed in the cpDNA and the incomplete lineage sorting in the nuclear genome support the hypothesis of a recent divergence in this subgenus.

2.4.2 Cultivated versus Wild Compartment of Grape

Even if comparisons of genetic diversity in cultivated and wild compartment of grape are still limited, the picture arising today is of a low but clear differentiation of cultivars and wild grape based either on cp-markers (Arroyo-Garcia et al. 2006; Grassi et al. 2006), nSSR (Snoussi et al. 2004; Grassi et al. 2008) or both (Grassi et al. 2003b; Sefc et al. 2003). The wild individuals also cluster according to their populations (Grassi et al. 2008). The histories of both compartments are also different and as a consequence linkage disequilibrium is more important in cultivated grape (Barnaud et al. 2006) than in wild individuals (Barnaud et al. 2010). Nevertheless, a low level of pollen mediated gene flow between both compartments was recorded in two locations in France (Di Vecchi-Staraz et al. 2009) and is more than likely to happen elsewhere having strong consequences on the differentiation of the compartments.

2.4.3 Structuration of the Diversity for *V. vinifera*

Even if a small part of the available *V. vinifera* diversity has been analyzed and compared so far, we nevertheless have some idea of the pattern of diversity structure in *V. vinifera*. Indeed, a clear differentiation of cultivars was observed according to their geographic origin (Sefc et al. 1999; Aradhya et al. 2003; Arroyo-Garcia et al. 2006; Di Vecchi-Staraz et al. 2007) or genetic differentiation between table versus wine cultivars (Heuertz et al. 2008; Le Cunff et al. 2008). Many parentage relationships have also been revealed in grape (Table 2-6). In this case, up to 60 markers were necessary since this latter analysis requires more information for accurate identification of parents but use of well dispersed SSR markers can increase the power of the analysis (Di Vecchi- Staraz et al. 2007; Boursiquot et al. 2009). These parentage relationships can also explain grape structuring (Di Vecchi-Staraz et al. 2007; Santana et al. 2008; Salmaso et al. 2008).

Table 2-6 Example of parentage relationships revealed for well known cultivars using SSR markers.

Cultivar name	SSR	cult	Article
American cultivars (Criollas)	9 + 25	79 acc 19 cvs	Tapia et al. 2007
Cabernet-Sauvignon	24	51	Bowers and Meredith 1997
Cornalin du Valais	50	400	Vouillamoz et al. 2003
Corrinto bianco = Pedro Ximenes seedless	25		Vargas et al. 2007
Diverse from Austria	24	52	Sefc et al. 1997
	32	52	Sefc et al. 1998b
Diverse parentage	15	24	Bautista et al. 2008
Fercal (Rootstock)	20	5 (2853)	Laucou et al. 2008
Malvasia familly	20	70 (2305)	Lacombe et al. 2007
Merlot, Cot, Carmenère	20 +35	2305	Boursiquot et al. 2009
Müller-Thurgau	6	5	Dettweiler et al. 2000
Muscat fleur d'oranger	9 27	-	Schneider et al. 2008
Parraleta/Graciano	6		Montaner et al. 2004
Petite Syrah = Durif	8 +17	6	Meredith et al. 1999
Pinot group	17 +15	300	Bowers et al. 1999b
	12 +34	800	Regner et al. 2000
Plavac Mali	13	12	Maletic et al. 2004
Raetica	60	151 (1700)	Voulliamoz et al. 2007b
Sangiovese	20 18	78 (2786)	Di Vecchi Staraz et al. 2007
	10 +40	180	Vouillamoz et al. 2007a
Sangiovese, Garganega	36	668	Crespan et al. 2008
Syrah	32	51	Bowers et al. 1998
	60	89	Vouillamoz and Grando 2006
Zinfandel	13	33	Calo et al. 2008

2.5 Concluding Remarks and Prospects

The natural diversity in the genus *Vitis* and more specifically for *V. vinifera* is important: grape is a diverse crop both at the morphological and molecular level, even if it is still underused. The genetic diversity, mostly recorded by SSR markers and only recently by SNP, is close to that found in maize or rice. The heterozygosity of grape is very high, thus rendering the analysis and use of this diversity more difficult. Nevertheless, it greatly varies from cultivar to cultivar (from 7 to 100%) and recent development may help to develop homozygous varietes for functional genomics. *V.v. sylvestris* as well as wild *Vitis* species tend to display less heterozygosity.

Finally, an interesting feature for future genomics works is the knowledge of long genealogies in grape (Bower et al. 1999b; Boursiquot et al. 2004), which might be useful for deciphering complex traits using the IBD method as already done for complex traits in animal sciences (Meuwissen et al. 2002).

Diversity in important genes linked to agronomical traits is still poorly documented in this species and may be of interest in helping the grape community to understand the functioning of grape genome, the use of naturally occuring variation being quite complementary to the use of mutants (either natural or induced) especially in the case of null or weak allele, lethality or redundancy (Alonso-Blanco and Koornneef 2000).

The exploitation of this source of variation will increase and become more systematic and efficient with the development of high-throughput automated marker technologies such as Illumina® marker sets (Gunderson et al. 2005). The naturally occurring diversity in grape could thus be used in association studies (see Chapter 4). The identification of selective sweeps along the genome as already performed for several plant species (Palaisa et al. 2003; Burke et al. 2005; Wright et al. 2005) and in *Drosophila* (Schlenke and Begun 2004) and humans (Sabeti et al. 2002; Akey et al. 2004) will also be possible by the use of naturally occuring variation. It will result in the identification of loci and alleles affecting important adaptive traits that have been selected during domestication and/or human selection of grape since domestication.

Finally, the knowledge of alleles affecting important traits and the availability of the genetic resources will also speed up the development of new cultivars tolerant or resistant to most diseases, adapted to the changing environment and of optimal quality.

References

Agaoglu YS, Ergul A, Aras S (2004) Molecular characterization of salt stress in grapevine cultivars (*V. vinifera* L.) and rootstocks. *Vitis* 43: 107–110.

Agati G, Meyer S, Matteini P, Cerovic ZG (2007) Assessement of anthocyanins in Grape (*V. vinifera* L.) berries using non invasive chlorophyll fluorescence methods. J Agri Food Chem 55: 1053–1061.

Akey JM, Eberle MA, Rieder MJ, Carlson CS, Shriver MD, Nickerson DA, Kruglyak L (2004) Population history and natural selection shape patterns of genetic variation in 132 genes. Plos Biol 2: 1591–1599.

Almadanim MC, Baleiras-Couto MM, Pereira HS, Carneiro LC, Fevereiro P, Eiras-Dias JE, Morais-Cecilio L, Viegas W, Veloso MM (2007) Genetic diversity of the grapevine (*V. vinifera* L.) cultivars most utilized for wine production in Portugal. Vitis 46: 116–119.

Alonso-Blanco C, Koornneef M (2000) Naturally occurring variation in Arabidopsis: an underexploited resource for plant genetics Trends Plant Sci 5: 22–29.

Aradhya MK, Dangl GS, Prins BH, Boursiquot JM, Walker MA, Meredith CP, Simon CJ (2003) Genetic structure and differentiation in cultivated grape, *Vitis vinifera* L. Genet Res 81: 179–192.

Arnold C, Rossetto M, Mcnally J, Henry RJ (2002) The application of SSRs characterized for grape (*V. vinifera*) to conservation studies in Vitaceae. Am J Bot 89: 22–28.

Arrigo N, Arnold C (2007) Naturalised *Vitis* rootstocks in Europe and consequences to native wild grapevine. PLoS ONE 2(6): e521.

Arroyo-Garcia R, Lefort F, De Andres MT, Ibanez J, Borrego J, Jouve N, Cabello F, Martinez-Zapater JM (2002) Chloroplast microsatellite polymorphisms in *Vitis* species. Genome 45: 1142–1149.

Arroyo-Garcia R, Ruiz-Garcia L, Bolling L, Ocete R, Lopez MA, Arnold C, Ergul A, Soylemezoglu G, Uzun HI, Cabello F, Ibanez J, Aradhya MK, Atanassov A, Atanassov I, Balint S, Cenis JL, Costantini L, Goris-Lavets S, Grando MS, Klein BY, Mcgovern PE, Merdinoglu D, Pejic I, Pelsy F, Primikirios N, Risovannaya V, Roubelakis-Angelakis KA, Snoussi H, Sotiri P, Tamhankar S, This P, Troshin L, Malpica JM, Lefort F, Martinez-Zapater JM (2006) Multiple origins of cultivated grapevine (*Vitis vinifera* L. ssp. *sativa*) based on chloroplast DNA polymorphisms. Mol Ecol 15: 3707–3714.

Asensio ML, E Valdes, F Cabello (2002) Characterisation of some Spanish white grapevine cultivars by morphology and amino acid analysis. Sci Hort 93: 289–299.

Baneh HD, Grassi F, Mohammadi A, Nazemieh A, De Mattia F, Imazio S, Labra M (2007) The use of AFLP and morphological markers to study Iranian grapevine germplasm to avoid genetic erosion. J Hort Sci Biotechnol 82: 745–752.

Barbeau G, Bournand S, Champenois R, Bouvet MH, Blin A, Cosneau M (2004) The behaviour of four red grapevine varieties of Val de Loire according to climatic variables. J Int Sci Vigne Vin 38: 35–40.

Barnaud A, Lacombe T, Doligez A (2006) Linkage disequilibrium in cultivated grapevine, *Vitis vinifera* L. Theor Appl Genet 112: 708–716.

Barnaud A, Laucou V, This P, Lacombe T, Doligez A (2010) Linkage disequilibrium in wild European grapevine, *Vitis vinifera* L. ssp. *silvestris.* Heredity 104: 431–437.

Baumgartner K, Rizzo DM (2006) Relative resistance of grapevine rootstocks to *Armillaria* root disease. Am J Enol Vitic 57: 408–414.

Bautista J, Dangl GS, Yang J, Reisch B, Stover E (2008) Use of genetic markers to assess pedigrees of grape cultivars and breeding program selections. Am J Enol Vitic 59: 248–254.

Boccacci P, Marinoni DT, Gambino G, Botta R, Schneider A (2005) Genetic characterization of endangered grape cultivars of Reggio Emilia province. Am J Enol Vitic 56: 411–416.

Bonomelli A, Mercier L, Franchel K, Baillieul F, Benizri E, Mauro MC (2004) Response of grapevine defenses to UV-C exposure. Am J Enol Vitic 55: 51–59.

Boso S, Kassemeyer HH (2008) Different susceptibility of European grapevine cultivars for downy mildew. Vitis 47: 39–49.

Boss PK, Davies CC, Robinson SP (1996) Anthocyanin composition and anthocyanin pathway gene expression in grapevine sports differing in berry colour. Aust J Grape Wine Res 2: 163–170.

Boss PK, Thomas MR (2002) Association of dwarfism and floral induction with a grape 'green revolution' mutation. Nature 416: 847–850.

Bouquet A, Boursiquot JM (1999) La sauvegarde des vieux cépages et la création de variétés nouvelles: une démarche conjointe pour concilier tradition et innovation en France. Bull OIV 825–826: 755–761.

Boursiquot JM, Dessup M, Rennes C (1995) Distribution of the Main Phenological, Agronomical and Technological Characters of *Vitis vinifera* L. Vitis 34: 31–35.

Boursiquot JM, This P (1996) Les nouvelles techniques utilisées en Ampélographie: informatique et marquage. J Int Sci Vigne Vin n° Hors Série, 13–23.

Boursiquot JM, Lacombe T, Bowers J, Meredith CP (2004) Le Gouais, un cépage clé du patrimoine viticole européen. Bull OIV 875–876: 5–19.

Boursiquot JM, Lacombe T, Laucou V, Julliard S, Perrin FX, Lanier N, Legrand D, Meredith CP, This P (2009) Parentage of Merlot and related winegrape cultivars of southwestern France: discovery of the missing link. Aust J Grape Wine Res 15: 144–155.

Bowers J, Dangl GS, Vignani R, Meredith CP (1996) Isolation and characterization of new polymorphic simple sequence repeat loci in grape (*Vitis vinifera* L.). Genome 39: 628–633.

Bowers JE, Meredith CP (1997) The parentage of a classic wine grape, Cabernet Sauvignon. Nat Genet 16: 84–87.

Bowers JE, Siret R, This P, Boursiquot JM, Meredith CP (1998) A single pair of parents proposed for a group of grapevine varieties in northeastern France. Acta Hort 528: 129–133.

Bowers JE, Dangl GS, Meredith CP (1999a) Development and characterization of additional microsatellite DNA markers for grape. Am J Enol Vitic 50: 243–246.

Bowers JE, Boursiquot JM, This P, Chu K, Johansson H, Meredith CP (1999b) Historical genetics: The parentage of Chardonnay, Gamay, and other wine grapes of northeastern France. Science 285: 1562–1565.

Braun P, Schmid J (1999) Sap flow measurements in grapevines (*Vitis vinifera* L.)—1 Stem morphology and use of the heat balance method. Plant Soil 215: 39–45.

Bronner A, Oliveira J (1991) Création et étude de lignées chez Pinot Noir (*V. vinifera* L.). J Int Vigne Vin 25: 133–148.

Bryan GJ, McNicoll J, Ramsay G, Meyers RC, De Jong WS (1999) Polymorphic simple sequence repeat markers in chloroplast genomes of Solanaceous plants. Theor App Genet 99, 859–867.

Bureau SM, Razungles AJ, Baumes RL (2000) The aroma of Muscat of Frontignan grapes: effect of the light environment of vine or bunch on volatiles and glycoconjugates. J Sci Food Agri 80: 2012–2020.

Burke JM, Knapp SJ, Rieseberg LH (2005) Genetic consequences of selection during the evolution of cultivated sunflower. Genetics 171: 1933–1940.

Cadle-Davidson L (2008) Monitoring Pathogenesis of Natural Botrytis Cinerea Infections in Developing Grape Berries. Am J Enol Vitic 59: 387–395.

Cadle-Davidson M, Owens C (2008) Genomic amplification of the Gret1 retroelement in white-fruited accessions of wild *Vitis* and interspecific hybrids. Theor Appl Genet 116: 1079–1094.

Calo A, Costacurta A, Maras V, Meneghetti S, Crespan M (2008) Molecular correlation of Zinfandel (Primitivo) with Austrian, Croatian, Hungarian cultivars and Kratosija, an additional synonym. Am J Enol Vitic 59: 205–209.

Castillo-Munoz N, Gomez-Alonso S, Garcia-Romero E, Hermosin-Gutierrez I (2007) Flavonol profiles of *Vitis vinifera* red grapes and their single-cultivar wines. J Agri Food Chem 55: 992–1002.

Chaïb J, Torregrosa L, Mackenzie D, Corena P, Bouquet A, Thomas MR (2010) The grape microvine—a model system for rapid forward and reverse genetics of grapevines. Plant J 62: 1083–1092.

Chatelet P, Laucou V, Fernandez L, Sreekantan L, Lacombe T, Martinez-Zapater JM, Thomas MR, Torregrosa L (2007) Characterization of *V. vinifera* L. somatic variants exhibiting abnormal flower development patterns. J Exp Bot 58: 4107–4118.

Chatelet P, Chabrillange N, Péros JP, Ortigosa P, Payrière A, Lacombe T, This P, Dussert S (2009) Cryopreservation in the Vitaceae: seed cryopreservation is effective for wild species accessions. 1st International Symposium on Cryopreservation in Horticultural Species, Leuven, Belgium 5–8 April 2009.

Ching A, Caldwell KS, Jung M, Dolan M, Smith OS, Tingey S, Morgante M, Rafalski AJ (2002) SNP frequency, haplotype structure, linkage disequilibrium in elite maize inbred lines. BMC Genet 3: 19.

Christen D, Schonmann S, Jermini M, Strasser RJ, Defago G (2007) Characterization and early detection of grapevine (*Vitis vinifera*) stress responses to esca disease by *in situ* chlorophyll fluorescence and comparison with drought stress. Environ Exp Bot 60: 504–514.

Chung SM, Staub JE (2003) The development and evaluation of consensus chloroplast primer pairs that possess highly variable sequence regions in a diverse array of plant taxa. Theor Appl Genet 107: 757–767.

Cipriani G, Marrazzo MT, Di Gaspero G, Pfeiffer A, Morgante M, Testolin R (2008) A set of microsatellite markers with long core repeat optimized for grape (*Vitis* spp.) genotyping. BMC Plant Biol 8.

Cooke D, Steward WP, Gescher AJ, Marczylo T (2005) Anthocyans from fruits and vegetables—Does bright colour signal cancer chemopreventive activity? Eur J Cancer 41: 1931–1940.

Cortell JM, Halbleib M, Gallagher AV, Righetti TL, Kennedy JA (2005) Influence of vine vigor on grape (*V. vinifera* L. cv Pinot noir), wine proanthocyanidins. J Agri Food Chem 53: 5798–5808.

Costantini L, Battilana J, Lamaj F, Fanizza G, Grando MS (2008) Berry and phenology-related traits in grapevine (*V. vinifera* L.): From Quantitative Trait Loci to underlying genes. BMC Plant Biol 8.

Crespan M, Calo A, Giannetto S, Sparacio A, Storchi P, Costacurta A (2008) 'Sangiovese' and 'Garganega' are two key varieties of the Italian grapevine assortment evolution. Vitis 47: 97–104.

Cunha J, Baleiras-Couto M, Cunha JP, Banza J, Soveral A, Carneiro LC, Eiras-Dias JE (2007a) Characterization of Portuguese populations of *Vitis vinifera* L. ssp. *sylvestris* (Gmelin) Hegi. Genet Resour Crop Evol 54: 981–988.

Cunha J, Cunha JP, Carneiro LC, Fevereiro P, Eiras-Dias JE (2007b) Carpometric measurements of seeds in the identification of wild ancestors of *Vitis vinifera* L. Ciencia Tec Vitivin 22: 29–34.

da Silva FCC, Viana AP, Da Silva MGO, De Oliveira JG, Gomes A (2008) Chemical characterization and phenological determination of grapes cultivated in the north area of Rio de Janeiro state. Rev Brasileira Fruticult 30: 38–42.

De Andres MT, Cabezas JA, Cervera MT, Borrego J, Martinez-Zapater JM, Jouve N (2007) Molecular characterization of grapevine rootstocks maintained in germplasm collections. Am J Enol Vitic 58: 75–86.

Debolt S, Cook DR, Ford CM (2006) L-Tartaric acid synthesis from vitamin C in higher plants. Proc Natl Acad Sci USA 103: 5608–5613.

Decroocq V, Fave MG, Hagen L, Bordenave L, Decroocq S (2003) Development and transferability of apricot and grape EST microsatellite markers across taxa. Theor Appl Genet 106: 912–922.

Dettweiler E, Jung A, Zyprian E, Töpfer R (2000) Grapevine cultivar Müller-Thurgau and its true to type descent. Vitis 39: 63–65.

Di Gaspero G, Peterlunger E, Testolin R, Edwards KJ, Cipriani G (2000) Conservation of microsatellite loci within the genus *Vitis*. Theor Appl Genet 101: 301–308.

Di Gaspero G, Cipriani G, Marrazzo MT, Andreetta D, Castro MJP, Peterlunger E, Testolin R (2005) Isolation of (AC) n-microsatellites in *V. vinifera* L. and analysis of genetic background in grapevines under marker assisted selection. Mol Breed 15: 11–20.

Di Vecchi-Staraz M, Bandinelli R, Boselli M, This P, Boursiquot JM, Laucou V, Lacombe T, Varès D (2007) Genetic structuring and parentage analysis for evolutionary studies in grapevine: Kin group and origin of the cultivar Sangiovese revealed. J Am Soc Hort Sci 132: 514–524.

Di Vecchi-Staraz M, Laucou V, Bruno G, Lacombe T, Gerber S, Bourse T, Boselli M, This P (2009) Low Level of Pollen-Mediated Gene Flow from Cultivated to Wild Grapevine: Consequences for the Evolution of the Endangered Subspecies *Vitis vinifera* L. subsp. *Sylvestris*. J Hered 100: 66–75.

Diez-Navajas AM, Greif C, Poutaraud A, Merdinoglu D (2007) Two simplified fluorescent staining techniques to observe infection structures of the oomycete Plasmopara viticola in grapevine leaf tissues. Micron 38: 680–683.

Duchêne E, Jaegli N, Salber R, Gaudillere JP (2003) Effects of ripening conditions on the following season's growth and yield components for Pinot noir and Gewürztraminer grapevines (*V. vinifera* L.) in a controlled environment. J Int Vigne Vin 37: 39–49.

Duchêne E, Legras JL, Karst F, Merdinoglu D, Claudel P, Jaegli N, Pelsy F (2009) Variation of linalool and geraniol content within two pairs of aromatic and non-aromatic grapevine clones. Aust J Grape Wine Res 15: 120–130.

Ebadi A, Moghadam JE, Fatahi R (2009) Evaluation of 22 populations achieved from controlled crossing between some seeded x seedless grapevine cultivars. Sci Hort 119: 371–376.

Ercisli S, Orhan E, Hizarci Y, Yildirim N, Agar G (2008) Genetic diversity in grapevine germplasm resources in the Coruh valley revealed by RAPD markers. Biochem Genet 46: 590–597.

Ergul A, Kazan K, Aras S, Cevik V, Celik H, Soylemezoglu G (2006) AFLP analysis of genetic variation within the two economically important Anatolian grapevine (*V. vinifera* L.) varietal groups. Genome 49: 467–475.

Fernandez L, Romieu C, Moing A, Bouquet A, Maucourt M, Thomas MR, Torregrosa L (2006) The grapevine fleshless berry mutation a unique genotype to investigate differences between fleshy and nonfleshy fruit. Plant Physiol 140: 537–547.

Fernandez L, Torregrosa L, Segura V, Bouquet A, Martínez Zapater JM (2010) Transposon induced gene activation as a mechanism generating cluster shape somatic variation in grapevine. The Plant Journal 61: 545–557.

Fernandez MP, Nunez Y, Ponz F, Hernaiz S, Gallego FJ, Ibanez J (2008) Characterization of sequence polymorphisms from microsatellite flanking regions in *Vitis* spp. Mol Breed 22: 455–465.

Fernandez-Gonzalez M, Mena A, Izquierdo P, Martinez J (2007) Genetic characterization of grapevine (*V. vinifera* L.) cultivars from Castilla La Mancha (Spain) using microsatellite markers. Vitis 46: 126–130.

Fritschi FB, Lin H, Walker MA (2007) Xylella fastidiosa population dynamics in grapevine genotypes differing in susceptibility to Pierce's disease. Am J Enol Vitic 58: 326–332.

Fritschi FB, Lin H, Walker MA (2008) Scanning electron Microscopy reveals different response pattern of four Vitis genotypes to Xylella fastidiosa infection. Plant Dis 92: 276–286.

Galbacs Z, Molnar S, Halasz G, Kozma P, Hoffman S, Kovacs L, Veres A, Galli Z, Szoke A, Heszky L, Kiss E (2009) Identification of grapevine cultivars using microsatellite-based DNA barcodes. Vitis 48: 17–24.

Galet P (2000) Dictionnaire encyclopédique des cépages. Hachette, Paris, France.

Gao LZ, Zhang CH, Li DY, Pan DJ, Jia JZ, Dong YS (2006) Genetic diversity within Oryza rufipogon germplasms preserved in Chinese field gene banks of wild rice as revealed by microsatellite markers. Biodiver Conserv 15: 4059–4077.

Gee CT, Gadoury DM, Cadle-Davidson L (2008) Ontogenic resistance to *Uncinula necator* varies by genotype and tissue type in a diverse collection of *Vitis* spp. Plant Dis 92: 1067–1073.
Gerrath JM, Posluszny U (2007) Shoot architecture in the *Vitaceae.* Can J Bot 85: 691–700.
Goldstein DG, Pollock DD (1997) Launching microsatellites: a review of mutation processes and methods of phylogenetic inference. J Hered 88: 335–342.
Gomez-Alonso S, Fernandez-Gonzalez M, Martinez AMJ, Garcia-Romero E (2007) Anthocyanin profile of Spanish *Vitis vinifera* L. red grape varieties in danger of extinction. Austr J Grape Wine Res 13: 150–156.
Goto-Yamamoto N, Mouri H, Azumi M, Edwards KJ (2006) Development of grape microsatellite markers and microsatellite analysis including oriental cultivars. Am J Enol Vitic 57: 105–108.
Grant DC, Helleur RJ (2008) Rapid screening of anthocyanins in berry samples by surfactant-mediated matrix-assisted laser desorption/ionization time-of-flight mass spectrometry. Rapid Com Mass Spectrom 22: 156–164.
Grassi F, Imazio S, Failla O, Scienza A, Rubio RO, Lopez MA, Sala F, Labra M (2003a) Genetic isolation and diffusion of wild grapevine Italian and Spanish populations as estimated by nuclear and chloroplast SSR analysis. Plant Biol 5: 608–614.
Grassi F, Labra M, Imazio S, Spada A, Sgorbati S, Scienza A, Sala F (2003b) Evidence of a secondary grapevine domestication centre detected by SSR analysis. Theor Appl Genet 107: 1315–1320.
Grassi F, Labra M, Imazio S, Rubio RO, Failla O, Scienza A, Sala F (2006) Phylogeographical structure and conservation genetics of wild grapevine. Conserv Genet 7: 837–845.
Grassi F, De Mattia F, Zecca G, Sala F, Labra M (2008) Historical isolation and Quaternary range expansion of divergent lineages in wild grapevine. Biol J Linn Soc 95: 611–619.
Guerrero RF, Liazid A, Palma M, Puertas B, Gonzalez-Barrio R, Gil-Izquierdo A, Garcia-Barroso C, Cantos-Villar E (2009) Phenolic characterisation of red grapes autochthonous to Andalusia. Food Chem 112: 949–955.
Gunderson KL, Steemers FJ, Lee G, Mendoza LG, Chee MS (2005) A genome-wide scalable SNP genotyping assay using microarray technology. Nat Genet 37: 549–554.
Halushka MK, Fan JB, Bentley K, Hsie L, Shen NP, Weder A, Cooper R, Lipshutz R, Chakravarti A (1999) Patterns of single-nucleotide polymorphisms in candidate genes for blood-pressure homeostasis. Nat Genet 22: 239–247.
Harbertson JF, Hodgins RE, Thurston LN, Schaffer LJ, Reid MS, Landon JL, Ross CF, Adams DO (2008) Variability of tannin concentration in red wines. Am J Enol Vitic 59: 210–214.
Heinonen M (2007) Antioxidant activity and antimicrobial effect of berry phenolics—a Finnish perspective. Mol Nutr Food Res 51: 684–691.
Heuertz M, Goryslavets S, Hausman JF, Risovanna V (2008) Characterization of grapevine accessions from Ukraine using microsatellite markers. Am J Enol Vitic 59: 169–178.
Imazio S, Labra M, Grassi F, Scienza A, Failla O (2006) Chloroplast microsatellites to investigate the origin of grapevine. Genet Resour Crop Evol 53: 1003–1011.
Ingrouille MJ, Chase MW, Fay MF, Bowman D, Van der Bank M, Bruijn DE (2002) Systematics of Vitacea from the viewpoint of plastid rbcl DNA sequence data. Bot J Linn Soc 2002, 138: 421–432.
Iriti M, Rossoni M, Faoro F (2006) Melatonin content in grape: myth or panacea? J Sci Food Agri 86: 1432–1438.
Jaillon O, Aury JM, Noel B, Policriti A, Clepet C, Casagrande A, Choisne N, Aubourg S, Vitulo N, Jubin C, Vezzi A, Legeai F, Hugueney P, Dasilva C, Horner D, Mica E, Jublot D, Poulain J, Bruyere C, Billault A, Segurens B, Gouyvenoux M, Ugarte E, Cattonaro F, Anthouard V, Vico V, Del Fabbro C, Alaux M, Di Gaspero G, Dumas V, Felice N, Paillard S, Juman I, Moroldo M, Scalabrin S, Canaguier A, Le Clainche I, Malacrida G, Durand E, Pesole G, Laucou V, Chatelet P, Merdinoglu D, Delledonne M, Pezzotti M, Lecharny A, Scarpelli C, Artiguenave F, Pe ME, Valle G, Morgante M, Caboche M, Adam-Blondon AF, Weissenbach J, Quetier F, Wincker P, The French-Italian Consortium (2007) The

grapevine genome sequence suggests ancestral hexaploidization in major angiosperm phyla. Nature 449: 463–U465.

Jansen RK, Kaittanis C, Saski C, Lee SB, Tomkins J, Alverson AJ, Daniell H (2006) Phylogenetic analyses of *Vitis* (*Vitaceae*) based on complete chloroplast genome sequences: effects of taxon sampling and phylogenetic methods on resolving relationships among rosids. BMC Evol Biol 6: 32.

Jones GV, Davis RE (2000) Climate influences on grapevine phenology, grape composition, and wine production and quality for Bordeaux, France. Am J Enol Vitic 51: 249–261.

Jones HG, Stoll M, Santos T, De Sousa C, Chaves MM, Grant OM (2002) Use of infrared thermography for monitoring stomatal closure in the field: application to grapevine. J Exp Bot 53: 2249–2260.

Jung SM, Yun HK, Ma KB, Song GC, Park SJ, Kim JB (2008) Influential factors for identifying levels of resistance and *in vitro* screening system against gray mold in grapevines. Kor J Hort Sci Technnol 26: 432–439.

Kallithraka S, Mohdaly AAA, Makris DP, Kefalas P (2005) Determination of major anthocyanin pigments in Hellenic native grape varieties (*Vitis vinifera* sp.): Association with antiradical activity. J Food Comp Anal 18: 375–386.

Karatas H, Degirmenci D, Velasco R, Vezzulli S, Bodur C, Agaoglu YS (2007) Microsatellite fingerprinting of homonymous grapevine (*V. vinifera* L.) varieties in neighboring regions of South-East Turkey. Sci Hort 114: 164–169.

Karatas H, Agaoglu YS (2008) Genetic diversity among Turkish local grape accessions (*V. vinifera* L.) using RAPD markers. Hereditas 145: 58–63.

Kellow AV, Mcdonald G, Corrie AM, Van Heeswijck R (2002) *In vitro* assessment of grapevine resistance to two populations of phylloxera from Australian vineyards. Austr J Grape Wine Res 8: 109–116.

Klein BY, Ben-Yair C, Bar-Gal GK, Greenblatt CL (2008) Microsatellite genotyping of cultivars of the Holy Land grapevine, *V. vinifera* ssp. sativa (Vitaceae). Bot J Linn Soc 156: 513–521.

Kobayashi S, Goto-Yamamoto N, Hirochika H (2004) Retrotransposon-induced mutations in grape skin color. Science 304: 982–982.

Korbuly J, Pedryc A, Jahnke GG (2004) Evaluation of frost resistance of traditional and newly bred Hungarian wine-grape cultivars. Acta Hort: 337–341.

Kortekamp A, Zyprian E (2003) Characterization of Plasmopara-resistance in grapevine using *in vitro* plants. J Plant Physiol 160: 1393–1400.

Ksouri R, Debez A, Mahmoudi H, Ouerghi Z, Gharsalli M, Lachaal M (2007) Genotypic variability within Tunisian grapevine varieties (*Vitis vinifera* L.) facing bicarbonate-induced iron deficiency. Plant Physiol Biochem 45: 315–322.

Labate JA, Robertson LD, Wu FN, Tanksley SD, Baldo AM (2009) EST, COSII, and arbitrary gene markers give similar estimates of nucleotide diversity in cultivated tomato (Solanum lycopersicum L.). Theor Appl Genet 118: 1005–1014.

Labra M, Imazio S, Grassi F, Rossoni M, Sala F (2004) Vine-1 retrotransposon-based sequence-specific amplified polymorphism for *V. vinifera* L. genotyping. Plant Breed 123: 180–185.

Lacombe T, Laucou V, Di Vecchi M, Bordenave L, Bourse T, Siret R, David J, Boursiquot JM, Bronner A, Merdinoglu D, This P (2003) Inventory and characterization of *Vitis vinifera* ssp. silvestris in France. Acta Hort 553–557.

Lacombe T, Boursiquot JM, Laucou V, Dechesne F, Vares D, This P (2007) Relationships and genetic diversity within the accessions related to malvasia held in the Domaine de Vassal grape germplasm repository. Am J Enol Vitic 58: 124–131.

Laiadi Z, Bentchikou MM, Bravo G, Cabello F, Martínez-Zapater JM (2009) Molecular identification and genetic relationships of Algerian grapevine cultivars maintained at the germplasm collection of Skikda (Algeria). Vitis 48: 25–32.

Lamboy WF, Alpha CG (1998) Using simple sequence repeats (SSRs) for DNA fingerprinting germplasm accessions of grape (*Vitis* L) species. J Am Soc Hort Sci, 123: 182–188.

Laucou V, Boursiquot JM, Lacombe T, Bordenave L, Decroocq S, Ollat N (2008) Parentage of grapevine rootstock 'Fercal' finally elucidated. Vitis 47: 163–167.
Laucou V, Lacombe T, Dechesne F, Siret R, Bruno JP, Dessup M, Dessup T., Ortigosa P, Parra P, Roux C, Santoni S, Varès D., Péros JP, Boursiquot JM, This P (2010) High throughput analysis of grape genetic diversity as a tool for germplasm collection management. Theor Appl Genet 112 DOI 10.1007/s00122-010-1527-y.
Lebon G, Duchene E, Brun O, Clement C (2005) Phenology of flowering and starch accumulation in grape (*Vitis vinifera* L.) cuttings and vines. Ann Bot 95: 943–948.
LeCunff L, Fournier-Level A, Laucou V, Vezzulli S, Lacombe T, Adam-Blondon AF, Boursiquot JM, This P (2008) Construction of nested genetic core collections to optimize the exploitation of natural diversity in *V vinifera* L. subsp. sativa. BMC Plant Biol 8: 31.
Lefort F, Roubelakis-Angelakis K (2002) Assessing the identity of grapevine plants from vineyards from Crete and Samos by microsatellite profiling. J Int Vigne Vin 36: 177–183.
Lefort F, Kyvelos CJ, Zervou M, Edwards KJ, Roubelakis-Angelakis KA (2002) Characterization of new microsatellite loci from *V vinifera* and their conservation in some *Vitis* species and hybrids. Mol Ecol Notes 2: 20–21.
Lefort F, Pelsy F, Schehrer L, Scott KD, Merdinoglu D (2003) Assessment of two highly polymorphic microsatellite loci in 103 accessions of *Vitis* species. J Int Sci Vigne Vin 37: 67–75.
Levadoux L, Boubals D, Rives M (1962) Le genre *Vitis* et ses espèces. Ann Amélior Plantes 12: 19–44.
Li D, Wan YZ, Wang YJ, He PC (2008) Relatedness of resistance to anthracnose and to white rot in Chinese wild grapes. Vitis 47: 213–215.
Liang ZC, Wu BH, Fan PG, Yang CX, Duan W, Zheng XB, Liu CY, Li SH (2008) Anthocyanin composition and content in grape berry skin in *Vitis* germplasm. Food Chem 111: 837–844.
Lijavetzky D, Ruiz-Garcia L, Cabezas JA, De Andres MT, Bravo G, Ibanez A, Carreno J, Cabello F, Ibanez J, Martinez-Zapater JM (2006) Molecular genetics of berry colour variation in table grape. Mol Genet Genom 276: 427–435.
Lijavetzky D, Cabezas J, Ibanez A, Rodriguez V, Martinez-Zapater JM (2007) High throughput SNP discovery and genotyping in grapevine (*V. vinifera* L.) by combining a re-sequencing approach and SNPlex technology. BMC Genom 8: 424.
Lin H, Walker MA (1998) Identifying grape rootstocks with simple sequence repeat (SSR) DNA markers. Am J Enol Vitic 49: 403–407.
Liu SM, Sykes SR, Clingeleffer PR (2003) A method using leafed single-node cuttings to evaluate downy mildew resistance in grapevine. Vitis 42: 173–180.
Lopes MS, Sefc KM, Dias EE, Steinkellner H, Machado MLD, Machado AD (1999) The use of microsatellites for germplasm management in a Portuguese grapevine collection. Theor Appl Genet 99: 733–739.
Lopes MS, Dos Santos MR, Dias JEE, Mendonca D, Machado AD (2006) Discrimination of Portuguese grapevines based on microsatellite markers. J Biotechnol 127: 34–44.
Lopez-Miranda S, Yuste J (2004) Influence of flowers per cluster, fruit-set and berry weight on cluster weight in Verdejo grapevine (*V vinifera* L.). J Int Vigne Vin 38: 41–47.
Louarn G, Guedon Y, Lecoeur J, Lebon E (2007) Quantitative analysis of the phenotypic variability of shoot architecture in two grapevine (*Vitis vinifera*) cultivars. Ann Bot 99: 425–437.
Lu J, Ren ZB, Cousins P (2008) Evaluation of grape rootstocks for resistance to Pierce's Disease and Adaptation to North Florida Environment. Proceedings of the International Symposium on Enhancing Economic and Environmental Sustainability of Fruit Production in a Global Economy, Leuven Belgium. Acta Hort 772: 257–261.
Mackay JF, Wright CD, Bonfiglioli RG (2008) A new approach to varietal identification in plants by microsatellite high resolution melting analysis: application to the verification of grapevine and olive cultivars. Plant Methods 4: 8.

Mahmoodzadeh H, Nazemieh A, Majidi I, Paygami I, Khalighi A (2004) Evaluation of crown gall resistance in *Vitis vinifera* and hybrids of *Vitis* spp. Vitis 43: 75–79.

Maletic E, Pejic I, Kontic JK, Piljac J, Dangl GS, Vokurka A, Lacombe T, Mirosevic N, Meredith CP (2004) Zinfandel, Dobricic, and Plavac mali: The genetic relationship among three cultivars of the Dalmatian coast of Croatia. Am J Enol Vitic 55: 174–180.

Mancuso S (1999) Elliptic Fourier Analysis (EFA) and Artificial Neural Networks (ANNs) for the identification of grapevine (*Vitis vinifera* L.) genotypes. Vitis 38: 73–77.

Maria F, Villota PD, Tortola MET (1994) Palynological study of the pollen grain of *vitis-vinifera* l. cultivars—some aspects of sculpturing and pollination. Vitis 33: 57–61.

Martin JP, Borrego J, Cabello F, Ortiz JM (2003) Characterization of Spanish grapevine cultivar diversity using sequence-tagged microsatellite site markers. Genome 46: 10–18.

Martin JP, Santiago JL, Pinto-Carnide O, Leal F, Martinez MD, Ortiz JM (2006) Determination of relationships among autochthonous grapevine varieties (*V. vinifera* L.) in the Northwest of the Iberian Peninsula by using microsatellite markers. Genet Resour Crop Evol 53: 1255–1261.

Martinez L, Cavagnaro P, Masuelli R, Rodriguez J (2003) Evaluation of diversity among Argentine grapevine (*V. vinifera* L.) varieties using morphological data and AFLP markers. Eur J Biotechnol 6: 244–253.

Martinez LE, Cavagnaro PF, Masuelli RW, Zuniga M (2006) SSR-based assessment of genetic diversity in South American *V. vinifera* varieties. Plant Sci 170: 1036–1044.

Martinez MC, Grenan S (1999) A graphic reconstruction method of an average vine leaf. Agronomie 19: 491–507.

Matsumoto T, Sakai A (2003) Cryopreservation of axillary shoot tips of *in vitro*-grown grape (*Vitis*) by a two-step vitrification protocol. Euphytica 131: 299–304.

Mattivi F, Guzzon R, Vrhovsek U, Stefanini M, Velasco R (2006) Metabolite profiling of grape: Flavonols and anthocyanins. J Agri Food Chem 54: 7692–7702.

Mattivi F, Vrhovsek U, Masuero D, Trainotti D (2009) Differences in the amount and structure of extractable skin and seed tannins amongst red grape varieties. Aust J Grape Wine Res 15: 27–35.

Maul E, Topfer R, Eibach R (2008) *Vitis* international variety catalogue: *http://www.vivc.bafz.de*

Mazhar H, Basha SM, Lu J (2002) Variation in berry protein composition of Muscadine cultivars. Am J Enol Vitic 53: 87–91.

Mcghie TK, MC Walton (2007) The bioavailability and absorption of anthocyanins: Towards a better understanding. Mol Nutr Food Res 51: 702–713.

McGovern PE (2003) Ancient wine: the search for the origins of viniculture. Princeton Univ Press, Princeton, New Jersey, USA.

Medrano H, Escalona JM, Cifre J, Bota J, Flexas J (2003) A ten-year study on the physiology of two Spanish grapevine cultivars under field conditions: effects of water availability from leaf photosynthesis to grape yield and quality. Funct Plant Biol 30: 607–619.

Merdinoglu D, Butterlin G, Bevilacqua L, Chiquet V, Adam-Blondon AF, Decroocq S (2005) Development and characterization of a large set of microsatellite markers in grapevine (*V. vinifera* L.) suitable for multiplex PCR. Mol Breed 15: 349–366.

Meredith CP, Bowers JE, Riaz S, Handley V, Bandman EB, Dangl GS (1999) The identity and parentage of the variety known in California as Petite Sirah. Am J Enol Vitic 50: 236–242.

Meuwissen THE, Karlsen A, Lien S, Olsaker I, Goddard ME (2002) Fine Mapping of a Quantitative Trait Locus for Twinning Rate Using Combined Linkage and Linkage Disequilibrium Mapping. Genetics 161: 373–379.

Moisy C, Garrison KE, Meredith CP, Pelsy F (2008) Characterization of ten novel Ty1/copia-like retrotransposon families of the grapevine genome. BMC Genom 9: 469.

Montaner C, Martin JP, Casanova J, Marti C, Badia D, Cabello F, Ortiz JM (2004) Application of microsatellite markers for the characterization of 'Parraleta': an autochthonous Spanish grapevine cultivar. Sci Hort 101: 343–347.

Montealegre RR, Peces RR, Vozmediano JLC, Gascuena JM, Romero EG (2006) Phenolic compounds in skins and seeds of ten grape (*Vitis vinifera*) varieties grown in a warm climate. J Food Comp Anal 19: 687–693.

Moore MO, Giannasi DE (1994) Foliar Flavonoids of Eastern North-American Vitis (Vitaceae) North of Mexico. Plant Syst Evol 193: 21–36.

Muganu M, Dangl G, Aradhya M, Frediani M, Scossa A, Stover E (2009) Ampelographic and DNA Characterization of Local Grapevine Accessions of the Tuscia Area (Latium, Italy). Am J Enol Vitic 60: 110–115.

Mullins MG, Bouquet A, Williams LE (1992) The grapevine and its relatives In: Biology of the Grapevine Cambridge Univ Press, Cambridge, UK, pp 17–27.

Myles S, Chia JM, Hurwitz B, Simon CJ, Zhong GY, Buckler ES, Ware D (2010) Rapid genomic characterization of the genus *Vitis*. PlosOne 5: e8219.

Nakagawa S, Horiuchi S, Matsui H, Yuda E, Yamada S, Murai Y, Komatsu H (1991) Distribution and Leaf Morphology of Wild Grapes Native to Japan. J Jpn Soc Hort Sci 60: 31–39.

Ocete, R, Lopez MA, Gallardo A, Arnold C (2008) Comparative analysis of wild and cultivated grapevine (*Vitis vinifera*) in the Basque Region of Spain and France. Agri Ecosys Environ 123: 95–98.

Office International de la vigne et du Vin (2008) 2nd edn of the OIV Descriptor List for Grape Varieties and *Vitis* Species. OIV, Paris, France.

Ollat N, Laborde W, Neveux M, Diakou-Verdin P, Renaud C, Moing A (2003) Organic acid metabolism in roots of various grapevine (*Vitis*) rootstocks submitted to iron deficiency and bicarbonate nutrition. J Plant Nutr 26: 2165–2176.

Olmo HP (1995) The origin, domestication of the vinifera grape In: PE Mc Govern , SJ Fleming, SH Katz (eds) The Origins and Ancient History of Wine. Gordon and Breach Publishers, Philadelphia, USA, pp 23–30.

Olmstead RG, Palmer JD (1994) Chloroplast DNA systematics: a review of methods and data analysis. Am J Bot 81: 1205–1224.

Ortega-Regules A, Romero-Cascales I, Lopez-Roca JM, Ros-Garcia JM, Gomez-Plaza E (2006) Anthocyanin fingerprint of grapes: environmental and genetic variations. J Sci Food Agri 86: 1460–1467.

Ortega-Regules A, Ros-Garcia JM, Bautista-Ortin AB, Lopez-Roca JM, Gomez-Plaza E (2008) Differences in morphology and composition of skin and pulp cell walls from grapes (*Vitis vinifera* L.): technological implications. Eur Food Res Technol 227: 223–231.

Ortiz JM, Martin JP, Borrego J, Chavez J, Rodriguez I, Munoz G, Cabello F (2004) Molecular and morphological characterization of a *Vitis* gene bank for the establishment of a base collection. Genet Resour Crop Evol 51: 403–409.

Palaisa KA, Morgante M, Williams M, Rafalski A (2003) Contrasting effects of selection on sequence diversity and linkage disequilibrium at two phytoene synthase loci. Plant Cell 15: 1795–1806.

Palaisa KA, Morgante M, Tingey S, Rafalski A (2004) Long-range patterns of diversity and linkage disequilibrium surrounding the maize Y1 gene are indicative of an asymmetric selective sweep. Proc Natl Acad Sci USA 101: 9885–9890.

Pastrana-Bonilla E, Akoh CC, Sellappan S, Krewer G (2003) Phenolic content and antioxidant capacity of muscadine grapes. J Agri Food Chem 51: 5497–5503.

Pavek DS, Lamboy WF, Garvey EJ (2003) Selecting *in situ* conservation sites for grape genetic resources in the USA. Genet Resour Crop Evol 50: 165–173.

Pellegrino A, Lebon E, Simonneau T, Wery J (2005) Towards a simple indicator of water stress in grapevine (*Vitis vinifera* L.) based on the differential sensitivities of vegetative growth components. Austr J Grape Wine Res 11: 306–315.

Pelsy F (2007) Untranslated leader region polymorphism of Tvv1, a retrotransposon family, is a novel marker useful for analyzing genetic diversity and relatedness in the genus *Vitis*. Theor Appl Genet 116: 15–27.

Peng JH, Bai Y, Haley SD, Lapitan NLV (2009) Microsatellite-based molecular diversity of bread wheat germplasm and association mapping of wheat resistance to the Russian wheat aphid. Genetica 135: 95–122.

Péros JP, Nguyen TH, Troulet C, Michel-Romiti C, Notteghem JL (2006) Assessment of powdery mildew resistance of grape and Erysiphe necator pathogenicity using a laboratory assay. Vitis 45: 29–36.

Péros JP, Berger G, Portemont A, Boursiquot JM, Lacombe T (2010) Genetic variation and biogeography of the disjunct Vitis subgenus(Vitaceae). Journal of Biogeography (doi 10.1111/j.1365-2699.2010.02410.x)

Perret C, Pezet R, Tabacchi R (2003a) Fractionation of grape tannins and analysis by matrix-assisted laser desorption/ionisation time-of-flight mass spectrometry. Phytochem Analys 14: 202–208.

Perret C, Pezet R, Tabacchi R (2003b) Qualitative analysis of grapevine tannins by mass spectrometry and their inhibitory effect on stilbene oxidase of Botrytis cinerea. Chimia 57: 607–610.

Peterlunger E, Di Gaspero G, Cipriani G, Sivilotti P, Zulini L, Marrazzo M T, Andreetta D, Testolin R (2003) Breeding strategy for the introgression of disease resistance genes into European grapevine. Acta Hort 603: 665–670.

Pico FX, Mendez-Vigo B, Martinez-Zapater JM, Alonso-Blanco C (2008) Natural genetic variation of arabidopsis thaliana is geographically structured in the Iberian peninsula. Genetics 180: 1009–1021.

Pindo M, Vezzulli S, Coppola G, Cartwright DA, Zharkikh A, Velasco R, Troggio M (2008) SNP high-throughput screening in grapevine using the SNPlex (TM) genotyping system. BMC Plant Biol 8.

Plessis P, Leddet C, Collas A, Dereuddre J (1993) Cryopreservation of *Vitis vinifera* L. cv. chardonnay shoot tips by encapsulation-dehydration—effects of pretreatment, cooling and postculture conditions. Cryo-Letters 14: 309–320.

Pomar F, Novo M, Masa A (2005) Varietal differences among the anthocyanin profiles of 50 red table grape cultivars studied by high performance liquid chromatography J Chrom 1094: 34–41.

Pouget R (1990) Histoire de la lutte contre le Phylloxéra de la vigne en France. INRA, OIV (eds), Editions Quae, Paris, France, pp 1–157.

Poutaraud A, Latouche G, Martins S, Meyer S, Merdinoglu D, Cerovic ZG (2007) Fast and local assessment of stilbene content in grapevine leaf by *in vivo* fluorometry. J Agri Food Chem 55: 4913–4920.

Provan J, Soranzo N, Wilson NJ, McNicol JW, Forrest GI, Cottrell J, Powell W (1998) Genepool variation in Caledonian and European Scots pine (*Pinus sylvestris* L.) revealed by chloroplast simple sequence repeats. Proc Roy Soc London B Biol Sci 265: 1697–1705.

Provan J (2000) Novel chloroplast microsatellites reveal cytoplasmic variation in *Arabidopsis thaliana*. Mol Ecol 9: 2183–2185.

Provan J, Powell W, Hollingsworth PM (2001) Chloroplast microsatellites: new tools for studies in plant ecology and evolution. Trends Ecol Evol 16: 142–147.

Ranc N, Munos S, Santoni S, Causse M (2008) A clarified position for *Solanum lycopersicum* var. cerasiforme in the evolutionary history of tomatoes (*solanaceae*). BMC Plant Biol 8.

Regner F, Stadlbauer A, Eisenheld C, Kaserer H (2000) Genetic relationships among Pinots and related cultivars. Am J Enol Vitic 51: 7–14.

Rivera D, Miralles B, Obon C, Carreno E, Palazon JA (2007) Multivariate analysis of *Vitis* subgenus *Vitis* seed morphology. Vitis 46: 158–167.

Romero-Cascales I, Fernandez-Fernandez JI, Lapez-Roca JM, Gamez-Plaza E (2005) The maceration process during winemaking extraction of anthocyanins from grape skins into wine. Eur Food Res Technnol 221: 163–167.

Rossetto M, Jackes BR, Scott KD, Henry RJ (2001) Intergeneric relationships in the Australian Vitaceae: new evidence from cpDNA analysis. Genet Resour Crop Evol 48: 307–314.

Rossetto M, Jackes BR, Scott KD, Henry RJ (2002a) Is the genus *Cissus* (Vitaceae) monophyletic? Evidence from plastid and nuclear ribosomal DNA. Syst Bot 27: 522–533.
Rossetto M, Mcnally J, Henry RJ (2002b) Evaluating the potential of SSR: flanking regions for examining taxonomic relationships in the Vitaceae. Theor Appl Genet 104: 61–66.
Roussel V, Leisova L, Exbrayat F, Stehno Z, Balfourier F (2005) SSR allelic diversity changes in 480 European bread wheat varieties released from 1840 to 2000. Theor Appl Genet 111: 162–170.
Roytchev V, Terziisky D, Dimova D, Karageorgiev S (1994) Scanning electron-microscopy study of pollen morphology in seedless grape (*Vitis vinifera* L.) cultivars. Vitis 33: 105–108.
Sabeti PC, Reich DE, Higgins JM, Levine HZ, Richter DJ, Schaffner SF, Gabriel SB, Platko JV, Patterson NJ, McDonald GJ, Ackerman HC, Campbell SJ, Altshuler D, Cooper R, Kwiatkowski D, Ward R, Lander ES (2002) Detecting recent positive selection in the human genome from haplotype structure. Nature 419: 832–837.
Sadras VO, Collins M, Soar CJ (2008) Modelling variety-dependent dynamics of soluble solids and water in berries of Vitis vinifera. Aust J Grape Wine Res 14: 250–259.
Salmaso M, Faes G, Segala C, Stefanini M, Salakhutdinov L, Zyprian E, Toepfer R, Grando MS, Velasco R (2004) Genome diversity and gene haplotypes in the grapevine (*V. vinifera* L.), as revealed by single nucleotide polymorphisms. Mol Breed 14: 385–395.
Salmaso M, Valle RD, Lucchin M (2008) Gene pool variation and phylogenetic relationships of an indigenous northeast Italian grapevine collection revealed by nuclear and chloroplast SSRs. Genome 51: 838–855.
Sanchez-Escribano EM, Martin JR, Carreno J, Cenis JL (1999) Use of sequence-tagged microsatellite site markers for characterizing table grape cultivars. Genome 42: 87–93.
Santana J C, Hidalgo E, De Lucas AI, Recio P, Ortiz JM, Martin JP, Yuste J, Arranz C, Rubio JA (2008) Identification and relationships of accessions grown in the grapevine (*V. vinifera* L.) Germplasm Bank of Castilla y Leon (Spain) and the varieties authorized in the VQPRD areas of the region by SSR-marker analysis. Genet Resour Crop Evol 55: 573–583.
Santiago JL, Boso S, Martin JP, Ortiz JM, Martinez MC (2005) Characterisation and identification of grapevine cultivars (*Vitis vinifera* L.) from northwestern Spain using microsatellite markers and ampelometric methods. Vitis 44: 67–72.
Santos C, Fragoeiro S, Phillips A (2005) Physiological response of grapevine cultivars and a rootstock to infection with Phaeoacremonium and Phaeomoniella isolates: an *in vitro* approach using plants and calluses. Sci Hort 103: 187–198.
Schlenke TA, Begun DJ (2004) Strong selective sweep associated with a transposon insertion in Drosophila simulans. Proc Natl Acad Sci USA 101: 1626–1631.
Schmid KJ, Sorensen TR, Stracke R, Torjek O, Altmann T, Mitchell- Olds T, Weisshaar B (2003) Large-scale identification and analysis of genome-wide single-nucleotide polymorphisms for mapping in *Arabidopsis thaliana.* Genome Res 13: 1250–1257.
Schneider A, Marinoni DT, Crespan M (2008) Genetics and ampelography trace the origin of Muscat fleur d'oranger. Am J Enol Vitic 59: 200–204.
Schultz HR (2003) Differences in hydraulic architecture account for near-isohydric and anisohydric behaviour of two field-grown *Vitis vinifera* L. cultivars during drought. Plant Cell Environ 26: 1393–1405.
Scott KD, Eggler P, Seaton G, Rossetto M, Ablett EM, Lee LS, Henry RJ (2000) Analysis of SSRs derived from grape ESTs. Theor Appl Genet 100: 723–726.
Sefc KM, Steinkellner H, Wagner HW, Glössl J, Regner F (1997) Application of microsatellite markers to parentage studies in grapevine. Vitis 36: 179–183.
Sefc KM, Regner F, Glössl J, Steinkellner H (1998a) Genotyping of grapevine and rootstock cultivars using microsatellite markers. Vitis 37: 15–20.
Sefc KM, Steinkellner H, Glössl J, Kampfer S, Regner F (1998b) Reconstruction of a grapevine pedigree by microsatellite analysis. Theor Appl Genet 97: 227–231.
Sefc KM, Regner F, Turetschek E, Glossl J, Steinkellner H (1999) Identification of microsatellite sequences in *Vitis riparia* and their applicability for genotyping of different *Vitis* species. Genome 42: 367–373.

Sefc KM, Lopes MS, Lefort F, Botta R, Roubelakis-Angelakis KA, Ibanez J, Pejic I, Wagner HW, Glossl J, Steinkellner H (2000) Microsatellite variability in grapevine cultivars from different European regions and evaluation of assignment testing to assess the geographic origin of cultivars. Theor Appl Genet 100: 498–505.
Sefc KM, Steinkellner H, Lefort F, Botta R, Machado AD, Borrego J, Maletic E, Glossl J (2003) Evaluation of the genetic contribution of local wild vines to European grapevine cultivars. Am J Enol Vitic 54: 15–21.
Shellie KC (2007) Viticultural performance of red and white wine grape cultivars in southwestern Idaho HortTechnology 17: 595–603.
Silva-Contreras C, Silva-Robledo H, Selles-Von Schouwen G, Ferreyra-Espada R (2008) Growth relationships between trunk and grape berry using linear variable differential sensors in *Vitis vinifera*. Agrociencia 42: 903–912.
Sinclair C, Hoffmann AA (2003) Monitoring salt stress in grapevines: are measures of plant trait variability useful? J Appl Ecol 40: 928–937.
Snoussi H, Harbi Ben Slimane H, Ruiz-Garcia L, Martinez-Zapater JM, Arroyo-Garcia R (2004) Genetic relationship among cultivated and wild grapevine accessions from Tunisia. Genome 47: 1211–1219.
Soejima A, Wen J (2006) Phylogenetic analysis of the grape family (Vitaceae) based on three chloroplast markers. Am J Bot 2006, 93: 278–287.
Somers DJ, Kirkpatrick R, Moniwa M, Walsh A (2003) Mining single-nucleotide polymorphisms from hexaploid wheat ESTs. Genome 46(3): 431–437.
Soylemezoglu G, Agaoglu YS, Uzun HI (2001) Ampelographic characteristics and isozymic analysis of Vitis vinifera spp. Sylvestris Gmel in southwestern Turkey. Biotechnol Biotechnol Equip 15: 106–113.
Stajner N, Korosec-Koruza Z, Rusian D, Javornik B (2008) Microsatellite genotyping of old Slovenian grapevine varieties (*V. vinifera* L.) of the Primorje (coastal) winegrowing region. *Vitis* 47: 201–204.
Stamatiadis S, Taskos D, Tsadilas C, Christofides C, Tsadila E, Schepers JS (2006) Relation of ground-sensor canopy reflectance to biomass production and grape color in two Merlot vineyards. Am J Enol Vitic 57: 415–422.
Steele MR, Gitelson AA, Rundquist DC (2008) A comparison of two techniques for nondestructive measurement of chlorophyll content in grapevine leaves. Agron J 100: 779–782.
Stefanini M, Porro D, Policarpo M, Colugnati G, Crespan G, Iacono F (2007) Employment of stability indexes to assess interaction genotype environment (G x E) in *Vitis vinifera*. Proc Int Workshop on Advances in Grapevine and Wine Research, October 2007, Venosa, Italy, pp 85–90.
Storchi P, Pieri M, Valentini P, Bucelli P, Faviere V, Giannetti F (2007) Evaluation of indigenous germplasm through "Environment x Genotype" interaction. Proc Int Workshop on Advances in Grapevine and Wine Research, October 2007, Venosa, Italy, pp 91–96.
Strefeler MS, Weeden NF, Reisch BI (1992) Inheritance of chloroplast DNA in two fullsib *Vitis* populations. Vitis 31: 183–187.
Stringer SJ, Marshall DA, Sampson BJ, Spiers JM (2008) Performance of muscadine grape cultivars in southern Mississippi. HortTechnology 18: 726–733.
Tapia AM, Cabezas JA, Cabello F, Lacombe T, Martinez-Zapater JM, Hinrichsen P, Cervera MT (2007) Determining the Spanish origin of representative ancient American grapevine varieties. Am J Enol Vitic 58: 242–251.
Tenaillon MI, Sawkins MC, Long AD, Gaut RL, Doebley JF, Gaut BS (2001) Patterns of DNA sequence polymorphism along chromosome 1 of maize (*Zea mays* ssp. *mays* L.). Proc Natl Acad Sci USA 98: 9161–9166.
Tesic D, Woolley DJ, Hewett EW, Martin DJ (2002) Environmental effects on cv Cabernet Sauvignon (*V. vinifera* L.) grown in Hawke's Bay, New Zealand. 1 Phenology and characterisation of viticultural environments. Aust J Grape Wine Res 8: 15–26.

This P, Roux C, Parra P, Siret R, Bourse T, Adam AF, Yvon M, Lacombe T, David J, Boursiquot JM (2001) Characterization of genetic diversity in a population of wild grapes from Pic Saint Loup area and it's relationship with the cultivated grapes. Genet Select Evol 33: S289–S304.

This P, Jung A, Boccacci P, Borrego J, Botta R, Costantini L, Crespan M, Dangl GS, Eisenheld C, Ferreira-Monteiro F, Grando S, Ibanez J, Lacombe T, Laucou V, Magalhaes R, Meredith CP, Milani N, Peterlunger E, Regner F, Zulini L, Maul E (2004) Development of a standard set of microsatellite reference alleles for identification of grape cultivars. Theor Appl Genet 109: 1448–1458.

This P, Lacombe T, Thomas MR (2006a) Historical origins and genetic diversity of wine grapes. Trends Genet 22: 511–519.

This P, Lacombe T, Laucou V, Siret R, Moreau F, Vares D (2006b) Grape and Wine varietal authentication by DNA analysis In: S Ebeler, G R Takeoka, P Wintherhalter (eds)] Food and Wine authentication, ACS Symposium Series ACS, Washington, DC, USA, pp 207–228.

This P, Lacombe T, Cadle-Davidson M, Owens CL (2007) Wine grape (*V. vinifera* L.) color associates with allelic variation in the domestication gene VvmybA1. Theor Appl Genet 114: 723–730.

Thomas MR, Cain P, Scott NS (1994) DNA Typing of Grapevines—a Universal Methodology and Database for Describing Cultivars and Evaluating Genetic Relatedness. Plant Mol Biol 25: 939–949.

Tomazic I, Korosec-Koruza Z (2003) Validity of phyllometric parameters used to differentiate local *Vitis vinifera* L. cultivars. Genet Resour Crop Evol 50: 773–778.

Troggio M, Malacarne G, Vezzulli S, Faes G, Salmaso M, Velasco R (2008) Comparison of different methods for SNP detection in grapevine. Vitis 47: 21–30.

Ulanovsky S, Gogorcena Y, De Toda FM, Ortiz JM (2002) Use of molecular markers in detection of synonymies and homonymies in grapevines (*V. vinifera* L.). Sci Hort 92: 241–254.

Upadhyay A, Saboji MD, Reddy S, Deokar K, Karibasappa GS (2007) AFLP and SSR marker analysis of grape rootstocks in Indian grape germplasm. Sci Hort 112: 176–183.

Valsesia G, Gobbin D, Patocchi A, Vecchione A, Pertot I, Gessler C (2005) Development of a high-throughput method for quantification of Plasmopara viticola DNA in grapevine leaves by means of quantitative real-time polymerase chain reaction. Phytopathology 95: 672–678.

Vargas AM, Velez MD, De Andres MT, Laucou V, Lacombe T, Boursiquot JM, Borrego J, Ibanez J (2007) Corinto bianco: A seedless mutant of Pedro Ximenes. Am J Enol Vitic 58: 540–543.

Velasco R, Zharkikh A, Troggio M, Cartwright DA, Cestaro A, Pruss D, Pindo M, Fitzgerald LM, Vezzulli S, Reid J, Malacarne G, Iliev D, Coppola G, Wardell B, Micheletti D, Macalma T, Facci M, Mitchell JT, Perazzolli M, Eldredge G, Gatto P, Oyzerski R, Moretto M, Gutin N, Stefanini M, Chen Y, Segala C, Davenport C, Dematte L, Mraz A, Battilana J, Stormo K, Costa F, Tao Q, Si-Ammour A, Harkins T, Lackey A, Perbost C, Taillon B, Stella A, Solovyev V, Fawcett JA, Sterck L, Vandepoele K, Grando SM, Toppo S, Moser C, Lanchbury J, Bogden R, Skolnick M, Sgaramella V, Bhatnagar SK, Fontana P, Gutin A, Van De Peer Y, Salamini F, Viola R (2007) A high quality draft consensus sequence of the genome of a heterozygous grapevine variety. PLoS ONE 2: e1326.

Vezzulli S, Micheletti D, Riaz S, Pindo M, Viola R, This P, Walker MA, Troggio M, Velasco R (2008) A SNP transferability survey within the genus *Vitis*. BMC Plant Biol 8.

Vigouroux Y, Mitchell S, Matsuoka Y, Hamblin M, Kresovich S, Smith JSC, Jaqueth J, Smith OS, Doebley J (2005) An analysis of genetic diversity across the maize genome using microsatellites. Genetics 169: 1617–1630.

Vouillamoz JF, Grando MS (2006) Genealogy of wine grape cultivars: 'Pinot' is related to 'Syrah'. Heredity 97: 102–110.

Vouillamoz J, Maigre D, Meredith C P (2003) Microsatellite analysis of ancient alpine grape cultivars: pedigree reconstruction of *Vitis vinifera* L. 'Cornalin du Valais'. Theor Appl Genet 107: 448–454.

Vouillamoz JF, Monaco A, Costantini L, Stefanini M, Scienza A, Grando MS (2007a) The parentage of 'Sangiovese', the most important Italian wine grape. *Vitis* 46: 19–22.

Vouillamoz JF, Schneider A, Grando MS (2007b) Microsatellite analysis of Alpine grape cultivars (*V. vinifera* L.): alleged descendants of Pliny the Elder's Raetica are genetically related. Genet Resour Crop Evol 54: 1095–1104.

Walker RR, Blackmore DH, Clingeleffer PR, Correll RL (2002) Rootstock effects on salt tolerance of irrigated field-grown grapevines (*Vitis vinifera* L. cv. Sultana) 1. Yield and vigour inter-relationships. Aust J Grape Wine Res 8: 3–14.

Walker RR, Blackmore DH, Clingeleffer PR, Correll RL (2004) Rootstock effects on salt tolerance of irrigated field-grown grapevines (*Vitis vinifera* L. cv. Sultana) 2. Ion concentrations in leaves, juice. Aust J of Grape Wine Res 10: 90–99.

Walker RR, Blackmore DH, Clingeleffer PR, Tarr CR (2007) Rootstock effects on salt tolerance of irrigated field-grown grapevines (*Vitis vinifera* L. cv. Sultana) 3. Fresh fruit composition and dried grape quality. Aust J Grape Wine Res 13: 130–141.

Wan YZ, Schwaninger H, He PC, Wang YJ (2007) Comparison of resistance to powdery mildew and downy mildew in Chinese wild grapes. Vitis 46: 132–136.

Wan YZ, Wang YJ, Li D, He PC (2008) Evaluation of agronomic traits in Chinese wild grapes and screening superior accessions for use in a breeding program. Vitis 47: 153–158.

Wang QC, Mawassi M, Sahar N, Li P, Violeta CT, Gafny R, Sela I, Tanne E, Perl A (2004) Cryopreservation of grapevine (*Vitis* spp.) embryogenic cell suspensions by encapsulation-vitrification. Plant Cell Tiss Org Cult 77: 267–275.

Weising K, Gardner RC (1999) A set of conserved PCR primers for the analysis of simple sequence repeat polymorphisms in chloroplast genomes of dicotyledoneous angiosperms. Genome 42: 9–19.

Wen J, Nie ZL, Soejima A, Meng Y (2007) Phylogeny of *Vitaceae* based on the nuclear GAI1 gene sequences. Can J Bot 85: 731–745.

Wright SI, Bi IV, Schroeder SG, Yamasaki M, Doebley JF, Mcmullen MD, Gaut BS (2005) The effects of artificial selection of the maize genome. Science 308: 1310–1314.

Wycislo AP, Clark JR, Karcher DE (2008) Fruit shape analysis of Vitis using digital photography. HortScience 43: 677–680.

Yamamoto T, Satoh H (1994) Relationship among berry cracking susceptibility, berry morphology and skin stress-distribution in several grape cultivars. J Jpn Soc Hort Sci 63: 247–256.

Yamane T, Shibayama K, Hamana Y, Yakushiji H (2009) Response of container-grown girdled grapevines to short-term water-deficit stress. Am J Enol Vitic 60: 50–56.

Yun HK, Park KS, Roh JH, Choi YJ, Jeong SB (2007) Developing a screening system for resistance to anthracnose in grapevines using culture filtrates from *Elsinoe ampelina*. J Hort Sci Biotechnol 82: 360–364.

Zdunic G, Pejic I, Kontic JK, Vukicevic D, Vokurka A, Pezo I, Maletic E (2008) Comparison of genetic and morphological data for inferring similarity among native Dalmatian (Croatia) grapevine cultivars (*Vitis vinifera* L.). J Food Agri Env 6: 333–336.

Zhang XK, Walker RR, Stevens RM, Prior LD (2002) Yield-salinity relationships of different grapevine (*V. vinifera* L.) scion-rootstock combinations. Aust J Grape Wine Res 8: 150–156.

Zhu YL, Song QJ, Hyten DL, Van Tassell CP, Matukumalli LK, Grimm DR, Hyatt SM, Fickus EW, Young ND, Cregan PB (2003) Single-nucleotide polymorphisms in soybean. Genetics 163: 1123–1134.

Zoghlami N, Riahi L, Laucou V, Lacombe T, Mliki A, Ghorbel A, This P (2009) Origin and genetic diversity of Tunisian grapes as revealed by microsatellite markers. Sci Hort 120: 479–486.

Zohary D (1995) Domestication of the Grapevine *Vitis vinifera* L. in the Near East. In: PE Mc Govern, SJ Fleming, SH Katz (eds) The Origins and Ancient History of Wine. Gordon and Breach Sciences Publisher, New York, USA, pp 23–30.

3

Origins and Consequences of Somatic Variation in Grapevine

Torregrosa L,[1,*] *Fernandez L,*[2] *Bouquet A,*[1] *Boursiquot J-M,*[3] *Pelsy F*[4] and *Martínez-Zapater JM*[2]

ABSTRACT

Somatic variation results from mutation or epimutation events that first takes place in a single cell. In a multi-cellular organism, for the variant phenotype to be observed at the plant level, the mutated cell has to "colonize" a cell layer (either L1 or L2) of an organ. Once a SAM is periclinaly colonized in either the L1, the L2 or both cell layers, the mutation can be transmitted by bud propagation or through sexual reproduction if present in the L2 (the cell layer giving rise to gametes) and not lethal for the gametophyte. In this chapter, we will focus on the sources of somatic variations that have been observed in grapevine, on the possible mechanisms involved, and on the chimeric states derived from the concurrent existence of genetically different cell lines in the same plant. Then, we will describe some of the somatic variants characterized in grapevine and will end with the exploitation of somatic variation for clonal selection and propagation.

Keywords: grapevine, phenotypic diversity, somatic variants, chimeras, mutants, vegetative propagation

[1]UMR INRA 1334 CIRAD Montpellier SupAgro AGAP, 2 Place P. Viala 34060 Montpellier Cedex, France.
[2]Instituto de Ciencias de la Vid y del Vino (ICVV), (CSIC, Universidad de La Rioja, Gobierno de La Rioja), C/Madre de Dios 51, 26006 Logroño, Spain.
[3]UMT Géno-Vigne, 2 Place P. Viala 34060 Montpellier Cedex, France.
[4]INRA, UMR SVQV INRA-ULP Strasbourg, 28 rue de Herrlisheim, 68021 Colmar Cedex, France.
*Corresponding author: *laurent.torregrosa@supagro.inra.fr*

3.1 Introduction

Somatic variation contributes to phenotypic variation in multi-cellular organisms. In plants, which lack a germ line, variation affecting somatic cells can be transmitted to the next sexual generation (Walbot and Evans 2003). The shoot apical meristem (SAM), which directs the growth of the plant shoot, produces vegetative structures, i.e., leaves and lateral shoots, until environmental and developmental signals trigger the switch to reproduction (Bäurle and Dean 2006) and initiates the development of reproductive structures. This lack of germ line can generate potential problems and benefits. On one hand, gametes could carry mutations accumulated during somatic growth. A problem that is partially overcome by the selection against cell lineages carrying deleterious mutations, which are impaired in growth and development and so are less likely to contribute to gamete formation, and by the selection during the haploid gametophytic phase which is characteristic of plants (Pineda-Krch and Fagerstrom 1999; Walbot and Evans 2003). On the other hand, somatic variation contributes to the generation of genetic variation and this contribution can be very relevant in perennial species (Antolin and Strobeck 1985). In species amenable to vegetative reproduction like grapevine, somatic variation can eventually give rise to independent plant variants without going through the gametophytic phase. Somatic variation has a strong relevance in fruit trees where vegetative reproduction is commonly used to propagate those relevant phenotypes appearing as spontaneous sports. In grapevine, somatic variation has been the only source of genetic variation used by breeders to improve the phenotypic features of classical cultivars (This et al. 2006). This is why, somatic variation deserves special consideration in a book of grapevine genomics as a source of phenotypic variation of relevance in the genetic improvement of the species as well as in functional analysis.

Somatic variation results from mutation or epimutation events taking place in a single cell belonging to a specific cell layer (L1 and L2 layers are distinguished in grapevine shoot apical meristems, Thompson and Olmo 1963). These somatic events will commonly affect one of the two alleles of the gene and therefore a derived phenotypic effect will only be expected for gain-of-function mutations or when loss-of-function mutations result in haploinsufficiency. Furthermore, for the mutant phenotype to be observed at the plant level, the mutant cell has to "colonize" the corresponding cell layer (either L1 or L2) in at least one SAM. Once a bud meristem is colonized in either the L1, the L2 or both cell layers, the mutation will be transmitted by bud propagation and also through sexual reproduction if present in the L2, the cell layer giving rise to gametes.

In this chapter, we will focus on the sources of somatic variations that have been observed in grapevine, the possible mechanisms involved, and then the chimeric states derived from the concurrent existence of genetically different cell lines in the same plant. Then, we will describe some of the grapevine somatic variants characterized in this species and will end with a description of the exploitation of somatic variation for clonal selection.

3.2 Sources of Somatic Variant Phenotypes

3.2.1 Spontaneous Variation (Sports)

Despite the use of a conservative strategy to maintain and propagate plants identical to the parental type, somatic variations can arise in vineyards or germplasm stock centers. When a spontaneous morpho-physiological variation appears on a shoot (bud sport), it can be established as a whole plant by vegetative propagation. Some cultivars, such as Pinot Noir, are known to be more prone to exhibit somatic polymorphisms than others and there are known Pinot Noir somatic variants for pigmentation of the skin (e.g., Pinot Noir, Pinot Gris and Pinot Blanc), pigmentation of the flesh (Pinot Teinturier), yield (Pinot Fin), leaf morphology (Pinot Meunier), shoot height (Pinot Droit) or absence of wax on berries (Pinot Moure). The basis for this high frequency of somatic variants could be genetic (e.g., a more unstable genetic background) or simply reflect a longer history of cultivation or a larger extension of growth. In fact, there are many examples of other spontaneous variations including berry color or flavor, early or late ripening, size and compactness of bunches, canopy growth or yield that have been identified in multiple cultivars (Table 3-1).

3.2.2 Artificial Mutagenesis

Mutagenic treatments have been used to increase the frequency of somatic variation and to generate agronomically relevant phenotypes in a given cultivar genetic background. In perennial fruit crops, induced mutations have been particularly investigated in apple, pear, and peach, but much less in grapevine (Khawale et al. 2007). Nevertheless, both physical (X-, UV- or γ-rays) or chemical [colchicine, ethylmethane sulfonate (EMS) or ethidium bromide(EB)] mutagens have been experimented. The organs of choice for mutagenesis experiments are buds that are treated either *in situ* or *in vitro* and the result is the appearance of phenotypic alterations.

Among physical mutagens, X-rays (Pratt 1959) and γ-rays have been used on buds of *V. vinifera* (Botta et al. 1987; Kuksova et al. 1997; Charbaji and Nabulsi 1999) or *Vitis* hybrids including rootstocks (Da Silva 1995; Charbaji and Nabulsi 1999). Botta et al. (1987), who irradiated five cultivars, described some negative effects on vegetative organs (growth, chlorophyll) and reproductive organs (berry set). However, some interesting phenotypes could

Table 3-1 Somatic variants described in grapevine.

Variant name	WT background	Phenotype	Reference
Small-berry Sultana	Sultana Seedless	Small berries	Olmo (1934)
Seeded Panariti	Panariti Seedless	Seedeed berries	Harmon and Snyder (1936)
Polyploid variants	Various	Polyploid traits	Olmo (1935, 1943, 1952)
Bruce's sport	Sultana	Low PPO	Antcliff and Webster (1962)
Gora Chirine	Sultana	Low berry acidity	Boubals et al. (1971)
Patricia Branca	Patricia (IAC 871-41)	White berries	Terra et al. (1984)
Rosinha Seedless	Niagara red	Seedless	Pires et al. (1988)
Rubi Italia	Italia	Light pink berries	Kishino and Mashima (1980)
Benitaka	Italia	Pink berries	Sousa (1996)
Red Chardonnay	Chardonnay	Red berries	Boss et al. (1996a)
Pink Sultana	Sultana	Pink berries	Boss et al. (1996a)
Bronze CS	Cab. Sauvignon	Bronze berries	Boss et al. (1996a)
Brazil	Italia	Black berries	Gonçalves (1995)
53XX lines	Gravesac	Vigour variations	Da Silva et al. (2000)
Cioutat	Chasselas	Extremely lobed blade	Galet (2000)
Freaux, De Chaudenay	Gamay	Red flesh	Galet (2000)
Early Flame	Flame seedless	Shorter cycle	Scott et al. (2001)
Pinot Meunier	Pinot Noir	Hairy epidermis	Franks et al. (2002)
Binova	SO4	Hermaphroditism	Wolf et al. (2003)
Bicchieri	Italia	Red berries	Fanizza et al. (2003)
La Notte	Italia	Longer berries	Fanizza et al. (2003)
Dipinto	Italia	Shorter cycle	Fanizza et al. (2003)
Mutated Italia	Italia	Cluster/berry size Short cycle	Fanizza et al. (2003)
Ruby Okuyama	Italia	Red berries	Kobayashi et al. (2004)
Flame Muscat	Muscat of Alexandria	Red berries	Kobayashi et al. (2004)
Fleshless berry	Ugni Blanc	No flesh	Fernandez et al. (2006a)
Malian/Shalistin	Cab-Sauvignon	Bronze/white berries	Walker et al. (2006)
Pinot Gris	Pinot Noir	Pink berries	Walker et al. (2006)
RRM	Carignan	Ramified bunch	Chatelet et al. (2007)
UFC	Bouchalès	Unfused carpels	Chatelet et al. (2007)
CLS	Mouvèdre	Carpel-less	Chatelet et al. (2007)
MPW	Gamay, Pinot	Multiple perianth whorls	Chatelet et al. (2007)
Star flowers	Syrah, Chardonnay	Calyptra anomalies	Longbottom et al. (2008)
Perfect-flowered	110R, 140Ru, ARG1	Hermaphroditism	Bouquet (2008)
Pink/green variants	Sangiovese	Colour variations	Ramazzoti et al. (2008)
Bequignol variant	Bequignol red	Variegated white/red berries	Delrot (2008)
Monococco Sultana	Sultana	Seeded berries	*http://bioweb.ensam.inra.fr/*
Red sultana	Sultana	Red berries	*http://bioweb.ensam.inra.fr/*

also be identified: a low berry dropping-type in Dolcetto, a polyploid Barbera with large berries less susceptible to *Botrytis* infection, a non-polyploid Delight mutant with large berries and a seedless Queen of Vineyards somatic variants. Kuksova et al. (1997) also reported ploidy changes in plants regenerated through somatic embryogenesis from irradiated embryogenic calli of cultivar Podarok Magaracha. The effect of chemical mutagens (EMS and EB) on buds of cultivar Pusa Seedless was studied under *in vitro* culture by Khawale et al. (2007). Furthermore, colchicine has also been used as a polyploidization agent with variable results on buds (Dermen 1954), embryogenic calluses (Kuksova et al. 1997) or single somatic embryos (Yang et al. 2006). Whichever organ is used in experiments (seed, bud or embryogenic tissue), mutagenic treatments result in chimerism (Alleweldt and Possingham 1988). However, the occurrence of chimeras was reduced when mutagenesis was combined with a highly regenerative procedure such as somatic embryogenesis (Kuksova et al. 1997).

3.2.3 Somaclonal Variation

The terms "somaclone" and "somaclonal variation" describe "the plants derived from any form of cell culture" and "the variation displayed amongst such plants", respectively (Larkin and Scowcroft 1981). Somaclonal variation can arise from pre-existing mutant somatic cells proliferating and/or regenerating in culture or from mutations or epimutations induced by the tissue culture procedures themselves (Karp 1991; Kaeppler et al. 2000). The potential of somaclonal variation in plant breeding was claimed earlier (Hammerschlag 1992) but few cultivars have been developed so far. In grapevine, somaclonal variation was proposed as a way to amplify clonal variability (Mullins 1987). But its successful exploitation depends upon the availability of *in vitro* screening procedures for abiotic or biotic traits, e.g., tolerance to NaCl (Lebrun et al. 1985) or *Eutypa lata* (Mauro 1986; Soulié et al. 1993).

Somaclonal variation is frequently observed among grapevine plants regenerated through somatic embryogenesis (Martinelli and Gribaudo 2001). Bouquet (1989) described a wide range of traits showing somaclonal variation, e.g., chlorophyll deficiencies, morphogenetic development, leaf shape, flower type and susceptibility to *Xylophyllus ampelinus*. The most spectacular variation observed was the conversion of male flowers into perfect-flowered flowers reported in Aramon x *rupestris* hybrids (Mullins 1987; Bouquet 2008) but also in various other male rootstocks (Bouquet 2008). In field trials, Mowbray et al. (1985) observed variations in berry size and sugar concentration among somaclones of Seyval. Mullins (1987) also reported that somaclones of Cabernet-Sauvignon were highly variable in growth and cropping, but phenotypes became similar with time. Variation in leaf shape was reported among a hundred somaclones of Grenache (Martinez et al. 1997). Considerable variability in vigor, cropping level, cluster size, berry weight, sugar concentration and total acidity, was observed on these somaclones (A Bouquet, unpublished results). Similar variability was also observed among somaclones of Gamay (Desperrier et al. 2003). But so far, no grapevine cultivar derived from somaclonal variation has been released. Finally, Kuksova et al. (1997) analyzed the ploidy level of somaclonal plants of Podarok Magaracha and found six tetraploid lines. Molecular polymorphisms have also been detected among somaclonal when using molecular markers such as random amplified polymorphic DNA (RAPD) (Schneider et al. 1996), amplified fragment length polymorphism (AFLP) (Popescu et al. 2002) or methylation-sensitive amplification polymorphism (MSAP).

Somaclonal variation could mainly result from the existing chimerism among cultivars L1 and L2 cell layers (Franks et al. 2002). In fact, the majority of the plants regenerated from somatic embryogenesis seem to derive from the L1 layer (Hocquigny et al. 2003). Furthermore, this type of variation could also be generated by changes in methylation (Schellenbaum et al. 2008) or ploidy levels.

3.3 Mechanisms Causing Somatic Variation

Several mechanisms, alone or in combination, may be at the origin of the somatic phenotype variation observed in grapevine: i.e., genetic alterations, epigenetic differences or changes in pathogen content.

3.3.1 Genetic Alterations

Genetic alterations causing somatic variation can range from major genetic changes such as whole genome duplications to minor point mutations. These alterations may result in loss of function (e.g., white leaf variegation) or gain of function (e.g., berry pigmentation); however, many of them will not result in an identified phenotype either because they are silent mutations, the mutated allele behaves as recessive or due to genetic redundancy (Crespan 2004). Among the major genetic changes, variation of ploidy level was found at the origin of variation for berry size in either *V. vinifera* or *V. labrusca*: e.g., the Muscat Cannon Hall variant derived from Muscat of Alexandria, the Early Giant from Campbell, the Wallis Giant from Concord or the Otsubu Niagara from Niagara (Olmo 1943). However, more frequently, a phenotypic change results from minor genetic modifications: e.g., insertions or deletions (INDELs), variations of SSR length or single nucleotide mutations resulting in SNPs. A single base substitution can be responsible for a huge somatic variation. This was demonstrated in Pinot Meunier, which is a well known periclinal somatic variant of Pinot Noir having tomentose vegetative organs. The regeneration of non-chimeric plants from Pinot Meunier L1 cell layer (Franks et al. 2002) revealed additional phenotypic effects like an extreme dwarfism and the production of inflorescences instead of tendrils. These variations were found genetically heritable and associated with a point mutation in *VvGAI1*, a gene involved in gibberelic acid signaling (Boss and Thomas 2002). The mutation causes amino acid substitution from leucine to histidine within the DELLA domain of the protein which impairs its function (L Torregrosa and MR Thomas, unpublished) and results in the critical Pinot L1 mutant phenotype.

Perhaps the best studied somatic varaitions in grapevine are those that affect the berry color. The lack of anthocyanin pigment biosynthesis was shown to be due to the insertion of *Gret1* (a 10,422 bp long retrotransposon) in the promoter of *VvmybA1* (Kobayashi et al. 2004). Further work showed that the berry color locus comprises another *VvmybA* gene, *VvmybA2* which is adjacent to *VvmybA1* (Walker et al. 2007). In the white allele, *VvmybA2* is inactivated by two non-conservative mutations so that the lack of transcription factor prevents anthocyanin biosynthesis. Most varieties with colored berries were found heterozygous at the berry color locus (Lijavetzky et al. 2006) but some cultivars such as Alphonse-Lavallée or Malbec carry two functional alleles (Spiegel-Roy et al. 1980; Madero et al. 1986). White berry somatic phenotypes observed in colored berry cultivars appear in heterozygous genotypes and result from deletions of functional *MybA* alleles (Walker et al. 2006; Yakushiji et al. 2006). On the other hand, somatic variation in white cultivars generating colored berries (generally pink or red) were shown to result from the excision of *Gret1* through LTR recombination (Kobayashi et al. 2004). Finally, the reiterated production of reproductive meristems observed in the RRM somatic variant of cultivar Carignan, has been shown to result from insertion of a functional class II transposable element (Fernandez et al. 2010), as described below. Thus, retrotransposons and transposons are also responsible for part of the somatic variation observed in grapevine. Three retrotransposons were characterized before the release of the grapevine genome sequence: *Vine-1* (Verriès et al. 2000), *Tvv1* (Pelsy and Merdinoglu 2002) and *Gret1* (Kobayashi et al. 2004). However, the release of grapevine genome sequence (Jaillon et al. 2007) has now provided the possibility to perform extensive census of transposable elements which increases the chance to predict the role of transposition in grapevine somatic variation (Benjak et al. 2008; Moisy et al. 2008).

3.3.2 Epigenetic Mechanisms

Epigenetic variation at the chromosomal level comprises two general mechanisms involved in regulating gene expression. One is chromatin configuration, which can be modulated by reversible histone modifications such as acetylation (Jarillo et al. 2009) affecting the spatial and temporal expression of genes. The second is cytosine methylation, a reversible covalent modification of DNA that seems to have dual roles in regulating gene expression and silent transposable elements or foreign DNA. Research on epigenetic mechanisms is scarce in grapevine. DNA methylation was shown to increase in ribosomal RNA genes of Sultanina *in vitro* plantlets (Harding et al. 1996) as well as in the T-DNA of transgenic plants. In this way, the T-DNA of Sultana transgenic plants showing reduced GUS expression had higher levels of methylation than plants with high GUS expression (Franks et al. 1998). Similarly, methylation of the *CP-GFLV* gene was detected in one transgenic plant of Nebbiolo with a low level of mRNA expression (Gambino et al. 2005). In somatic variation, differences among clones of cultivar Traminer were detected using methylation MSAP analysis (Imazio et al. 2002), however no phenotypic somatic variation has so far been linked with a specific sequence methylation in grapevine.

3.3.3 *Changes in Pathogen Contents*

The sanitary status of cultivar clones can also be a source of phenotypic variation (Mannini et al. 1999). Infection of plants with viruses, bacteria or phytoplasms can alter their morphology and physiology. Viroids have also been shown to be widespread in grapevine cultivars (Semancik et al. 1987), although the evaluation of the effect of viroids on morphological traits, fitness and fruit quality, depends on the availability of viroid-free grapes (Duran-Vila et al. 1988). Morpho-physiological changes (Borgo and Angelini 2002) observed among different cultivar clones could be associated with general changes of gene expression caused by virus infections (Espinosa et al. 2007). In any case, this source of phenotypic variation is generally minimized in grapevine as clonal selection aims to release only healthy clones (Golino and Wolpert 2003), discarding or curing clones infected with microorganisms.

3.4 Chimerism

Higher plants have a stratified apical meristems containing up to three layers of dividing cells, each cell layer remaining developmentally independent from the adjacent layers and giving rise to different plant tissues (Neilson-Jones 1969). For this reason, when a mutation arises in a SAM cell, during plant development, the mutant cell proliferates by mitosis producing a mutated sector. Then, somatic chimerism can be defined as a type of tissue organization that contains several cell genotypes. Chimeras are common in plants (Marcotrigiano 1997) and frequently maintained as samples. In grapevine, chimerism was first determined in sports showing variation for cell ploidy level (Einset and Pratt 1954). It has been proposed that *Vitis* SAM have two different cell layers (Thompson and Olmo 1963), L1 forming epidermis and L2 making up most of the other parts of the plant including mesophyll cells and gametes. Here chimerism is considered at the tissue level, excluding the case of grafted plants which results from an artificial association of two partners.

3.4.1 *Different Types of Chimeras*

According to their spatial patterning, chimeras can be classified as sectorial or periclinal (Fig. 3-1). In sectorial (syn. mericlinal or anticlinal) chimera, the phenotype differs from cell to cell within the same cell layer whereas in periclinal chimera, cell layers are genetically distinct (Dermen 1960). However, as somatic variation is a single cell event, generally a chimera is initially established at both the anticlinal and periclinal level. Only mutations taking place in an initial cell of the SAM can be uniformly distributed to whole organs. Green color variegations of the leaves are the best-known plant chimeras (Fig. 3-2). Chimerism can also modify various other traits, e.g. trichome density (Boss and Thomas 2002) or fruit development (Fernandez et al. 2006a). The phenotypic effect of a given somatic mutation depends on its location. For instance, the mutation *Vvgai1* results in very different phenotypes when the mutation in only present in L1 cell layer or in all cell layers (Boss and Thomas 2002).

3.4.2 *Occurrence and Detection of the Chimeric Status*

There is no information on the relative frequency of somatic variation in annual and perennial plants. However, lifespan of annuals is short and the regular formation of zygotic embryos reduces the possibility of somatic variation transmission to only those occurring in L2 cells. In perennial plants

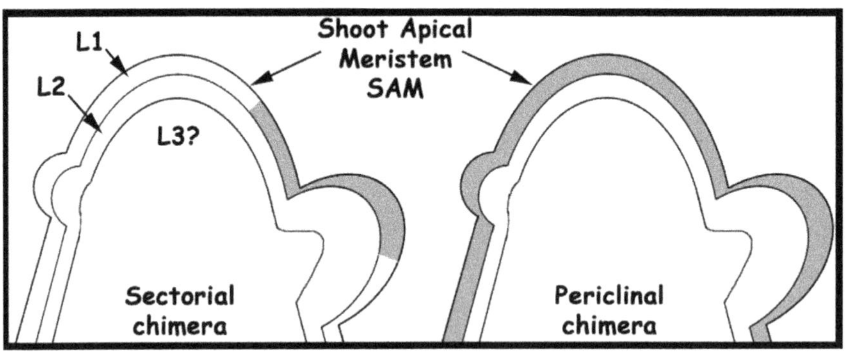

Figure 3-1 Diagram of the two types of chimera usually found in grapevine.

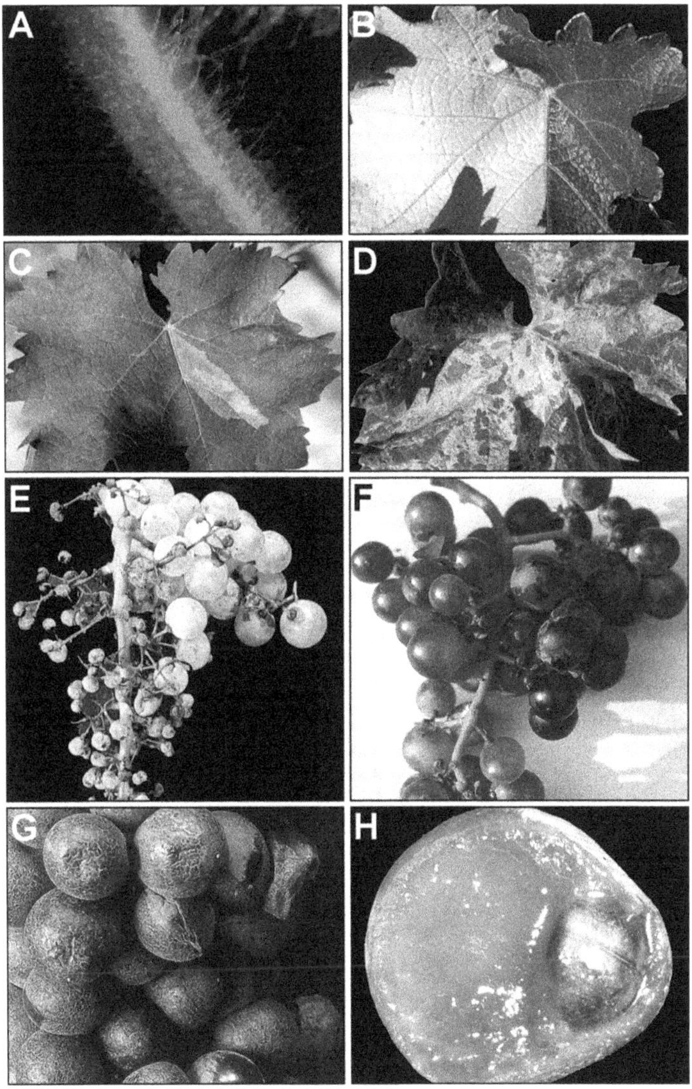

Figure 3-2 Some examples of grapevine chimeras (A) Transgenic root expressing localized GFP ectopic expression (B) Leaf with half of the blade showing chlorophyll deficiency (C) Leaf blade showing three different somatic phenotypes (wild type, chlorophyll deficiency in either on cell layer, chlorophyll deficiency in L1 and L2 cell layer) (D) Patchy chlorophyll deficiency variegations (E) Chimeric variation of FLB phenotype at bunch level (F) Somatic variation of berry color development at bunch level (G) Suberification of berry skin showing somatic variations at bunch and berry levels (unsuberified sector within suberified berry), (H) Ovary of the FLB somatic variant showing a reversion to normal development in one of the two carpels.

Color image of this figure appears in the color plate section at the end of the book.

that develop during a long period of time, genetic or epigenetic variations can independently accumulate in the different cell layers. As a result, during plant development, different tissues can have slightly different genotypes. Consequently, in perennials, chimerism is not only potentially more frequent but also much more complex than in annual species. Somatic chimerism seems to be quite common in grapevine where it has been observed in many genotypes such as Chardonnay (Bertsch et al. 2005), Greco di Tufo and Corvina Veronese (Crespan 2004), Primitivo (Franks et al. 2002) or Cabernet-Sauvignon (Walker et al. 2006). SSR screening within several other cultivars including Pinot clones as Pinot Gris, Pinot Noir, Pinot Blanc, Pinot Meunier, Pinot Moure and varieties such as, Cabernet Franc, Chenin, Grolleau, Cabernet Sauvignon, Riesling and Savagnin revealed a frequent prevalence of three and sometimes four alleles per microsatellite locus (Riaz et al. 2002; Hocquigny et al. 2004; Pelsy et al. unpublished).

3.4.3 Stability and Propagation of Chimeras

The case of Pinot Meunier which has been propagated for centuries, demonstrates the stability of chimeric status in grapevine. However, some chimeras may show instability. For instance, the reversion from Brazil cultivar (dark-red berries) to Benitaka (pink berries) was estimated to be 28.57% by Oliveira-Collet et al. (2005). Similarly, reversion from Pinot Meunier to Pinot Noir is also commonly observed (Franks et al. 2002; Rose 2005). Pinot Gris also exhibited some instability, displaying colored berries with white variegation (Hocquigny et al. 2004), clusters with white berries or canes with white berries. Malian was also found to produce canes with white fruits at a frequency of 1–2 per thousand vines per year (Walker et al. 2006). Similarly, Fernandez et al. (2006a) who characterized the *fleshless* berry somatic variant observed reversions to the wild type occurring at a rate of 10^{-3} to 10^{-4}.

Theoretically, a chimera can return to the original phenotype if the mutated gene allele reverts to the wild-type stage or if some cellular rearrangements occur during organogenesis. The invasion by cells from the inner layers into the outer L1 layer is thought to be the most frequent cause of reversion because of the low level of division organization in L2 layer (Stewart and Dermen 1970). It can occur as a result of cell or tissue damage either by insect or mechanical disruption. Such cellular rearrangements could have occurred to generate white berries in Pinot Gris, in agreement with the white phenotype of the L2 layer. Examination of the Pinot Blanc skin confirmed that the epidermal and other skin cells are not colored in this cultivar (Walker et al. 2006). In the same way, the white-berried Shalistan, whose epidermal and other skin cells are not colored could be derived from Malian by the displacement of L2 mutated cell towards the L1 layer. The invasion of the L2 cell layer by cells from the outer L1 layer seems to be more rare probably due to the stability of the anticlinal cell divisions.

Somatic mutations can be propagated to the whole plant by mutant cell invasion of non-mutant layers and can be sexually transmitted if located in L2 cell layer (Neilson-Jones 1969). When located in the L1, a somatic variation can be transmitted via adventitious organogenesis (Skene and Barlass 1983) or somatic embryogenesis (Franks et al. 2002) because regeneration techniques allow the recovery of whole plants from individual somatic cells. However, the only way to propagate a chimeric state is the formation of plants from established buds through the use of cuttings, layering or grafting.

3.5 Somatic Variants of Economic and Academic Interest

Somatic variants potentially represent a commercial interest if they improve an important agronomical trait (e.g., height, yield, pathogen susceptibility, etc.) or allow product innovation. For instance, berry color somatic variants of Cabernet-Sauvignon (Malian and Shalistin) gave the opportunity to produce some unique wines in South Australia (*http://www.cleggettwines.com.au/*). However, many polymorphisms were only maintained as samples in germplasm collections. Some of these variants have been described (Viala and Vermorel 1901–1910; Olmo 1935; Snyder and Harmon 1934; Scherz 1940; Einset and Pratt 1954; Sousa 1959; Breider 1962; Thompson and Olmo 1963; Ourecky et al. 1967; Kishino and Mashima 1980; Terra et al. 1984; Scott et al. 2001; Fanizza et al. 2003; Fernandez et al. 2006a; Chatelet et al. 2007). However, most grapevine somatic variants have not been described in detail and their genetics and nature of the mechanism underlying phenotypic changes remain unknown. Here, we will briefly describe those somatic variants showing alterations in plant development and berry traits that have been more deeply analyzed given their conspicuous phenotypes.

3.5.1 Somatic Variants for Vegetative Development

There is a large set of somatic variants that have been found to be associated with either general developmental modifications or more subtle phenotypes (Table 3-1). Some of these phenotypes are related to phenology timing of vegetative or reproductive phase transitions. These modifications are particularly valuable for table grape production as earliness or lateness releases competition within the production peaks and allows higher market prices (Scott et al. 2001). Most often, somatic variation modifies a particular trait of organ morphology, such as the Meunier phenotype identified in several cultivars (e.g., Grenache, Pinot Noir and Alicante-Bouschet) or the Cioutat somatic variation identified in Chasselas. Some of these variations, like Lledoner Pelut or the Pinot Meunier were largely used by the wine industry. The molecular determinism of the Pinot Meunier variant which is a major cultivar of the Champagne region was associated with the presence at the heterozygous state of the *Vvgai1* mutation. However, the genetic and molecular mechanisms causing Meunier phenotype in other cultivars remain unknown.

The determinism of chlorophyll variegation that appears in some grapevine somatic variants (Fig. 3-2) was examined by Reisch and Watson (1984) leading to the identification of a single recessive gene. This phenotype could be useful for ornamental use or for scientific research, e.g., to score cell layer fate during organ and tissue ontogenesis.

Disease susceptibility was also found to be subjected to somatic variation. For instance, three classes of foliar susceptibility to downy mildew have been reported in the *V. vinifera* cv. Albariño (Boso and Kassemeyer 2008). The underlying mechanism of these phenotypes could be related to trichoma density on the abaxial leaf side (Loureiro et al. 1998). Investigation of the Syrah decline identified considerable susceptibility variations among clones (Renault-Spilmont et al. 2007).

3.5.2 Somatic Variants for Reproductive Traits

In grapevine, inflorescences are initiated in latent buds during the vegetative cycle and remain indeterminate until the proleptic axis develops. Although grapevine reproductive biology has been carefully described in different genotypes (see Carmona et al. 2008 for a review), the molecular mechanisms of flowering induction and reproductive organ initiation need to be elucidated. Somatic variants have provided the strongest genetic evidence available on the participation of specific genes in these developmental processes. In this way, the somatic mutation present in the L1 cell layer of Pinot Meunier described previously, affects reproductive development when it is also present in the L2 cell layer. In mutant plants carrying this mutation, the anlagen opposite to the leaf primordia gives rise to inflorescence meristems in place of tendril primordia (Franks et al. 2002). The dominant negative mutation in the *VvGAI1* gene of Pinot Meunier prevents GA response indicating a major role for GAs in the inhibition of inflorescence differentiation (Boss and Thomas 2002).

Variation for inflorescence size and branching is commonly observed among cultivars (Shavrukov et al. 2003). These features are critical for final compactness of the cluster, which is directly linked to *Botrytis* susceptibility. Somatic variation for this trait has been observed among cultivars such as Chardonnay, where the French clone 548 exhibited lower bunch compactness than regular clones (Maigre 2005). More extreme phenotypes are observed in the reiterated reproductive meristems (RRM) variant of Carignan

(Fernandez et al. 2010) and the Mutated Branch variant of Italia (Fanizza et al. 2003), both developing very large and highly branched clusters. In RRM, all types of lateral organs are affected, with tendrils also being more branched and with a higher presence of intermediate structures between leaf and tendril or inflorescence. Genetic and molecular analyses have shown that the RRM phenotype is inherited as if caused by a single dominant mutation. Transcriptional profiling along early stages of RRM inflorescence development associated the mutant phenotype with overexpression of a few regulatory genes including *VvTFL1A* and molecular analyses have shown that the insertion of a class II transposable element in *VvTFL1A* promoter is the possible cause of *VvTFL1A* overexpression in an allele-specific cis-activation manner (Fernandez et al. 2010). In *Arabidopsis*, *TFL1* is required to maintain inflorescence meristem indeterminacy (Bradley et al. 1997) and *TFL1A* overexpression in the RRM plants is hypothesized to induce the reiteration of inflorescence meristems resulting in higher cluster branching and a delay in flower development. Similar phenotypic effects were produced by overexpression of *Arabidopsis TFL1* or *V. vinifera TFL1A* in transgenic *Arabidopsis* plants (Ratcliffe et al. 1998; Carmona et al. 2007).

Several somatic variants altered in flower development were described by Chatelet et al. (2007). Among them, the multiple perianth whorls (MPW) variants found in cultivars Gamay, Morrastel, and Pinot, are able to develop perianth organs, but are impaired in the specification of anther and carpel identity, coherent with the down-regulation of *VvMADS1*, they showed an *AG* homolog.. The Carpel-less (CLS) variant of cultivar Mourvedre exhibits a late formation of ovule-like structures in the absence of carpels, coherent with a delay in *VvMADS1* up-regulation. The Unfused carpels (UFC) variant of Bouchales exhibits an abnormal development of reproductive whorls with no effect on *VvMADS1*, *VvMADS2* (a *SEP* homolog), and *VvMADS3* (an *AGL13* homolog) expression. However, another study has identified a delayed overexpression of *VvMADS9* (*PI* homolog) in UFC (Sreekantan et al. 2006). Even if expression data of some MADS-box genes are correlated with flower alterations, the genetic control and the molecular defects of these somatic variants remain unknown. Finally, Star flower somatic variants have recently been described in Chardonnay and Shiraz (Longbottom et al. 2008). Flowers of these variants precociously open the petals from the top of the calyptra, before the anthers and pistil mature, resulting in male sterility. Ovaries of the Star Shiraz do not persist after flowering. However, Star flower Chardonnay variants produce abundant seedless berries. Therefore, either different mechanisms are affected in these somatic variants or phenotypic variation is due to additional genetic interactions in their genetic background. The genetic and molecular mechanisms are so far unknown.

3.5.3 Berry Quality Traits

Within *V. vinifera* there is a wide range of berry shape and size (Galet 2000) that can range from 1–2 g in berries for wine production to 5–10 g in table grape berries. Somatic variation of berry size can result from three different sources. Somatic variation for ploidy levels, putative pleiotropic effects of somatic variation for stenospermocarpic seedlessness and somatic mutations. Somatic ploidy variations were the first to be identified as the origin of berry size sports such as Muscat Cannon Hall, Black King or Benikawachi that were commercialy developed as table grape cultivars (Olmo 1943). Furthermore, the presence of a dominant mutation at the *SDI* locus in cultivar Sultanina or Thompson Seedless causes stenospermocarpy (Lahogue et al. 1998) and reduced fruit size. This mutation has been sexually transmitted and is now the reason for seedlessness in most table grape cultivars. Finally, six somatic variants for berry size were analyzed in terms of cell division and pericarp enlargement to understand the base of phenotypic variation (Fernandez et al. 2006b). Both, cell number and cell volume were affected in the small-sized berries of the *flb* somatic variant, whereas differences in fruit size in other small-size clones were always due to cell expansion. The *fleshless berry* somatic mutant was further characterized at either genetic and biochemical levels (Fernandez et al. 2006a, c). Transcriptome analyses (Fernandez et al. 2005, 2007) led to the identification of several putative regulators of berry early morphogenesis (*VvNap, VvYabby, VvHB13, VvPI, VvSpec* or *VvBurp*) that were underexpressed or overexpressed in *flb* variant. Some orthologs of these genes were also identified during the morphogenesis of other fleshly fuits (Lemaire-Chamley et al. 2005) showing critical function in the regulation of fruit size (Cong et al. 2008). However, the underlying ontogenic and molecular mechanisms of grapevine berry growth regulation in this somatic variant is still unknown.

Regarding berry color, the basic mechanisms of fruit coloration are now established for several plants (Allan et al. 2008) but the fine tuning of this trait is still poorly understood. In grapevine, somatic changes in either berry skin or flesh pigmentation are widely-known, probably because they are easy to identify (Viala and Vermorel 1901–1910; Müller-Stoll 1950; Galet 2000). Variations from white to pink or red berries and from black to white have been identified in a large number of cultivars (Table 3-1). Furthermore, color somatic variations can affect either skin, flesh or both producing different results. For instance, the variants of Gamay Noir, Gamay Freaux and Gamay de Chaudenay develop colored flesh berries (Galet 2000) allowing the production of highly colored red wine. The first decisive progress on the genetic basis of berry color somatic variation was reported by Kobayashi et al. (2004) based on the characterization of two somatic variants, Ruby Okuyama and Flame Muscat, derived from Italia and Muscat of Alexandria respectively. Since then, using somatic polymorphisms, several studies investigated the specific function of structural genes (Ageorges et al. 2006; Castellarin and Di Gaspero 2007), the genetic and molecular mechanism of the pathway regulation (Lijavetzky et al. 2006; This et al. 2007; Walker et al. 2007) and the role of chimerism in the phenotype determination (Walker et al. 2006).

There are many other biochemical properties that are essential for wine or table grape quality in relation to the accumulation of specific metabolites (proanthocyanidins, organic acids, sugars, aromas, etc.) or the production of specific enzymes modifying their structure, many of which are affected by a wide range of somatic variations (Table 3-1). For instance, Gora Chirine is thought to be a somatic variation of Sultana producing acidless berries (Boubals et al. 1971). The decrease in acidity appears very early before véraison (Diakou et al. 1997). The lower TA of Gora Chirine juice could be attributed to lower concentrations of all three major acids, malic, tartaric and citric, and to a higher potassium concentration. At *véraison*, the malic acid concentration in the Gora Chirine berry juice was about seven times lower than for a standard *V. vinifera* cultivar. Interestingly, a precocious accumulation of hexoses compensates cell osmotic variation to support vacuolar expansion and berry growth. Bruce's sport is also another interesting somatic variant of Sultana which produces raisins with a lighter color than other seedless cultivars (Antcliff and Webster 1962). Bruce's sport shows a reduction of polyphenol oxidase (PPO) enzyme activity (Rathjen and Robinson 1992) that decreases browning during berry drying. The Bruce's sport grapes were found to be variegated and the green regions of skin had a similar PPO activity to the wild-type Sultana while the white regions had very low PPO activity.

There are also somatic variants differing in berry and/or wine aromas. For instance, two types of Chardonnay grown in Burgundy to produce wines are specified in the French official catalog as being either neutral or having a "muscat" aroma (Anonymous 2007). Savagnin variants, that are grown to produce either aromatic or neutral wines (Pacottet 1903), are classified as three distinct cultivars in the French official catalog (Anonymous 2007): Gewurztraminer, with pink berries and aromatic wine, Savagnin Rose with pink berries and neutral wine and Savagnin Blanc (also known as Traminer) with white berries and neutral wine. While comparing Savagnin Rose and Gewurztraminer, the aroma profile of Gewurztraminer was found to contain a higher geraniol content than in Savagnin Rose (Duchêne et al. 2009). Moreover, these aromatic traits were shown to be heritable as a single Mendelian trait in the self-progenies of Chardonnay and Gewurztraminer.

3.6 Clonal Selection

Initially, viticulturists used cuttings or layering for propagation. However, since the phylloxera crisis, genotypes are mainly propagated by grafting. Such ways of propagation used for a long time period, have resulted in phenotypic heterogeneity (Duchêne et al. 2009) that becomes a new source of variation for genetic improvement of grape cultivars through clonal selection. The aim of clonal selection is to take advantage of this somatic variability for grapevine improvement. While one of the precursors was Gustav Froelich in Germany, who began to multiply and to compare selected single vines of Sylvaner in 1878 (Schmid et al. 1995), the main clonal selection programs were developed during the second part of the 20th century. According to the definition of the International Organization of Vine and Wine, a clone can be regarded as the vegetative conformant to a vine chosen for its unquestionable identity, its morphological characters and its sanitary condition (Anonymous 2007).

3.6.1 Typology of Clonal Variation

The typology of clonal variation differs from one cultivar to another (Cid-Alvarez et al. 1998). A large range of trait variations is commonly observed during the clonal selection process affecting many different traits such as morphological traits: Density of prostrate (Aramon, Tannat) and erect hairs (Muscat à petits grains), pigmentation of leaf veins (Chenin) and nodes (Sauvignon); texture (Syrah), color (Muscat d'Alexandrie), brightness (Mauzac), number of lobes (Gamay), lateral sinuses depth (Pinot noir) of the leaf, opening and shape of the petiolar sinus base (Syrah), berry shape (Gamay, Muscat de Hambourg), berry skin thickness (Pinot); Physiological traits: Shoot attitude (Merlot, Pinot noir), vigor (Syrah), ripening time (Chardonnay); Production traits: Fruitfullness (Syrah), blossom drop (Grenache), cluster compactness (Viognier), millerandage and berry size (Chardonnay, Syrah); Berry quality traits: sugar content (Grenache), acidity (Syrah, Mourvèdre), anthocyanins (Cabernet franc), tannins (Cabernet-Sauvignon), and aromatic compounds (Muscat of Alexandria and Chardonnay).

3.6.2 Achievement of the Clonal Selection Programs

Several steps are necessary to achieve clonal selection on sanitary and genetic basis. The entire process is long, as a minimum of 10–15 years is required for completion. Depending on the country, sanitary or genetic approaches may be more or less favored, but due to the different possible origins of somatic variation (see Section 3.2), best results are obtained when both objectives are balanced (Mannini 2000). As an example, the pipeline of the French clonal selection program is given in Figure 3-3. The first step is to pick accessions in old vineyards or from repositories. Then, the program considers sanitary selection. The set of viruses that are scored by indexing, ELISA or PCR methods varies according to the institutions and/or the country rules. However, there is an agreement to only release clones free from Grapevine Fanleaf Virus (GFLV), Arabis Mosaic Virus (ArMV) and Grapevine Leafroll-associated Viruses (GLRaV), which are the most harmful viruses. In some cases, additional viruses or virus-like diseases such as Grapevine Fleck Virus (GFkV), Rupestris Stem Pitting (RSP), Kober Stem Grooving (KSG), Grapevine Corky Bark (GCB), LN33 Stem Grooving (LN33SG), Grapevine Vein Mosaic (GVM) or Grapevine Vein Necrosis (GVN) may also be investigated. If no healthy accession is found, *in vitro* sanitation techniques, i.e., thermotherapy, shoot tip meristem culture, embryogenesis or micro-grafting (Barlass et al. 1982; Duran-Vila et al. 1988; Goussard et al. 1991; Torres-Vinals et al. 2004) need to be performed. After such treatments, as the pathogen content can be related to the variation of the phenotype, conformity and behavior of the material need to be rechecked before further release.

With regards to the genetic selection part, the scheme is currently based on experimental field trials with adapted protocols (Boidron 1995). The aim is to evaluate agronomical and technological traits (for wine production) in order to select the most diverse and appropriate clones according to the needs of industry. The evaluation is preferably performed in the main culture area

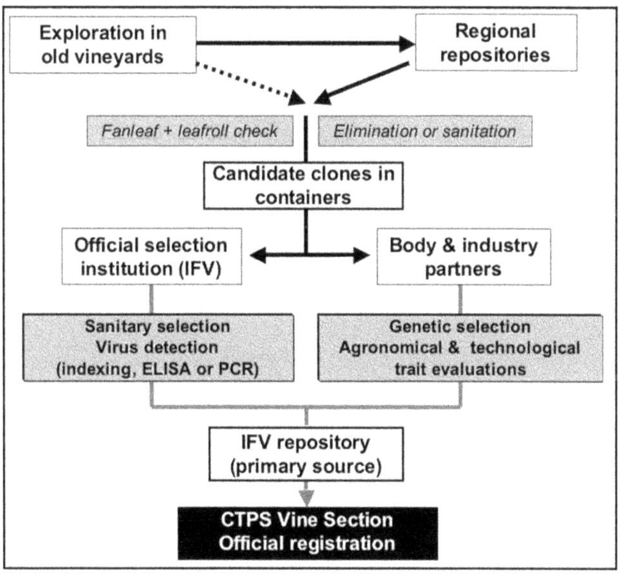

Figure 3-3 Description of the French clonal selection scheme (IFV: Institut Français de la Vigne et du Vin, CTPS: Comité Technique Permanent de la Sélection).

of the cultivar. This step is particularly important for the best accurate use of clonal material. But is very time-consuming as several years are required to get a good picture of the phenotypic characteristics of each selected clone. In the future, it is expected that tools for marker-assisted selection (MAS) could provide some help to carry out this decisive part of the selection.

3.6.3 Propagation and Certification

Clonal selection represents huge investments to preserve the integrity of the selected material and to provide the maximum guarantee to grapevine industry. To strictly control all propagation steps, certification procedures need to be developed. The tracking of clones has to be carefully performed at all stages of propagation to ensure the trueness-to-type and the sanitary status of the material. Regular sanitary checks have to be done in clonal germplasm and propagating material. All the propagation steps have to be clearly identified in order to avoid any confusion between the different categories of propagating material. In particular, it is crucial to restrict the number of propagation generations of mother vines and special attention has to be paid to holding the conditions of the primary source of clones.

3.7 Conclusions

Somatic variation in grapevine plays a major role as a source of phenotypic variation for clonal selection, representing the only approach allowing the genetic improvement of classical cultivars. Clonal selection is a very powerful tool to improve grapevine. Progress with the application of this technique proved very significant to improve culture conditions and output quality for both table grape or wine cultivars. However, as old vineyards become rare, preservation of the natural intra-varietal diversity is a stake and a responsibility for breeders and institutions interested in the use of somatic variation. Even if some biotechnologies (e.g., induced mutagenesis or somaclonal variation) may increase the frequency of variation, it is essential to maintain the largest natural diversity as possible, for instance in repositories, to be able to select new clonal properties in response to the evolution of viticulture challenges, like the reduction of chemical inputs, fall in cost pricing, adaptation to climatic changes or market demands.

The grapevine genome sequence (Jaillon et al. 2007) provides new information and tools to understand the origin of somatic variation at both genetic and molecular levels which is a prerequisite for its efficient use in grapevine breeding. Understanding the molecular genetics of somatic variation can help devise strategies to increase its frequency, develop new selections schemes or as a final application develop molecular markers for clone identification and protection. In addition, somatic variation proved useful to investigate gene biological function in grapevine. Careful study of the genetic control of somatic variant phenotype, identification of the involved genes and characterization of their nucleotide diversity in independent variants of similar phenotype can provide insights on the role of those genes in a specific trait. Furthermore, understanding the origin and effects of somatic variation can help in avoiding undesirable changes, such as morphological mutations, that can occur during genetic transformation of grapevine. Thus, grapevine somatic variants constitute unique resources for both functional genomics and breeding worth collecting and describing.

References

Ageorges A, Fernandez L, Vialet S, Merdinoglu D, Terrier N, Romieu C (2006) Four specific isogenes of the anthocyanin metabolic pathway are systematically co-expressed with the red colour of grape berries. Plant Sci 170: 372–383.

Allan AC, Hellens RP, Laing WA (2008) MYB transcription factors that colour our fruit. Trends Plant Sci 13: 99–102.

Alleweldt G, Possingham JV (1988) Progress in grapevine breeding. Theor Appl Genet 75: 669–673.

Anonymous (2007) IFV-INRA-Montpellier SupAgro-Vinifhlor. Catalogue des variétés et clones de vigne cultivés en France, 2ème édn. Institut Français de la Vigne et du Vin, Le Grau du Roi, France.

Antcliff AJ, Webster WJ (1962) Bruces Sport, a mutant of the Sultana. Aust J Exp Agri 2: 97–100.

Antolin MF, Strobeck C (1985) The population genetics of somatic mutations in plants. Am Nat 126: 52–62.

Barlass M, Skene KG, Woodham RD, Krake LR (1982) Regeneration of virus-free grapevines using *in vitro* apical culture. Ann Appl Biol 101: 291–295.

Bäurle I, Dean C (2006) The timing of developmental transitions in plants. Cell 125: 655–664.

Benjak A, Forneck A, Casacuberta JM (2008) Genome-wide analysis of the cut-and-paste transposons of grapevine. PLoS ONE 3: e3107.

Bertsch C, Kieffer F, Maillot P, Farine S, Butterlin G, Merdinoglu D, Walter B (2005) Genetic chimerism of *Vitis vinifera* cv. Chardonnay 96 is maintained through organogenesis but not somatic embryogenesis. BMC Plant Biol 5: 20.

Boidron R (1995) Clonal selection in France. Methods, organization and use. Proc 3rd Int Symp on Clonal Selection, June 20–21, ASEV, Portland, Oregon, USA, pp 1–7.

Borgo M, Angelini E (2002) Influence of grapevine leaf roll (GLRaV3) on Merlot cv. grape production. Bull OIV 859–860: 611–623.

Boso S, Kassemeyer HH (2008) Different susceptibility of European grapevine cultivars for downy mildew. Vitis 47: 39–49.

Boss PK, Thomas MR (2002) Association of dwarfism and floral induction with a grape 'green revolution' mutation. Nature 416: 847–850.

Boss PK, Davies C, Robinson SP (1996a) Anthocyanin composition and anthocyanin pathway gene expression in grapevine sports differing in berry skin colour. Aust J Grape Wine Res 2: 163–170.

Boss PK, Davies C, Robinson SP (1996b) Expression of anthocyanin biosynthesis pathway genes in red and white grapes. Plant Mol Biol 32: 565–569.

Botta R, Vellania R, Me G (1987) Grapevine breeding by gamma radiations: results and prospects. Acta Hort 224: 30–36.

Boubals D, Bourzeix M, Guitraud J (1971) Gora Chirine, variete de vigne iranienne à faible teneur en acides organiques dans les baies. Ann Amelior Plantes 21: 281–285.

Bouquet A (1989) *In vitro* culture for grapevine breeding: *In ovulo* embryo culture, somatic embryogenesis and somaclonal variation (in French). Quad Vitic Enol Univ Torino 13: 51–64.
Bouquet A (2008) Grapevine and its genetic diversity (a review). Progr Agric Vitic 125: 91–102.
Bradley D, Ratcliffe O, Vincent C, Carpenter R, Coen E (1997) Inflorescence commitment and architecture in Arabidopsis. Science 275: 80–83.
Breider H (1962) Spontane Blütenmutationen bei der Rebe. Züchter 32: 100–102.
Carmona MJ, Calonje M, Martinez-Zapater JM (2007) The FT/TFL1 gene family in grapevine. Plant Mol Biol 63: 637–650.
Carmona MJ, Chaib J, Martınez-Zapater JM, Thomas MR (2008) A molecular genetic perspective of reproductive development in grapevine. J Exp Bot 59: 2579–2596.
Castellarin SD, Di Gaspero G (2007) Transcriptional control of anthocyanin biosynthetic genes in extreme phenotypes for berry pigmentation of naturally occurring grapevines. BMC Plant Biol 7: 46.
Charbaji T, Nabulsi I (1999) Effect of low doses of gamma irradiation on *in vitro* growth of grapevine. Plant Cell Tiss Organ Cult 57: 129–132.
Chatelet P, Fernandez L, Laucou V, Sreekantan L, Lacombe T, Martinez-Zapater JM, Thoma MR, Torregrosa L (2007) Characterization of *Vitis vinifera* L. somatic variants with exhibiting abnormal flower development patterns. J Exp Bot 58: 4107–4118.
Cid-Alvarez N, Boursiquot JM, Martinez LR, Saa Otero MP (1998) Differentiation of clones of varieties of grapevines that are traditionally cultivated in NW Spain (Galicia) and elaboration of determination key for application of philometric measures. J Int Sci Vigne Vin 4: 183–191.
Cong B, Barrero LS, Tanksley SD (2008) Regulatory change in YABBY-like transcription factor led to evolution of extreme fruit size during tomato domestication. Nat Genet 40: 800–804.
Crespan M (2004) Evidence on the evolution of polymorphism of microsatellite markers in varieties of *Vitis vinifera* L. Theor Appl Genet 108: 231–237.
Da Silva AL (1995) *In vitro* mutagenesis on grapevine rootstocks: methodology of irradiation, isolation and characterisation of the mutants. PhD Thesis, Univ of Bordeaux, France.
Da Silva AL, Hariscain P, Ollat N, Doazan JP (2000) Comparative *in vitro* development of five grapevine rootstock varieties and mutants from the cultivar 'Gravesac'. Acta Hort 528: 351–357.
Delrot S (2008) Understanding the biology: Moving away from observation based science to predictive science (oral presentation). 8th Int Symp on Grape Physiology and Biotechnology, November 23–28 2008, Adelaide, Australia.
Dermen H (1954) Colchiploidy in grapes. J Hered 45: 159–172.
Dermen H (1960) Nature of plant sports. Am Hort Mag 39: 123–173.
Desperrier JM, Berger JL, Bessis R, Fournioux JC, Labroche C (2003) Directed clonal creation by somatic embryogenesis (in French). Bull OIV 871–872: 751–765.
Diakou P, Moing A, Svanella L, Ollat N, Rolin DB, Gaudillere M, Gaudillere JP (1997) Biochemical comparison of two grape varieties differing in juice acidity. Aust J Grape Wine Res 3: 117–126.
Duchêne E, Legras JL, Karst F, Merdinoglu D, Claudel P, Jaegli N, Pelsy F (2009) Variation of linalool and geraniol content within two pairs of aromatic and non aromatic grapevine clones. Aust J Grape Wine Res 15: 120–130.
Duran-Vila N, Juarez J, Arregui M (1988) Production of viroid-free grapevines by shoot tip culture. Am J Enol Vitic 39: 217–220.
Einset J, Pratt C (1954) Giant sport of grapes. Proc Am Soc Hort Sci 63: 251–256.
Espinosa C, Vega A, Medina C, Schlauc K, Cramer G, Arce-Johnson P (2007) Gene expression associated with compatible viral diseases in grapevine cultivars. Funct Integr Genom 7: 95–110.

Fanizza G, Chaabane R, Ricciardi L, Resta P (2003) Analysis of a spontaneous mutant and selected clones of cv. Italia (*Vitis vinifera*) by AFLP markers. Vitis 42: 27–30.
Fernandez L, Ageorges A, Torregrosa L (2005) A putative NAP homolog specifically expressed during grapevine flower and berry development. Vitis 44: 51–52.
Fernandez L, Doligez A, Lopez G, Thomas MR, Bouquet A, Torregrosa L (2006a) Somatic chimerism, genetic inheritance, and mapping of the fleshless berry (*flb*) mutation in grapevine (*Vitis vinifera* L.). Genome 49: 721–728.
Fernandez L, Pradal M, Lopez G, Berud F, Romieu C, Torregrosa L (2006b) Berry size variability in *Vitis vinifera* L. Vitis 45: 53–55.
Fernandez L, Romieu C, Moing A, Bouquet A, Maucourt M, Thomas MR, Torregrosa L (2006c) The grapevine fleshless berry mutation: A unique genotype to investigate differences between fleshy and non-fleshy fruit. Plant Physiol 140: 537–547.
Fernandez L, Torregrosa L, Terrier N, Sreekantan L, Grimplet J, Davies C, Thomas MR, Romieu, C, Ageorges A (2007) Identification of genes associated with flesh morphogenesis in grapevine (*V. vinifera* L.) berry. Plant Mol Biol 63: 307–323.
Fernandez L, Torregrosa L, Segura V, Bouquet A, Martinez-Zapater JM (2010) Transposon-induced gene activation as a mechanism generating cluster shape somatic variation in grapevine. Plant J 61: 545–557.
Franks T, He DG, Thomas MR (1998) Regeneration of transgenic *Vitis vinifera* L. sultana plants: genotypic and phenotypic analysis. Mol Breed 4: 321–333.
Franks T, Botta R, Thomas MR (2002) Chimerism in grapevines: implications for cultivar identity, ancestry and genetic improvement. Theor Appl Genet 104: 192–199.
Galet P (2000) Dictionnaire Encyclopédique des Cépages. Hachette pratique, Paris, France (ISBN: 9782012363311).
Gambino G, Gribaudo I, Leopold S, Schartl A, Laimer M (2005) Molecular characterization of grapevine plants transformed with GFLV resistance genes. Plant Cell Rep 24: 655–662.
Gonçalves JA (1995) Parana descobre nova variedade de uva. Folha de Sao Paulo, Agrofolha, Sao Paulo, 12 dezembro, p 1.
Golino D, Wolpert J (2003) Vine Selection and Clones. In: Wine Grape Varieties in California. Univ of California Agricultural and Natural Resources Publication, Oakland, CA, USA, pp 8–11.
Goussard PG, Wiid J, Kasdorf GGF (1991) The effectiveness of *in vitro* somatic embryogenesis in eliminating fanleaf virus and leafroll associated viruses from grapevines. S Afr J Enol Vitic 12: 77–81.
Hammerschlag FA (1992) Somaclonal variation. In: FA Hammerschlag, RE Litz (eds) Biotechnology of Perennial Fruit Crops, CABI Publ, Wallingford, Oxon, UK, pp 35–55.
Harding K, Benson EE, Roubelakis-Angelakis KA (1996) Methylated DNA changes associated with the initiation and maintenance of *Vitis vinifera in vitro* shoot and callus cultures : A possible mechanism for age-related changes. Vitis 35: 79–86.
Harmon FN, Snyder E (1936) A seeded mutation of the Panariti grape. J Hered 27: 77–78.
Hocquigny S, Merdinoglu D, Heloir MC, Pelsy F (2003) Chimerism and genetic diversity within the cultivar group of pinots. Acta Hort 603: 535–544.
Hocquigny S, Pelsy F, Dumas V, Kindt S, Heloir MC, Merdinoglu D (2004) Diversification within grapevine cultivars goes through chimeric states. Genome 47: 579–589.
Imazio S, Labra M, Grassi F, Winfield M, Bardini M, Scienza A (2002) Molecular tools for clone identification : the case of grapevine cultivar 'Traminer'. Plant Breed 121: 531–535.
Jaillon O, Aury JM, Noel B, Policriti A, Clepet C, Casagrande A, Choisne N, Aubourg S, Vitulo N, Jubin C, Vezzi A, Legeai F, Hugueney P, Dasilva C, Horner D, Mica E, Jublot J, Poulain J, Bruyere C, Billault A, Segurens B, Gouyvenoux M, Ugarte E, Cattonaro F, Anthouard V, Vico V, Del Fabbro C, Alaux M, Di Gaspero G, Dumas V, Felice N, Paillard S, Juman I, Moroldo M, Scalabrin S, Canaguier A, Le Clainche I, Malacrida G, Durand E, Pesole G, Laucou V, Chatelet P, Merdinoglu D, Delledonne M, Pezzotti M, LecharnyA, Scarpelli C, Artiguenave F, Pè ME, Valle G, Morgante M, Caboche M, Adam-Blondon AF, Weissenbach J, Quétier F, Wincker P (2007) The grapevine genome sequence suggests ancestral hexaploidization in major angiosperm phyla. Nature 449: 463–467.

Jarillo JA, Piñeiro M, Cubas P, Martinez-Zapater JM (2009) Chromatin remodeling in plant development. Int J Dev Biol 53: 1581–1596.
Kaeppler SM, Kaeppler HF, Rhee Y (2000) Epigenetic aspects of somaclonal variation in plants. Plant Mol Biol 43: 179–188.
Karp A (1991) On the current understanding of somaclonal variation. In: BJ Miflin (ed). Oxford Surveys of Plant Molecular and Cell Biology. Oxford Univ Press, UK, pp 1–58.
Khawale RA, Yerramilli V, Singh SK (2007) Molecular marker-assisted selection of *in vitro* chemical mutagen-induced grapevine mutants. Curr Science 92: 1056–1060.
Kishino AY, Mashima M (1980) Uva: *Vitis vinifera* L. In: Manual Agropecua Del Rio do Parana. IAPAR, Londrina, Brazil, pp 138–177.
Kobayashi S, Goto-Yamamoto N, Hirochika H (2004) Retrotransposon-induced mutations in grape skin color. Science 304: 982–982.
Kuksova VB, Piven NM, Gleba YY (1997) Somaclonal variation and *in vitro* induced mutagenesis in grapevine. Plant Cell Tiss Organ Cult 49: 17–27.
Lahogue F, This P, Bouquet A (1998) Identification of a codominant marker linked to the seedlessness character in grapevine. Theor Appl Genet 97: 950–959.
Larkin PJ, Scowcroft WR (1981) Somaclonal variation: a novel source of variability from cell culture for plant improvement. Theor Appl Genet 60: 197–214.
Lebrun L, Rajasekaran C, Mullins MG (1985) Selection *in vitro* for NaCl tolerance in *Vitis rupestris* Scheele. Ann Bot 5: 733–739.
Lemaire-Chamley M, Petit J, Garcia V, Just D, Baldet P, Germain V, Fagard M, Mouassite M, Cheniclet C, Rothan C (2005) Changes in Transcriptional Profiles Are Associated with Early Fruit Tissue Specialization in Tomato. Plant Physiol 139: 750–769.
Lijavetzky D, Ruiz-García L, Cabezas JA, De Andrés MT, Bravo G, Ibáñez A, Carreño J, Cabello F, Ibáñez J, Martínez-Zapater JM (2006) Molecular genetics of berry colour variation in table grape. Mol Genet Genom 276: 427–35.
Longbottom ML, Dry PR, Sedgley M (2008) Observations on the morphology and development of star flowers of *Vitis vinifera* L. cvs Chardonnay and Shiraz. Aust J Grape Wine Res 14: 203–210.
Loureiro MD, Martinez MC, Boursiquot JM, This P (1998) Molecular marker analysis of *Vitis vinifera* Albariño and some similar grapevine cultivars. J Am Soc Hort Sci 123: 842–848.
Madero J, Boubals D, Truel D (1986) Transmission héréditaire des principaux caractères des cépages Cabernet Franc, Cabernet Sauvignon et Merlot (*Vitis vinifera* L.). Vigne Vini 13 209–219.
Maigre D (2005) Evaluation de quelques clones de Chardonnay à Changins. Rev Suisse Vitic Arbor Hort 37: 97–101.
Mannini F (2000) Clonal selection in grapevine: interactions between genetic and sanitary strategies to improve propagation material. Acta Hort 528: 703–712.
Mannini F, Argamante N, Credi R (1999) Contribution of virus infections to clonal variability of some *Vitis vinifera* L. cultivars. Bull OIV 817–818: 146–160.
Marcotrigiano M (1997) Chimeras and variegation: patterns of deceit. Hort Sci 32: 773–784.
Martinelli L, Gribaudo I (2001) Somatic embryogenesis in grapevine. In: KA Roubelakis-Angelakis (ed) Molecular Biology and Biotechnology of the Grapevine. Kluwer Academic Publ, Dordrecht, The Netherlands, pp 327–351.
Martinez MC, Boursiquot JM, Grenan S, Boidron R (1997) Etude ampélométrique de feuilles adultes de somaclones du cv. Grenache N (*Vitis vinifera* L.). Can J Bot 75: 333–345.
Mauro MC (1986) Somatic embryogenesis in *Vitis vinifera* cv. Cabernet-Sauvignon. Application to the selection of clones resistant to *Eutypa lata* (in French). Thesis INP, Toulouse, France.
Moisy C, Garrison K, Meredith CP, Pelsy F (2008) Characterization of ten novel Ty1 copia-like retrotransposon families of the grapevine genome. BMC Genom 9: 469.
Mowbray GH, Sherrer CE, Desmonts MH (1985) Agronomic characteristics of vitroplants from somatic embryos. In: Amélioration de la vigne et culture *in vitro*. Colloque Moet-Hennessy, Paris, France, pp 121–126.
Müller-Stoll W (1950) Mutant colour changes in wine grapes. Der Züchter 20: 288–291.

Mullins MG (1987) Propagation and genetic improvement of temperate fruits: the role of tissue culture. In: CE Green , DA Somers ,WP Hackett, DD Bieshoer (eds) Plant Tissue and Cell Culture, Liss AR , New York, USA, pp 395–406.
Neilson-Jones W (1969) Plant Chimeras. Methuen (ed). Methuen and Co LTD, London, UK.
Oliveira-Collet SA, Collet MA, Machado FM (2005) Differential gene expression for isozymes in somatic mutants of *Vitis vinifera* L. (*Vitaceae*). Biochem Syst Ecol 33: 691–703.
Olmo HP(1934) Bud mutation in the *vinifera* grape: 1. Parthenocarpic sultanina. Proc Am Soc Hort Sci 31: 118–121.
Olmo HP (1935) Bud mutation in the *vinifera* grape: 2. Sultanina gigas. Proc Am Soc Hort Sci 33: 436–439.
Olmo HP (1943) Breeding new tetraploid grapes. Proc Am Soc Hort Sci 47: 225–227.
Olmo HP (1952) Breeding tetraploid grapes. Proc Am Soc Hort Sci 57: 284–290.
Ourecky DK, Pratt C, Einset J (1967) Fruiting behavior of large-berried and large-clustered sports of grapes. Proc Am Soc Hort Sci 91: 217–223.
Pacottet P (1903) Savagnins. In: P Viala, V Vermorel (eds) Ampélographie, Masson, Paris, France, pp 307–309.
Pelsy F, Merdinoglu F (2002) Complete sequence of Tvv1, a family of Ty1 copia-like retrotransposons of *Vitis vinifera* L. reconstituted by chromosome walking. Theor Appl Genet 105: 614–621.
Pineda-Krch M, Fagerström T (1999) On the potential for evolutionary change in meristematic cell lineages through intraorganismal selection. J Evol Biol 12: 681–688.
Pires EJP, Pommer CV, Passos IRS, Terra MM (1988) Seedless somatic mutant in the Niagara Red grape cultivars. Bragantia 47: 171–176.
Popescu CF, Falk A, Glimelius K (2002) Application of AFLPs to characterize somaclonal variation in anther-derived grapevines. Vitis 41: 177–182.
Pratt C (1959) Radiation damage in shoot apices of concord graope. Am J Bot 46: 103–109.
Ramazotti S, Filippetti I, Intrieri C (2008) Expression of genes associated with anthocyanin synthesis in red-purplish, pink, pinkish-green and green grape berries from mutated 'Sangiovese' biotypes. Vitis 47: 147–151.
Ratcliffe OJ, Amaya I, Vincent CA, Rothstein S, Carpenter R, Coen ES, Bradley DJ (1998) A common mechanism controls the life cycle and architecture of plants. Development 125: 1609–1615.
Rathjen AH, Robinson SP (1992) Aberrant proccessing of polyphenol oxidase in a variegated mutant. Plant Physiol 99: 1619–1625.
Reisch BI, Watson JP (1984) Inheritance of leaf variegation in *Vitis* species. J Heredity 75: 417–418.
Renault-Spilmont AS, Grenan S, Boursiquot JM (2007) Syrah decline in French vineyards: Rootstocks and Syrah clone impact, pathological and genetic studies. Proc Syrah Vine Health Symp, Nov 07, Davis, USA, pp 5–8.
Riaz S, Garrison KE, Dangl GS, Boursiquot JM, Meredith CP (2002) Genetic divergence and chimerism within ancient asexually propagated winegrape cultivars. J Am Soc Hort Sci 127: 508–514.
Rose B (2005) Mutation of Pinot Noir. Aust NZ Grapegrower Winemaker 87: 21–22.
Schellenbaum P, Mohler V, Wenzel G, Walter B (2008) Variation in DNA methylation patterns of grapevine somaclones (*Vitis vinifera* L.). BMC Plant Biol 8: 78.
Scherz W (1940) Die Mutationen der Rebe, ihre Bedeutung und Auswertung für die Züchtung. Wein Rebe 22: 73–86.
Schmid J, Ries R, Rühl EH (1995) Aims and achievements of clonal selection at Geisenheim. In: JA Rantz (ed). Proc Int Symp On Clonal Selection, Portland, Oregon, USA, pp 70–73.
Schneider S, Reustle G, Zyprian E (1996) Detection of somaclonal variation in grapevine regenerants from protoplasts by RAPD-PCR. Vitis 35: 99–100.
Scott KD, Ablet EM, Lee LS, Henry RJ (2001) AFLP markers distinguishing an early mutant of Flame Seedless grape. Euphytica 113: 245–249.
Semancik JS, Rivera-Bustamante R, Goheen AC (1987) Widespread occurrence of viroid-like RNAs in grapevines. Am J Enol Vitic 38: 35–40.

Shavrukov YN, Dry IB, Thomas MR (2003) Inflorescence and bunch architecture development in *Vitis vinifera* L. Aust J Grape Wine Res 10: 116–124.
Skene KGM, Barlass M (1983) Studies on the fragmented shoot apex of grapevine. IV. Separation of phenotypes in a periclinal chimera *in vitro*. J Exp Bot 34: 1271–1280.
Snyder E, Harmon FN (1934) Three mutations of *Vitis vinifera*. Proc Am Soc Hort Sci 29: 435–436.
Souliè O, Roustan JP, Fallot J (1993) Early *in vitro* selection of eutypine tolerant plantlets. Application in screening of *Vitis vinifera* cv. Ugni-Blanc somaclones. Vitis 32: 243–244.
Sousa JSI (1959) Mutações somáticas na videira Niagara. Bragantia 18: 377–415.
Sousa JSI(1996) Uvas para o Brasil (Rev. Aum.), 2nd edn. FEALQ, Piracicaba, Brazil.
Sparvoli F, Martin C, Scienza A, Gavazzi G, Tonelli C (1994) Cloning and molecular analysis of structural genes involved in flavonoid and stilbene biosynthesis. Plant Mol Biol 24: 743–755.
Spiegel-Roy P, Assaf R, Baron I (1980) Inheritance of some characters in progenies of *Vitis vinifera* from crosses with Dabouki and Alphonse Lavallee. Proc 3rd Int Symp on Grape Breeding, June 15–18, Davis, USA, pp 210–219.
Sreekantan L, Torregrosa L, Fernandez L, Thomas MR (2006) *VvMADS9*, a class B MADS-box gene involved in grapevine flowering, shows different expression patterns in mutants with abnormal petal and stamen structures. Funct Plant Biol. 33: 877–886.
Stewart RN, Dermen H (1970) Determination of number and mitotic of shoot initial apical cells by analysis of mericlinal chimeras. Am J Bot 57: 816–826.
Terra MM, Pires EJP, Ribeiro IJA, Passos IRS (1984) Patricia Branca: mutacao somatica na videira cultivar Patricia (IAC 871–41).
This P, Lacombe T, Thomas MR (2006) Historical origins and genetic diversity of wine grapes. Trends Genet 22: 511–519.
This P, Lacombe T, Cadle-Davidson M, Owens CL (2007) Wine grape (*Vitis vinifera* L.) color associates with allelic variation in the domestication gene *VvmybA1*. Theor Appl Genet 114: 723–730.
Thompson MM, Olmo HP (1963) Cytohistological studies of cytochimeric and tetraploid grapes. Am J Bot 50: 901–906.
Torres-Vinals M, Sabate-Casaseca S, Aktouche N, Grenan S, Lopez G, Porta-Falguera M, Torregrosa L. (2004) Large-scale production of somatic embryos as a source of hypocotyl explants for *Vitis vinifera* micrografting. Vitis 4: 163–168.
Verriès C, Bès C, This P, Tesnière C (2000) Cloning and characterization of *Vine-1*, a LTR-retrotransposon-like element in *Vitis vinifera* L. and other *Vitis* species. Genome 43: 366–376.
Viala P, Vermorel V (1901–1010). Traité général de viticulture. Ampelographie. Masson, Paris, France.
Walbot V, Evans MM (2003) Unique features of the plant life cycle and their consequences. Nat Rev Genet 4: 369–379.
Walker AR, Lee E, Robinson SP (2006) Two new grape cultivars, bud sports of Cabernet Sauvignon bearing pale-coloured berries, are the result of deletion of two regulatory genes of the berry color locus. Plant Mol Biol 62: 623–635.
Walker AR, Lee E, Bogs J, McDavid DAJ, Thomas MR, Robinson SP (2007) White grapes arose through mutation of two similar and adjacent regulatory genes. Plant J 49: 772–785.
Wolf T, Cabezas JA, Martinez-Zapater JM (2003) Genetic characterization of closely related rootstock varieties based on AFLP and SAMPL markers. Acta Hort 603: 291–300.
Yakushiji H, Kobayashi S, Goto-Yamamoto N, Tae Jeong S, Sueta T, Mitani N, Azuma A (2006) A skin color mutation of grapevine, from black-skinned Pinot Noir to white-skinned Pinot Blanc, is caused by a deletion of the functional *VvmybA1* allele. Biosci Biotechnol Biochem 70: 1506–1508.
Yang XM, Cao ZY, An LZ, Wang YM, Fang XW (2006) *In vitro* tetraploid induction via colchicine treatment from diploid somatic embryos in grapevine (*Vitis vinifera* L.). Euphytica 152: 217–224.

4

Linkage Disequilibrium and Prospects for Association Mapping in *Vitis*

Christopher L. Owens

ABSTRACT

Association mapping, also known as linkage disequilibrium (LD) mapping, is rapidly becoming an important method in the dissection of complex traits of crop plants. The potential of association mapping, especially the prospects of a greater level of map resolution without a concomitant increase in population size has led to interest in developing these methods and populations in many crop plants, particularly long-lived perennial plants such as grape. Association mapping in *Vitis* is currently in its infancy, although some examples with the testing of candidate-genes have been reported. Linkage disequilibrium levels in the cultivated grape appear often to be very low, and have been shown to decay to background levels within 3 kb. As the cost of genotyping and DNA sequencing continue to decline the ability to conduct genome-wide association screens in grape should become feasible and will likely be necessary to adequately genotype the entire genome.

Keywords: Genetic mapping, grape, LD analysis, quantitative traits

4.1 Introduction

Association mapping, also known as linkage disequilibrium (LD) mapping, is rapidly becoming an important method in the dissection of complex traits of crop plants. Highly successful in determining the genetic causes of human disease, association mapping is likely to become a common approach for elucidating the genetic mechanisms of many agriculturally important

USDA-ARS, Grape Genetics Research Unit, Cornell University, Geneva, NY, USA;
e-mail: *chris.owens@ars.usda.gov*

traits. The addition of association mapping to the suite of genetic mapping approaches, such as linkage mapping of quantitative trait loci (QTL), is likely to have a dramatic effect on the ability to identify the causal genes and alleles underlying many important traits in crop plants, including grape. Several recommended recent reviews explore many of the topics relevant to conducting association mapping in plants (Flint-Garcia et al. 2003; Yu et al. 2006; Zhu et al. 2008).

The genetic mapping of complexly inherited traits is based on the shared inheritance of variation in the DNA sequence and phenotypes without prior information about biological function. Genetic mapping has been conducted since the early part of the 20th century, long before the advent of molecular markers, but only since the 1980s have DNA based molecular markers been used to create increasingly dense linkage maps suitable for the dissection of complex traits in plants (Osborn et al. 1987; Paterson et al. 1988). Over time the nature of the markers employed has shifted from restriction fragment length polymorphisms (RFLPs) and PCR-based markers to SNPs and we are now entering the age of complete genome sequencing for large-scale polymorphism discovery.

From the earliest linkage analysis in fruit flies to the wide-spread adoption of DNA-based molecular markers in the 1980s, the genetic mapping of traits in grape (*Vitis vinifera* L. and other *Vitis* species) was largely unexplored. Since then, many genetic linkage maps have been constructed for grapevine since the first reported genetic map utilizing DNA based markers in the mid-1990s (Lodhi et al. 1995). Linkage maps have been constructed for *V. vinifera* intraspecific crosses (Adam-Blondon et al. 2004; Riaz et al. 2004; Fanizza et al. 2005; Doligez et al. 2006a; Salmaso et al. 2008; Vezzulli et al. 2008) as well as interspecific crosses utilizing *V. vinifera* (Grando et al. 2003; Marguerit et al. 2009), and more complex interspecific crosses as well as crosses between wild species (Doucleff et al. 2004; Fischer et al. 2004; Lodhi et al. 1995; Lowe and Walker 2006; Mandl et al. 2006; Riaz et al. 2006, 2008; Xu et al. 2008).

Most economically important traits in grape are believed to be complexly inherited and the majority of genetic mapping efforts have focused on quantitative traits, although some single-gene traits have been mapped as well. QTL analysis has been conducted for many traits in grape in many populations, including resistance to several pathogens, abiotic stress tolerance, and fruit quality traits (Dalbo et al. 2001; Fischer et al. 2004; Fanizza et al. 2005; Cabezas et al. 2006; Doligez et al. 2006b; Krivanek et al. 2006; Riaz et al. 2006; Mejia et al. 2007; Welter et al. 2007; Costantini et al. 2008; Xu et al. 2008; Battilana et al. 2009; Dêchene et al. 2009; Marguerit et al. 2009).

Due to the expense of maintaining extensive populations in the vineyard and the amount of time necessary for seedling vines to reach maturity, most

QTL analysis in grape have only limited replication and progeny size. The majority of these populations are of modest size, of generally under 200 individuals. The map resolution achieved in these circumstances is generally low. While numerous successes have been achieved in identifying genomic regions harboring QTL for many important traits, little or no validation across environments or between populations, or fine mapping of those loci has been accomplished to date.

The potential of association mapping, especially the prospects of a greater level of map resolution without a concomitant increase in population size has led to interest in developing these methods and populations in many crop plants, particularly long-lived perennial plants such as grape. Recently, dissection of complex traits through genetic association in populations has become increasingly useful in plants. Such studies have been conducted in human genetic research for decades, but have only recently begun to yield tremendous advances due to the dramatic increase in DNA based molecular markers available, such as single nucleotide polymorphisms (SNPs).

Contributing to increased interest in association mapping is the rapid decline in the cost of DNA sequencing and genotyping, which has made it possible to develop large numbers of DNA based markers, typically SNPs, within plant genomes. The genetic mapping of complex traits in plants, while providing many successes in identifying genomic regions of importance, has yielded very little success in elucidating the Mendelian factors, or specific genes underlying QTL. In many crop species association mapping offers the possibility of higher resolution mapping than QTL analysis of segregating populations. The promise of accelerating research progress by taking advantage of pre-existing germplasm collections or diverse pools of genotypes provides another major advantage for conducting association mapping in grape.

4.2 What is Association Mapping and What is Linkage Disequilibrium?

Like QTL analysis, association mapping seeks to study the co-inheritance of genetic variants (or markers) with phenotypic traits of interest. In QTL analysis mapping resolution is based on gametic recombination of sibling populations, typically only one generation in grape. The key difference in LD-based mapping, is that it is based on historical recombination events that have occurred in a group of unrelated individuals, and treats a collection of individuals as "a very large and very old extended family" (Altshuler et al. 2008). This approach is made possible because a grouping of unrelated individuals that vary in a phenotype of interest will also show differences in the underlying causal alleles. This simple relationship can be confounded by many factors as two groups of individuals varying in ancestry may differ in allele frequencies at many additional genetic loci (Rafalski and Ananiev 2009).

LD measures the degree of non-random association between alleles at different loci (Hartl and Clark 1997). Several statistics are used to measure LD, but the basic measurement of LD is determined by calculating the difference between observed haplotype frequency and that expected based on allele frequency (Flint-Garcia et al. 2003). The two most commonly used and related statistics are D' and r^2 (Lewontin 1964; Hill and Robertson 1968).

LD can be caused and influenced by several factors including: mutation, stochastic effects such as genetic drift or artifacts of sampling, population structure, and selection (Slatkin 2008). Mutation can lead to LD between two loci if widely divergent mutation rates exist, particularly in the case where one locus is monomorphic in a population. Population structure, or the differential relatedness among individuals within a sample, can cause problems in implementing a successful association mapping screen. Population structure can exist due to several factors: subpopulation structure, admixture, migration, and historical pedigree relationships. While taking these factors into consideration, LD due to linkage is of primary value in mapping complex traits, which measures the net effect of all recombination events since the origin of the alleles.

The decay of LD across a chromosome is of particular importance in developing an association genetics panel. The expectation is that a population undergoing random mating and of constant size will have mutation and drift continuously generating LD, while recombination removes it as a function of distance between loci (Falconer and Mackay 1996).

One of the main advantages of association mapping is the promise of improved resolution. Most grape QTL analyses have been conducted with small population sizes, with QTL resolved to approximately 10 cM or more. In those species in which LD decays rapidly, resolution achieved in association mapping can be under 1 cM and approach the level of only a few thousand base pairs. Another key difference between linkage analysis and association mapping is that the results of linkage analysis are typically only relevant to the same or genetically related populations, while association mapping provides information on a much wider germplasm base. In other words, the number of alleles sampled is much higher in association mapping, with a concomitant reduction in power for determining the effect per individual allele. For analysis of traits controlled by very low frequency alleles in a population, it may be best to construct specific bi-parental populations for analysis. For those traits that are rarely represented within the available germplasm collections, it may still be advisable to construct a segregating population specifically for the dissection of that trait.

It will be important to realize that positive results from an association mapping analysis is not conclusive proof of that gene's involvement in the trait of interest, although species with low levels of LD will limit the size of intervals possessing the underlying alleles. Additionally, negative results do not fully eliminate a gene as playing a functional role, a lack of association may be due to insufficient diversity within the population, or the presence of rare alleles, or alleles with small phenotypic effects (Rafalski and Ananiev 2009). Although association mapping has great promise, it is not a cure-all for poorly replicated or poorly phenotyped experiments, insufficient sample size, and confounded genetic and environmental interactions.

Association mapping seeks to uncover some of the key genomic regions and genes underlying important traits, but ultimately does not provide any functional context. Additional hypothesis testing or complimentary studies will still be required to provide support for the identification of causal alleles.

4.3 Germplasm for Association Mapping

The choice of individuals to utilize as a diversity panel for genetic association mapping is critical for success. The genetic diversity, extent of LD within the genome, and relatedness of the germplasm are all key determinants of mapping resolution (Flint-Garcia et al. 2003; Breseghello and Sorrells 2006; Yu et al. 2006; Zhu et al. 2008). Depending on the degree of population structure and familial relatedness, the possible germplasm will be more or less amenable to association mapping studies, with populations showing low levels of population structure and familial relatedness being ideal (Flint-Garcia et al. 2005; Yu and Buckler 2006; Yu et al. 2006). In grape, little is known about the genetic consequences of selection, breeding history, and population structure.

Grape, particularly *V. vinifera*, offers several unique characteristics that are likely to influence the selection of germplasm for association mapping. Thousands of cultivars of *V. vinifera* are recognized, with many of these cultivars having been cultivated for hundreds of years (Alleweldt et al. 1990; This et al. 2006). The effective population size of the cultivated genepool of *V. vinifera* has not been extensively explored. Close pedigree relationships are known for many genotypes, particularly several of the historical cultivars originating in Bordeaux and Burgundy (Bowers and Meredith 1997; Bowers et al. 1999). Due to the ease of asexual propagation, immortalized lines can easily be replicated in many locations.

Efforts to define core collections, or panels of genotypes that replicate the majority of diversity within the species will become valuable resources for use as diversity panels in association mapping (Le Cunff et al. 2008). Many countries for which viticulture is an important industry possess diverse germplasm collections that can be used as starting points for developing association mapping panels. Several advances have been made over the last several years characterizing germplasm collections with molecular markers, typically microsatellite markers, and developing core collections that represent significant proportions of the species-wide diversity (Aradhya et al. 2003; Le Cunff et al. 2008). The utility in developing common pools of accessions to be used by researchers in multiple locations is an important question to address as association mapping becomes more common in grape. Obstacles in importation of woody plant material may limit the ability to do so, but a large diverse panel of genotypes that is well replicated across environments will be highly valuable. Currently, most QTL mapping populations are unique to individual researcher's interests and are not distributed across laboratories. Existing germplasm collections have poor levels of replication of unique accessions. If diversity panels suitable for association mapping are identified, adequate statistical design will need to be addressed to increase the power to detect significant associations, such as the level of replication, or the use of incomplete blocks or augmented designs in the absence of full replication.

Genetic diversity of several large collections of grape germplasm has been conducted with small numbers of simple sequence repeat (SSR) loci, typically 20 or less (Lopes et al. 1999; Aradhya et al. 2003; Martin et al. 2003; Lopes et al. 2006; Le Cunff et al. 2008). These data can be used in the selection of a panel of individuals for association analysis, which will be augmented by the availability of additional genome sequence data and larger SNP data sets. Although greater numbers of biallelic SNP markers will be needed to adequately describe genetic diversity, as compared to SSR markers (Hamblin et al. 2007), the rapid decline in cost of sequencing and SNP genotyping will allow for substantial increases in the markers utilized in the described available genetic diversity and in defining groups of diverse individuals for association mapping.

The increased interest and likely adoption of association mapping efforts in grape suggest a continued importance of germplasm collections. As larger, diverse sets of accessions are used in genetic analyses, it becomes imperative that existing collections are well characterized. Outside of *V. vinifera*, most additional wild species of *Vitis* are not well represented in the world's germplasm collections. For example, there are typically well under 100 unique accessions of any given wild *Vitis* species in the US national germplasm collections. Some wild species are directly under threat due to habitat loss and global climate change. Efforts to preserve wild germplasm should be renewed to ensure sufficient genetic diversity is maintained.

4.4 Candidate Genes vs. Whole Genome Scans

The type and magnitude of genotyping available for association mapping is quickly changing in grape. Large scale genotyping and next-generation whole-genome sequencing are becoming a computational and data analysis challenge and less of an experimental obstacle. Due to the confounding risks caused by inherent population structure and to reduce the occurrence of false-positive associations, it is necessary to genotype a set of unlinked, selectively neutral background markers. The background markers are critical in preventing spurious associations due to population structure and estimating levels of familial relatedness (Pritchard et al. 2000a; Yu and Buckler 2006). Although, previously identified SSR markers available to the grape genetic community can be utilized as background markers, they are quickly being replaced by SNP markers, due to their higher genome density and amenability to high-throughput genotyping methods. However, due to their biallelic nature they are less informative than SSRs and require larger numbers of markers to adequately assess population structure and familial relationships (Hamblin et al. 2007).

Most association mapping efforts in plants have thus far focused on the testing of candidate-gene associations. In this scenario, biologically relevant candidates are chosen and tested for association with a trait of interest. For candidate-gene studies, SNPs or other sequence variants are initially discovered in a sub-set of individuals chosen to represent the allelic diversity present. For grape it should be feasible to utilize small sub-sets of diverse genotypes, as few as 24, to sample for SNP discovery in candidate genes of interest (Le Cunff et al. 2008). Once SNPs are identified they can be genotyped on the germplasm chosen to conduct the association analysis. Candidate gene approaches should have increased chances for success in cases where there is ample biochemical data on which to base the selection of candidate genes (Rafalski and Ananiev 2009). However, more complexly inherited traits, or for those for which there are poor biochemical support, it will be difficult to determine significant associations through candidate gene approaches. As more association mapping studies are conducted in plants, it will be valuable to learn the success rate of candidate-gene selection. Many examples from human genetics have demonstrated that the candidate-gene approach has been largely inefficient (Altshuler et al. 2008). For candidate gene testing to be successful, associated markers need to be in high LD with underlying causal alleles, but they also need to have comparable allele frequencies. The power of detection is a function of the QTL effect size, number of observations, and the LD between the QTL and marker.

Whole-genome scans, in which there is no *a priori* knowledge of the genetic control of a trait, are becoming increasingly feasible. Human geneticists are now routinely conducting whole-genome scans with SNP data sets in the millions. Once the genome-wide LD structure and haplotype diversity are better characterized in grape, it should be feasible to develop a strategy for genotyping an appropriate number and distribution of SNP markers to conduct a whole-genome scan in grape. However, whole-genome scans do impose two burdens on the analysis: higher genotyping cost, and a higher multiplicity of testing.

One potential application of association mapping in grape may be to test candidate genes in a map interval first identified through linkage or QTL analysis. The ability to increase map resolution in a segregating population through an increase of progeny number is challenging even in annual crops and will remain challenging in most cases for grape. The ability to examine specific genomic regions through the association mapping of the candidate genes in that interval may be one approach to develop hypotheses of genes underlying important QTL.

4.5 Phenotyping

Although the costs of sequencing and genotyping continue to decline rapidly, the cost of obtaining phenotypic data is not. It is intuitively obvious that the ability to conduct successful association analysis in grape will be dependent on the ability to collect high-quality phenotypic data. The establishment and maintenance of vineyards over time and space is expensive, yet it is necessary to maximize the long-term ability to dissect complexly inherited traits in grape. The additional challenge of measuring phenotypic traits on a large number of diverse accessions will require careful attention to experimental design. The incorporation of high-throughput data collection systems can be used to increase the speed of collection and reduce errors, such as bar-code scanners for certain phenotypes. However, the collection of high quality phenotypic data will remain a time consuming and expensive endeavor.

Due to the large number of markers required, most grape association mapping studies will be conducted with sets of candidate genes in the short term. If high quality, replicated phenotypic data is collected from such studies, these data can easily be analyzed in the future as the marker coverage approaches that are necessary to conduct a whole genome scan.

4.6 Statistical Analysis—Methodology

Significant advances have been made over the last few years in the statistical methodology used for association analysis. In the absence of population structure the statistical analysis of association could be conducted with common statistical methods. However, the risk of numerous spurious associations creates the need for additional methodology to control these confounding factors. The majority of the advances in statistical methodology have the intention of reducing spurious associations due to population structure or familial relatedness. Several methods are now routinely used to control for false positives, all of which rely on the separate genotyping of random markers. The genomic control (GC) and structured association (SA) methods use a set of random markers to control the impact of population structure. GC uses a set of random markers to estimate the degree that test statistics are inflated by population structure (Devlin and Roeder 1999). SA uses a set of markers that are randomly distributed across the genome to estimate population structure by allocating individuals to populations and then incorporates this estimate into the statistical test of association (Pritchard and Rosenberg 1999; Pritchard et al. 2000b; Falush et al. 2003). A unified mixed-model test that accounts for population structure and kinship simultaneously have also been recently developed (Yu et al. 2006). Principal component analysis (PCA) has also been used extensively to describe variation observed across all markers into component variables (Price et al. 2006; Zhao et al. 2007).

Zhu et al. (2008) recently reviewed the sample size and number of background markers used in published plant association mapping studies and reported that the typical number of individuals sampled was approximately 100, with a small number of studies using larger sample sizes. Population sizes of approximately 100 individuals would be consistent with typical population sizes for many of the published QTL analysis in grape, but due to the increase in genetic diversity sampled in an association diversity panel, sample sizes in this range will be unlikely to detect any but the most significant associations and that larger populations will be required to detect loci of moderate or small effect. Simulations to determine ideal number of background markers to genotype for conducting structured association mapping have estimated that over 1000 SNP markers would be needed to define population structure of a group of diverse maize lines. A much smaller number, approximately 100 markers, of multi-allelic SSRs were capable of estimating kinship of the same lines (Yu et al. 2009).

4.7 Association Mapping in Humans

Due to limitations in conducting linkage analysis in humans, particularly for complexly inherited traits that are difficult to study in multi-generational families, association mapping has been viewed as an alternative approach with great promise. As a result, association mapping studies have become increasingly powerful for detecting disease related genes in human genetic research. Association mapping in humans has been conducted since the middle of the 20th century. Early examples included examining the correlations between peptic ulcer disease and blood-group antigens (Aird et al. 1954), and the association of human leukocyte antigen with several immunologically mediated diseases (Tiwari and Terasaki 1985). Many subsequent studies showed poor replicability, thought to be due to several common problems: small sample-size, poorly matched control groups, and results confounded with population structure (Cardon and Bell 2001). An additional problem was that many early studies were limited to small numbers of candidate genes. Many of the recent advances in statistical analysis have sought to rectify many of these shortcomings. The lack of significant DNA based markers from which to conduct large scale studies was an additional limitation.

Over the last 10 years the number of SNPs identified in humans and the number of SNPs genotyped while conducting association mapping have grown considerably (Frazer et al. 2007). Development of these large datasets of genotyped SNPs has allowed conducting genome-wide association studies (GWAS) in humans in which the entire genome is scanned for significant associations. These studies have lead to the reports of many SNPs associated with common diseases, clinical conditions, and various individual traits. By 2008, over 150 relationships had been established between common SNPs and disease traits in humans (Altshuler et al. 2008).

4.8 Association Mapping in Plants

Association mapping studies are becoming more common in plants, with much of the early work being conducted in maize. In addition to maize, association mapping studies have been reported for *Arabidopsis*, wheat, barley, potato, rice, *Pinus taeda*, and sugarcane (Thornsberry et al. 2001; Aranzana et al. 2005; Breseghello and Sorrells 2006; Camus-Kulandaivelu et al. 2006; Kraakman et al. 2006; Wei et al. 2006; Gonzalez-Martinez et al. 2007; Malosetti et al. 2007; Belo et al. 2008; Harjes et al. 2008).

Typically, mapping of important traits in plants has been conducted through controlled crosses, which randomize and control for genetic background amongst closely related progeny. The heightened interest in association mapping in plants has been triggered by several factors: the rapid decline in the cost of sequencing and genotyping, allowing for the development of large numbers of genetic markers; improvements in statistical methodology that controls for spurious associations; the ability to test greater allelic diversity in an association panel; and the recognition that in many crops association mapping offers the potential of much greater mapping resolution compared to QTL analysis.

LD will vary in different plant populations based on the mating system, the degree of selection at the loci assayed, diversity within the population, and the population structure. Plants with a high selfing rate will have a much slower decay of LD than obligate out-crossing species. LD decay has been measured in several plant species, and seems largely effected by the mating system as it pertains to selfing rate as well as pollen and seed dispersal distances, and degree of selection of the population in question.

4.9 Association Mapping in *Vitis*

Linkage disequilibrium (LD) based association mapping is of interest to grape geneticists considering the potentially high resolution and the time saved associated with the utilization of existing germplasm collections (This et al. 2006). Development of appropriate populations for association mapping and expectations of achievable mapping resolution are highly dependent on estimates of genetic diversity and the decay of linkage disequilibrium within grapevine germplasm. These factors are significantly influenced by the current and historic levels of genetic diversity and the severity of the population bottleneck that would have occurred during grape domestication.

The domestication of *V. vinifera* is thought to have occurred approximately 6,000 to 8,000 years ago (Levadoux 1956; Zohary and Hopf 2000; McGovern 2003). There are several morphological and biochemical traits associated with the domestication of *V. vinifera* that were derived from the progenitor species *V. vinifera* subsp. *sylvestris*. Comparisons of chloroplast and nuclear markers between wild and cultivated grapes have been used to investigate aspects of grape domestication, particularly the timing and location of key domestication events (Arroyo-Garcia et al. 2006; De Mattia et al. 2008; Grassi et al. 2008), but the direct impact of domestication on genetic diversity and estimates of the severity of the domestication bottleneck have not been extensively investigated.

Significant haplotypic LD was observed over 30 cM in a *V. vinifera* core collection when estimating LD with 38 microsatellite markers genotyped on 5 linkage groups (Barnaud et al. 2005). Utilizing a subset of the pigmented accessions from the same core collection a much more rapid decay in LD at the single locus level was observed (This et al. 2007). In the most comprehensive study published to date, rapid decay of LD was observed in over 200 random loci, representing over 1 Mb of total sequence and over 1,500 SNPs within 11 genotypes (Lijavetzky et al. 2007). While focusing on coding sequence of individual genes, these studies suggest that LD decay is extremely rapid in grape and that LD will decay to background levels in only a small number of kilobases in many situations. This observation is supported by a recent work with thousands of SNPs identified by next-generation sequencing of 10 cultivated *V. vinifera* and 7 lines of wild *Vitis* species, in which LD levels were generally low, and decayed to background levels within 3 kb (Myles et al. 2010).

One likely consequence of the observed rapid decay of LD in *Vitis* is that additional whole-genome sequencing efforts will be required to provide sufficient marker coverage to conduct whole-genome scans. As sequencing costs continue to decline, it should be feasible to whole-genome sequence all individuals in an association diversity panel.

A similar discrepancy between LD estimates has been observed in maize, in which LD estimated by SSRs was greater than that measured with SNPs (Remington et al. 2001). In maize, it is thought this discrepancy is most likely due to differences in mutation rate of the marker types, and the resultant difference in the ages of the alleles being genotyped. Most SNPs genotyped in maize predate the divergence of cultivated maize from teosinte, the wild progenitor, and that perhaps the SSR loci more accurately reflect population structure following the domestication bottleneck. In grape, there is little available evidence comparing mutation rates between *V. vinifera* ssp. *sativa* and *V. vinifera* ssp. *sylvestris*. With the rapid decline in the cost of sequencing and genotyping, additional data will soon be generated that can provide clearer estimates of genomic patterns of LD in *V. vinifera*.

No large scale association mapping studies have currently been reported in *Vitis*. Structured association mapping has been used to test for genetic association of the gene *Vvmyba1* and fruit color in a diverse set of *V. vinifera* accessions (This et al. 2007). By genotyping the SNPs within 100 diverse accessions and utilizing 20 SSR markers for estimating population structure, it was possible to identify SNPs significantly associated with variation in berry color. Recently, an approach combining QTL mapping with LD mapping of candidate genes within a significant QTL for berry color identified several individual polymorphisms within a cluster of genes explaining 84% over observed variation (Fournier-Level et al. 2009).

Several key questions need to be addressed before association mapping gains wide adoption in grape. Several studies have sought to explore in greater depth the level of genetic diversity present in *Vitis*, but questions on the extent of and variation in genome-wide LD and population structure remain. Available evidence suggests that LD decay will be extremely rapid in *V. vinifera*, and presumably the other *Vitis* species. The high mapping resolution that would exist will require extensive marker data sets in order to conduct whole genome scans. Until large-scale marker data sets exist, association mapping of candidate genes should be a valuable approach to understanding specific traits.

Little is presently known about the patterns of LD across the genome in grape. Evidence for LD blocks due to variation in recombination rate has been observed in humans and *Arabidopsis* (Gabriel et al. 2002; Kim et al. 2007). It is thought that heterogeneity in recombination rate as well as random sampling of alleles and timing of mutation relative to recombination events leads to blocks. If blocks exist in grape it will be important to determine if the blocks arose from a consistent recombination process and thus likely to be consistent across populations or from other random processes.

Association mapping and QTL mapping have some complementary strengths that can be exploited through mapping approaches that combine a joint analysis of linkage and LD mapping. Several approaches have been suggested and have now been implemented in several organisms to accommodate the mating system and level of LD present in the species, including cattle, mouse, and maize (Meuwissen et al. 2002; Blott et al. 2003; Churchill et al. 2004; Yu et al. 2008). Establishment of similar populations in grape would be time-consuming and expensive to maintain, but may be worth considering as a means to reduce genotyping costs and to increase the power of analysis.

Association mapping is now being conducted on many crop plants as a method to dissect the genetics of complex traits. The high map resolution that can be attained and the ability to mine the rich allelic diversity present within the genepool make grape an attractive crop for association mapping. It is likely that over the coming years, association mapping of candidate genes or targeting of genomic regions will add to our understanding of the genetic architecture of important traits. As the cost of genotyping and DNA sequencing continue to decline the ability to conduct genome-wide association screens in grape should become feasible.

References

Adam-Blondon A-F, Roux C, Claux D, Butterlin G, Merdinoglu D, This P (2004) Mapping 245 SSR markers on the *Vitis vinifera* genome: a tool for grape genetics. Theor Appl Genet 109: 1017–1027.

Aird I, Bentall HH, Mehigan JA, Roberts JAF (1954) The blood groups in relation to peptic ulceration and carcinoma of colon, rectum, breast, and bronchus—an association between the Abo groups and peptic ulceration. Brit Med J 2: 320–321.

Alleweldt G, Spiegel-Roy P, Reisch B (1990) Grapes (*Vitis*). Acta Hort 290: 289–327.

Altshuler D, Daly MJ, Lander ES (2008) Genetic mapping in human disease. Science 322: 881–888.

Aradhya MK, Dangl GS, Prins BH, Boursiquot JM, Walker MA, Meredith CP, Simon CJ (2003) Genetic structure and differentiation in cultivated grape, *Vitis vinifera* L. Genet Res 81: 179–192.

Aranzana MJ, Kim S, Zhao KY, Bakker E, Horton M, Jakob K, Lister C, Molitor J, Shindo C, Tang CL, Toomajian C, Traw B, Zheng HG, Bergelson J, Dean C, Marjoram P, Nordborg N (2005) Genome-wide association mapping in Arabidopsis identifies previously known flowering time and pathogen resistance genes. PLoS Genet 1: 531–539.

Arroyo-Garcia R, Ruiz-Garcia L, Bolling L, Ocete R, Lopez MA, Arnold C, Ergul A, Soylemezoglu G, Uzun HI, Cabello F, Ibanez J, Aradhya MK, Atanassov A, Atanassov I, Balint S, Cenis JL, Costantini L, Goris-Lavets S, Grando MS, Klein BY, Mcgovern PE,

Merdinoglu D, Pejic I, Pelsy F, Primikirios N, Risovannaya V, Roubelakis-Angelakis KA, Snoussi H, Sotiri P, Tamhankar S, This P, Troshin L, Malpica JM, Lefort F, Martinez-Zapater JM (2006) Multiple origins of cultivated grapevine (*Vitis vinifera* L. ssp. *sativa*) based on chloroplast DNA polymorphisms. Mol Ecol 15: 3707–3714.

Barnaud A, Lacombe T, Doligez A (2005) Linkage disequilibrium in cultivated grapevine, *Vitis vinifera* L. Theor Appl Genet 112: 708–716.

Battilana J, Costantini L, Emanuelli F, Sevini F, Segala C, Moser S, Velasco R, Versini G, Grando MS (2009) The 1-deoxy-D: -xylulose 5-phosphate synthase gene co-localizes with a major QTL affecting monoterpene content in grapevine. Theor Appl Genet 118: 653–669.

Belo A, Zheng PZ, Luck S, Shen B, Meyer DJ, Li BL, Tingey S, Rafalski A (2008) Whole genome scan detects an allelic variant of fad2 associated with increased oleic acid levels in maize. Mol Genet Genom 279: 1–10.

Blott S, Kim JJ, Moisio S, Schmidt-Kuntzel A, Cornet A, Berzi P, Cambisano N, Ford C, Grisart B, Johnson D, Karim L, Simon P, Snell R, Spelman R, Wong J, Vilkki J, Georges M, Farnir F, Coppieters W (2003) Molecular dissection of a quantitative trait locus: A phenylalanine-to-tyrosine substitution in the transmembrane domain of the bovine growth hormone receptor is associated with a major effect on milk yield and composition. Genetics 163: 253–266.

Bowers JE, Meredith CP (1997) The parentage of a classic wine grape, Cabernet Sauvignon. Nat Genet 16: 84–87.

Bowers JE, Boursiquot JM, This P, Chu K, Johansson H, Meredith CP (1999) Historical genetics: The parentage of chardonnay, gamay, and other wine grapes of northeastern France. Science 285: 1562–1565.

Breseghello F, Sorrells ME (2006) Association mapping of kernel size and milling quality in wheat (*Triticum aestivum* L.) cultivars. Genetics 172: 1165–1177.

Cabezas JA, Cervera MT, Ruiz-Garcia, Carreño J, Martínez-Zapater JM (2006) A genetic analysis of seed and berry weight in grapevine. Genome 49: 1572–1585.

Camus-Kulandaivelu L, Veyrieras JN, Madur D, Combes V, Fourmann M, Barraud S, Dubreuil P, Gouesnard B, Manicacci D, Charcosset A (2006) Maize adaptation to temperate climate: Relationship between population structure and polymorphism in the Dwarf8 gene. Genetics 172: 2449–2463.

Cardon LR, Bell JI (2001) Association study designs for complex diseases. Nat Rev Genet 2: 91–99.

Churchill GA, Airey DC, Allayee H, Angel JM, Attie AD, Beatty J, Beavis WD, Belknap JK, Bennett B, Berrettini W, Bleich A, Bogue M, Broman KW, Buck KJ, Buckler E, Burmeister M, Chesler EJ, Cheverud JM, Clapcote S, Cook MN, Cox RD, Crabbe JC, Crusio WE, Darvasi A, Deschepper CF, Doerge RW, Farber CR, Forejt J, Gaile D, Garlow SJ, Geiger H, Gershenfeld H, Gordon T, Gu J, Gu W, de Haan G, Hayes NL, Heller C, Himmelbauer H, Hitzemann R, Hunter K, Hsu HC, Iraqi FA, Ivandic B, Jacob HJ, Jansen RC, Jepsen KJ, Johnson DK, Johnson TE, Kempermann G, Kendziorski C, Kotb M, Kooy RF, Llamas B, Lammert F, Lassalle JM, Lowenstein PR, Lu L, Lusis A, Manly KF, Marcucio R, Matthews D, Medrano JF, Miller DR, Mittleman G, Mock BA, Mogil JS, Montagutelli X, Morahan G, Morris DG, Mott R, Nadeau JH, Nagase H, Nowakowski RS, O'Hara BF, Osadchuk AV, Page GP, Paigen B, Paigen K, Palmer AA, Pan HJ, Peltonen-Palotie L, Peirce J, Pomp D, Pravenec M, Prows DR, Qi Z, Reeves RH, Roder J, Rosen GD, Schadt EE, Schalkwyk LC, Seltzer Z, Shimomura K, Shou S, Sillanpää MJ, Siracusa LD, Snoeck HW, Spearow JL, Svenson K, Tarantino LM, Threadgill D, Toth LA, Valdar W, de Villena FP, Warden C, Whatley S, Williams RW, Wiltshire T, Yi N, Zhang D, Zhang M, Zou F; Complex Trait Consortium (2004) The Collaborative Cross, a community resource for the genetic analysis of complex traits. Nat Genet 36: 1133–1137.

Costantini L, Battilana J, Lamaj F, Fanizza G, Grando MS (2008) Berry and phenology-related traits in grapevine (*Vitis vinifera* L.). From quantitative trait loci to underlying genes. BMC Plant Biol 8: 38.

Dalbo MA, Ye GN, Weeden NF, Wilcox WF, Reisch BI (2001) Marker-assisted selection for powdery mildew resistance in grape. J Am Soc Hort Sci 126: 83–89.
De Mattia F, Imazio S, Grassi F, Baneh HD, Scienza A, Labra M (2008) Study of genetic relationships between wild and domesticated grapevine distributed from Middle East regions to European countries. Rendiconti Lincei-Scienze Fisiche E Naturali 19: 223–240.
Devlin B, Roeder K (1999) Genomic control for association studies. Biometrics 55: 997–1004.
Doligez A, Adam-Blondon AF, Cipriani G, Di Gaspero G, Laucou V, Merdinoglu D, Meredith CP, Riaz S, Roux C, This P (2006a) An integrated SSR map of grapevine based on five mapping populations. Theor Appl Genet 113: 369–382.
Doligez A, Audiot E, Baumes R, This P (2006b). QTLs for Muscat flavor and monoterpenic odorant content in grapevine (*Vitis vinifera* L.). Mol Breed 18: 109–125.
Doucleff M, Jin Y, Gao F, Riaz S, Krivanek AF, Walker MA (2004) A genetic linkage map of grape, utilizing *Vitis rupestris* and *Vitis arizonica*. Theor Appl Genet 109: 1178–1187.
Duchêne E, Butterlin G, Claudel P, Dumas V, Jaegli N, Merdinoglu D (2009) A grapevine (*Vitis vinifera* L.) deoxy-D-xylulose synthase gene colocates with a major quantitative trait loci for terpenol content. Theor Appl Genet 118: 541–552.
Falconer DS, Mackay TFC (1996) Introduction to Quantitative Genetics. Longman Group, Harlow, UK.
Falush D, Stephens M, Pritchard JK (2003) Inference of population structure using multilocus genotype data: Linked loci and correlated allele frequencies. Genetics 164: 1567–1587.
Fanizza G, Lamaj F, Costantini L, Chaabane R, Grando MS (2005) QTL analysis for fruit yield components in table grapes (*Vitis vinifera*). Theor Appl Genet 111: 658–664.
Fischer B, Salakhutdinov I, Akkurt M, Eibach R, Edwards KJ, Töpfer R, Zyprian E (2004) Quantitative trait locus analysis of fungal disease resistance factors on a molecular map of grapevine. Theor Appl Genet 108: 505–515.
Flint-Garcia SA, Thornsberry JM, Buckler ES (2003) Structure of linkage disequilibrium in plants. Annu Rev Plant Biol 54: 357–374.
Flint-Garcia SA, Thuillet AC, Yu JM, Pressoir G, Romero SM, Mitchell SE, Doebley J, Kresovich S, Goodman MM, and Buckler ES (2005) Maize association population: a high-resolution platform for quantitative trait locus dissection. Plant J 44: 1054–1064.
Fournier-Level A, Le Cunff L, Gomez C, Doligez A, Ageorges A, Roux C, Betrand Y, Souquet JM, Cheynier V, This P (2009) Quantitative genetic bases of anthocyanin variation in grape (*Vitis vinifera* L. ssp. *sativa*) berry: A quantitative trait locus to quantitative nucleotide integrated study. Genetics 183: 1127–1139.
Frazer KA, Ballinger DG, Cox DR, Hinds DA, Stuve LL, Gibbs RA, Belmont JW, Boudreau A, Hardenbol P, Leal SM, Pasternak S, Wheeler DA, Willis TD, Yu F, Yang H, Zeng C, Gao Y, Hu H, Hu W, Li C, Lin W, Liu S, Pan H, Tang X, Wang J, Wang W, Yu J, Zhang B, Zhang Q, Zhao H, Zhou J, Gabriel SB, Barry R, Blumenstiel B, Camargo A, Defelice M, Faggart M, Goyette M, Gupta S, Moore J, Nguyen H, Onofrio RC, Parkin M, Roy J, Stahl E, Winchester E, Ziaugra L, Altshuler D, Shen Y, Yao Z, Huang W, Chu X, He Y, Jin L, Liu Y, Shen Y, Sun W, Wang H, Wang Y, Wang Y, Xiong X, Xu L, Waye MM, Tsui SK, Xue H, Wong JT, Galver LM, Fan JB, Gunderson K, Murray SS, Oliphant AR, Chee MS, Montpetit A, Chagnon F, Ferretti V, Leboeuf M, Olivier JF, Phillips MS, Roumy S, Sallée C, Verner A, Hudson TJ, Kwok PY, Cai D, Koboldt DC, Miller RD, Pawlikowska L, Taillon-Miller P, Xiao M, Tsui LC, Mak W, Song YQ, Tam PK, Nakamura Y, Kawaguchi T, Kitamoto T, Morizono T, Nagashima A, Ohnishi Y, Sekine A, Tanaka T, Tsunoda T, Deloukas P, Bird CP, Delgado M, Dermitzakis ET, Gwilliam R, Hunt S, Morrison J, Powell D, Stranger BE, Whittaker P, Bentley DR, Daly MJ, de Bakker PI, Barrett J, Chretien YR, Maller J, McCarroll S, Patterson N, Pe'er I, Price A, Purcell S, Richter DJ, Sabeti P, Saxena R, Schaffner SF, Sham PC, Varilly P, Altshuler D, Stein LD, Krishnan L, Smith AV, Tello-Ruiz MK, Thorisson GA, Chakravarti A, Chen PE, Cutler DJ, Kashuk CS, Lin S, Abecasis GR, Guan W, Li Y, Munro HM, Qin ZS, Thomas DJ, McVean G, Auton A, Bottolo L, Cardin N, Eyheramendy S, Freeman C, Marchini J, Myers S, Spencer C, Stephens M, Donnelly

P, Cardon LR, Clarke G, Evans DM, Morris AP, Weir BS, Tsunoda T, Mullikin JC, Sherry ST, Feolo M, Skol A, Zhang H, Zeng C, Zhao H, Matsuda I, Fukushima Y, Macer DR, Suda E, Rotimi CN, Adebamowo CA, Ajayi I, Aniagwu T, Marshall PA, Nkwodimmah C, Royal CD, Leppert MF, Dixon M, Peiffer A, Qiu R, Kent A, Kato K, Niikawa N, Adewole IF, Knoppers BM, Foster MW, Clayton EW, Watkin J, Gibbs RA, Belmont JW, Muzny D, Nazareth L, Sodergren E, Weinstock GM, Wheeler DA, Yakub I, Gabriel SB, Onofrio RC, Richter DJ, Ziaugra L, Birren BW, Daly MJ, Altshuler D, Wilson RK, Fulton LL, Rogers J, Burton J, Carter NP, Clee CM, Griffiths M, Jones MC, McLay K, Plumb RW, Ross MT, Sims SK, Willey DL, Chen Z, Han H, Kang L, Godbout M, Wallenburg JC, L'Archevêque P, Bellemare G, Saeki K, Wang H, An D, Fu H, Li Q, Wang Z, Wang R, Holden AL, Brooks LD, McEwen JE, Guyer MS, Wang VO, Peterson JL, Shi M, Spiegel J, Sung LM, Zacharia LF, Collins FS, Kennedy K, Jamieson R, Stewart J (2007) A second generation human haplotype map of over 3.1 million SNPs. Nature 449: 851–861.

Gabriel SB, Schaffner SF, Nguyen H, Moore JM, Roy J, Blumenstiel B, Higgins J, DeFelice M, Lochner A, Faggart M, Liu-Cordero SN, Rotimi C, Adeyemo A, Cooper R, Ward R, Lander ES, Daly MJ, Altshuler D (2002) The structure of haplotype blocks in the human genome. Science 296: 2225–2229.

Gonzalez-Martinez SC, Wheeler NC, Ersoz E, Nelson CD, Neale DB (2007) Association genetics in *Pinus taeda* L. I. Wood property traits. Genetics 175: 399–409.

Grando MS, Bellin D, Edwards KJ, Pozzi C, Stefanini M, Velasco R (2003) Molecular linkage maps of *Vitis vinifera* L. and *Vitis riparia* Mchx. Theor Appl Genet 106: 1213–1224.

Grassi F, De Mattia F, Zecca G, Sala F, Labra M (2008) Historical isolation and Quaternary range expansion of divergent lineages in wild grapevine. Biol J Linn Soc 95: 611–619.

Hamblin MT, Warburton ML, Buckler ES (2007) Empirical comparison of Simple Sequence Repeats and single nucleotide polymorphisms in assessment of maize diversity and relatedness. PLoS ONE 2: e1367.

Harjes CE, Rocheford TR, Bai L, Brutnell TP, Kandianis CB, Sowinski SG, Stapleton AE, Vallabhaneni R, Williams M, Wurtzel ET, Yan JB, Buckler ES (2008) Natural genetic variation in lycopene epsilon cyclase tapped for maize biofortification. Science 319: 330–333.

Hartl DL, Clark AG (1997) Principles of Population Genetics. 3rd edn. Sinauer Associates, Sunderland, Massachusetts, USA.

Hill WG, Robertson A (1968) Linkage Disequilibrium in Finite Populations. Theor Appl Genet 38: 226–231.

Kim S, Plagnol V, Hu TT, Toomajian C, Clark RM, Ossowski S, Ecker JR, Weigel D, Nordborg M (2007) Recombination and linkage disequilibrium in *Arabidopsis thaliana*. Nat Genet 39: 1151–1155.

Kraakman ATW, Martinez F, Mussiraliev B, van Eeuwijk FA, Niks RE (2006) Linkage disequilibrium mapping of morphological resistance, and other agronomically relevant traits in modern spring barley cultivars. Mol Breed 17: 41–58.

Krivanek AF, Riaz S, Walker MA (2006) Identification and molecular mapping of *PdR1*, a primary resistance gene to Pierce's disease in *Vitis*. Theor. Appl. Genet. 112: 1125–1131.

Le Cunff L, Fournier-Level A, Laucou V, Vezzulli S, Lacombe T, Adam-Blondon AF, Boursiquot JM, This P (2008) Construction of nested genetic core collections to optimize the exploitation of natural diversity in *V vinifera* L. subsp. Sativa. BMC Plant Biol 8: 31.

Levadoux L (1956) Les populations sauvages et cultivees de *Vitis vinifera* L. Ann Amelior Plant 6: 59– 118.

Lewontin RC (1964) Interaction of selection + linkage .I. General considerations—heterotic models. Genetics 49: 49–67.

Lijavetzky D, Cabezas J, Ibanez A, Rodriguez V, Martinez-Zapater JM (2007) High throughput SNP discovery and genotyping in grapevine (*V. vinifera* L.) by combining a re-sequencing approach and SNPlex technology. BMC Genom 8: 424.

Lodhi MA, Daly MJ, Ye GN, Weeden NF, Reisch BI (1995) A molecular marker based linkage map of *Vitis*. Genome 38: 786–794.

Lopes MS, Sefc KM, Dias EE, Steinkellner H, Machado MLD, Machado AD (1999) The use of microsatellites for germplasm management in a Portuguese grapevine collection. Theor Appl Genet 99: 733–739.

Lopes MS, Dos Santos MR, Dias JEE, Mendonca D, Machado AD (2006) Discrimination of Portuguese grapevines based on microsatellite markers. J Biotechnol 127: 34–44.

Lowe KM, Walker MA (2006) Genetic linkage map of the interspecific grape rootstock cross Ramsey (*Vitis champinii*) x Riparia Gloire (*Vitis riparia*). Theor Appl Genet 112: 1582–1592.

Malosetti M, van der Linden CG, Vosman B, van Eeuwijk FA (2007) A mixed-model approach to association mapping using pedigree information with an illustration of resistance to Phytophthora infestans in potato. Genetics 175: 879–889.

Mandl K, Santiago J-L, Hack R, Fardossi A, Regner F (2006) A genetic map of Welschriesling x Sirius for the identification of magnesium-deficiency by QTL analysis. Euphytica 149: 133–144.

Marguerit E, Boury C, Manicki A, Donnart M, Butterlin G, Némorin A, Wiedemann-Merdinoglu S, Merdinoglu D, Ollat N, Decroocq S (2009) Genetic dissection of sex determinism, inflorescence morphology and downy mildew resistance in grapevine. Theor Appl Genet 118: 1261–1278.

Martin JP, Borrego J, Cabello F, Ortiz JM (2003) Characterization of Spanish grapevine cultivar diversity using sequence-tagged microsatellite site markers. Genome 46: 10–18.

McGovern PE (2003) Ancient Wine: The Search for the Origins of Viticulture. Princeton Univ Press, Princeton, USA.

Mejía N, Gebauer M, Muñoz L, Hewstone N, Muñoz C, Hinrichsen P (2007) Identification of QTLs for seedlessness, berry size, and ripening date in a seedless x seedless table grape progeny. Am J Enol Vitic 58: 499–507.

Meuwissen THE, Karlsen A, Lien S, Olsaker I, Goddard ME (2002) Fine Mapping of a Quantitative Trait Locus for Twinning Rate Using Combined Linkage and Linkage Disequilibrium Mapping. Genetics 161: 373–379.

Myles S, Chia JM, Hurwitz B, Simon CJ, Zhong GY, Buckler ES, Ware D (2010) Rapid genomic characterization of the genus *Vitis*. PlosOne 5: e8219.

Osborn TC, Alexander DC, Fobes JF (1987) Identification of Restriction-Fragment-Length-Polymorphisms Linked to Genes-Controlling Soluble Solids Content in Tomato Fruit. Theor Appl Genet 73: 350–356.

Paterson AH, Lander ES, Hewitt JD, Peterson S, Lincoln SE, Tanksley SD (1988) Resolution of Quantitative Traits into Mendelian Factors by Using a Complete Linkage Map of Restriction Fragment Length Polymorphisms. Nature 335: 721–726.

Price AL, Patterson NJ, Plenge RM, Weinblatt ME, Shadick NA, Reich D (2006) Principal components analysis corrects for stratification in genome-wide association studies. Nature Genetics 38: 904–909.

Pritchard JK, Rosenberg NA (1999) Use of unlinked genetic markers to detect population stratification in association studies. Am J Hum Genet 65: 220–228.

Pritchard JK, Stephens M, Donnelly P (2000a) Inference of population structure using multilocus genotype data. Genetics 155: 945–959.

Pritchard JK, Stephens M, Rosenberg NA, Donnelly P (2000b) Association mapping in structured populations. American Journal of Human Genetics 67: 170–181.

Rafalski A, Ananiev E (2009) Genetic diversity, linkage disequilibrium and association mapping. In: JL Bennetzen, S Hake (eds) Handbook of Maize: Genetics and Genomics. Springer, New York, USA, pp 201–219.

Remington DL, Thornsberry JM, Matsuoka Y, Wilson LM, Whitt SR, Doebley J, Kresovich S, Goodman MM, Buckler ES (2001) Structure of linkage disequilibrium and phenotypic associations in the maize genome. Proc Natl Acad Sci USA 98: 11479–11484.

Riaz S, Dangl GS, Edwards KJ, Meredith CP (2004) A microsatellite marker based framework linkage map of *Vitis vinifera* L. Theor Appl Genet 108: 864–872.

Riaz S, Krivanek AF, Xu K, Walker MA (2006) Refined mapping of the Pierce's disease resistance locus, PdR1, and sex on the extended genetic map of *Vitis rupestris* x *V. arizonica*. Theor Appl Genet 113: 1317–1329.

Riaz S, Tenscher AC, Rubin J, Graziani R, Pao SS, Walker MA (2008) Fine-scale genetic mapping of two Pierce's disease resistance loci and a major segregation distortion region on chromosome 14 of grape. Theor Appl Genet 117: 671–681.

Salmaso M, Malacarne G, Troggio M, Faes G, Stefanini M, Grando MS, Velasco R (2008) A grapevine (*Vitis vinifera* L.) genetic map integrating the position of 139 expressed genes. Theor Appl Genet 116: 1129–1143.

Slatkin M (2008) Linkage disequilibrium—understanding the evolutionary past and mapping the medical future. Nat Rev Genet 9: 477–485.

This P, Lacombe T, Thomas MR (2006) Historical origins and genetic diversity of wine grapes. Trends Genet 22: 511–519.

This P, Lacombe T, Cadle-Davidson M, Owens CL (2007) Wine grape (*V. vinifera* L.) color associates with allelic variation in the domestication gene VvmybA1. Theor Appl Genet 114: 723–730.

Thornsberry JM, Goodman MM, Doebley J, Kresovich S, Nielsen D, Buckler ES (2001) *Dwarf8* polymorphisms associate with variation in flowering time. Nat Genet 28: 286–289.

Tiwari JL, Terasaki PI (1985) Hla and Disease Associations. Springer, New York, USA.

Vezzulli S, Troggio M, Coppola G, Jermakow A, Cartwright D, Zharkikh A, Stefanini M, Grando MS, Viola R, Adam-Blondon AF, Thomas M, This P, Velasco R (2008) A reference integrated map for cultivated grapevine (*Vitis vinifera* L.) from three crosses, based on 283 SSR and 501 SNP-based markers. Theor Appl Genet 117: 499–511.

Wei XM, Jackson PA, McIntyre CL, Aitken KS, Croft B (2006) Associations between DNA markers and resistance to diseases in sugarcane and effects of population substructure. Theor Appl Genet 114: 155–164.

Welter LJ, Göktürk-Baydar N, Akkurt M, Maul E, Eibach R, Toepfer R, Zyprian E (2007) Genetic mapping and localization of quantitative trait loci affecting fungal disease resistance and leaf morphology in grapevine (*Vitis vinifera* L.). Mol Breed 20: 359–374.

Xu K, Riaz S, Roncoroni NC, Jin Y, Hu R, Zhou R, Walker MA (2008) Genetic and QTL analysis of resistance to *Xiphinema index* in a grapevine cross. Theor Appl Genet 116: 305–311.

Yu JM, Buckler ES (2006) Genetic association mapping and genome organization of maize. Curr Opin Biotechnol 17: 155–160.

Yu JM, Pressoir G, Briggs WH, Bi IV, Yamasaki M, Doebley JF, McMullen MD, Gaut BS, Nielsen DM, Holland JB, Kresovich S, Buckler ES (2006) A unified mixed-model method for association mapping that accounts for multiple levels of relatedness. Nat Genet 38: 203–208.

Yu JM, Holland JB, McMullen MD, Buckler ES (2008) Genetic design and statistical power of nested association mapping in maize. Genetics 178: 539–551.

Yu JM, Zhang ZW, Zhu CS, Tabanao DA, Pressoir G, Tuinstra MR, Kresovich S, Todhunter RJ, Buckler ES (2009) Simulation appraisal of the adequacy of number of background markers for relationship estimation in association mapping. Plant Genome 2: 63–77.

Zhao KY, Aranzana MJ, Kim S, Lister C, Shindo C, Tang CL, Toomajian C, Zheng HG, Dean C, Marjoram P, Nordborg M (2007) An Arabidopsis example of association mapping in structured samples. PLoS Genet 3.

Zhu CS, Gore M, Buckler ES, Yu JM (2008) Status and prospects of association mapping in plants. Plant Genome 1: 5–20.

Zohary D, Hopf M (2000) Domestication of Plants in the Old World. Oxford Univ Press, London, UK.

5

Molecular Linkage Maps: Strategies, Resources and Achievements

Guido Cipriani,[1,*] *Gabriele Di Gaspero*,[1,2] *Aurélie Canaguier*,[3] *Julie Jusseaume*,[3] *Johan Tassin*,[4] *Arnaud Lemainque*,[4,6] *Vincent Thareau*,[5] *Anne-Françoise Adam-Blondon*[3] and *Raffaele Testolin*[1,2]

ABSTRACT

The development of genetic maps in grapevine started in the late 90s. It greatly benefited from the development of SSR markers by an international consortium, leading to the construction of integrated genetic maps. Software was improved for facilitating map construction in full sib families. A review of the different strategies that were developed for mapping in grapevine, including mapping populations, software and markers is provided here together with a discussion on their interest and limitations. Thanks to all these resources and experience, genetic mapping is now easy to handle in grapevine and is now widely used for QTL detection, map-based cloning, comparative mapping across species and genome sequence anchoring.

Keywords: Genetic map, SSR, SNP, integrated map, full-sib family, heterozygosity

[1]Dipartimento di Scienze Agrarie e Ambientali, University of Udine, via delle scienze 208, 33100 Udine, Italy.
[2]Istituto di Genomica Applicata, Parco Scientifico e Tecnologico, via Jacopo Linussio 51, 33100, Udine, Italy.
[3]UMR INRA UEVE ERL CNRS Génomique Végétale, 2 rue Gaston Crémieux, BP 5708, 91 057 Evry cedex, France.
[4]Centre National de Génotypage, CNG-IG CEA, 2 rue Gaston Crémieux CP 5721, 91057 Evry Cedex, France.
[5]IBP, UMR CNRS Université Paris-Sud, bat 630, 91405 Orsay cedex, France.
[6]Present address: IMAGENE, Bat Genavenir 6, Genopole campus 1, 5 rue Henri Desbruères, 91 030 Evry cedex, France.
*Corresponding author: *guido.cipriani@uniud.it*

5.1 Introduction

Genetic maps based on molecular markers emerged on the scene of plant genetics in the 1980s (Beckmann and Soller 1983; Tanksley et al. 1989) and in the course of time became indispensable tools for grapevine geneticists.

In perennials, the development of the first genetic maps occurred later than in annuals due to the higher difficulty of generating large progeny from controlled crosses (e.g., Tulsieram et al. 1992) and to the difficulty of extracting large amounts of high quality DNA, which was required for the production of the first-generation molecular markers, such as restriction fragment length polymorphism (RFLP) markers (e.g., Lodhi et al. 1994). The first grapevine linkage map was constructed in 1995 (Lodhi et al. 1995), while the first genetic maps in tomato and maize were published in 1986 and 1988, respectively (Bernatzky and Tanksley 1986; Helentjaris et al. 1988).

Genetic mapping provided significant benefit to the grapevine community only after the development of molecular markers based on polymerase chain reaction (PCR). In parallel, improvement of statistical methods and software required for handling segregation data in various genetic contexts (full-sib families, back-crosses, F_2) were regularly achieved.

In the following paragraphs we will review how and when molecular markers were introduced in grapevine genetics, the evolution of grapevine linkage maps, as well as the algorithms and mapping populations used.

5.2 Evolution of Molecular Markers for Genetic Mapping: from RFLPs to SNPs

The development and use of molecular markers in grape followed the development of molecular biology that began in the 1970s. Initially, markers were used for the analysis of genetic diversity in the grapevine germplasm (see Chapter 2).

5.2.1 Isozymes

Isoenzymes became very popular in the 1980s as markers capable of discriminating grape accessions and to confirm parentage (Loukas et al. 1983; Arulsekar and Parfitt 1986). Segregation data provided by this kind of biochemical marker can be found in the first grapevine linkage maps (Lodhi et al. 1995; Dalbò et al. 2000; Doligez et al. 2002; Grando et al. 2003).

5.2.2 RFLP

The first kind of DNA-based molecular markers that came into play for genetic mapping was RFLP marker (Beckmann and Soller 1983). RFLP markers have never received the same attention from grape breeders as they got for instance in humans and in other crop species. This was mainly due to the high cost of using this technique, the large amount of DNA required, and the necessity to access and store large collections of probes. Most of these probes were heterologous, that is developed from the DNA of other plant species. In spite of these technical constraints, RFLP contributed to the construction of the earliest grape maps (Lodhi et al. 1995; Dalbò et al. 2000; Doligez et al. 2002; Grando et al. 2003).

5.2.3 *RAPD, AFLP, SAMPLE, and S-SAP*

The first revolution in grape molecular genetics took place with the advent of the PCR technique and the use of arbitrarily designed primers that do not require *a priori* knowledge of the DNA sequence of the species of interest (Williams et al. 1990). Random amplified polymorphic DNA (RAPD; Williams et al. 1990) markers were easy to produce, and their cost was affordable in terms of labor and investment, even in small laboratories.

In grapevine, RAPD markers were used in several maps (Lodhi et al. 1995; Lahogue et al. 1998; Dalbò et al. 2000), but they were very quickly replaced by another type of marker produced with a combination of endonucleases and arbitrarily designed primers: the amplified fragment length polymorphism (AFLP) marker (Vos et al. 1995). AFLPs were more reproducible and yielded a greater number of polymorphic loci per experiment, with a single gel displaying as many as one hundred different bands per lane, a dozen of which could segregate in a mapping population. AFLP-based linkage maps were produced between 2002 and 2007 (Doligez et al. 2002; Grando et al. 2003; Doucleff et al. 2004; Fischer et al. 2004; Riaz et al. 2004; Cabezas et al. 2006; Troggio et al. 2007).

Both RAPD and AFLP markers have several drawbacks, such as dominance, reproducibility issues, and low transportability across distantly related genotypes. The latter is a severe limitation, as geneticists and especially breeders frequently need to share knowledge acquired in a particular mapping population with others.

Two marker classes, based on the AFLP technology and thus showing the same major drawbacks, deserve mentioning, although their use was mostly aimed at distinguishing grape accessions rather than at producing linkage maps: selective amplification of polymorphic loci (SAMPL), which combines the AFLP technology with the abundance and polymorphism of simple sequence repeat (SSR) markers, and sequence-specific amplified polymorphism (S-SAP), in which the AFLP technology is combined with the retrotransposon long terminal repeat (LTR) (Pelsy et al. 2003; Labra et al. 2004; Pelsy 2007). A single paper reports the use of SAMPL and S-SAP markers in grapevine mapping (Cabezas et al. 2006).

5.2.4 SSR

RAPD, AFLP, and their derived classes of markers were never adopted in humans, where genetic maps were initially produced using RFLP and then microsatellite or SSR markers. A highly saturated linkage map based on 5,264 SSRs was published in 1996 to anchor the human physical map in the frame of the Human Genome Sequencing project (Dib et al. 1996).

Microsatellite repeats are also abundant in plant genomes, and show a preferential distribution to non-repetitive regions (Morgante et al. 2002). They are amplified as single loci in diploid genomes, they are codominant, and are highly polymorphic in grapevine (see for, e.g., Aradhya et al. 2003). In addition, they have conserved flanking regions that make them transferable to related species (Moore et al. 1991), which is particularly true within the Vitaceae family (Sefc et al. 1999; Di Gaspero et al. 2000; Decroocq et al. 2003). All of these positive features made them the ideal markers for the construction of linkage maps that are transferable from one cross to another.

The SSR markers were first isolated in grapevine in the early 1990s (Thomas and Scott 1993; Bowers et al. 1996; Bowers et al. 1999). During the International Conference on Grape Genetics and Breeding in 1998 held at Montpellier (France), Mark Thomas, Carol Meredith, and Patrice This launched the *Vitis* Microsatellite Consortium (VMC), a project aiming to develop a large number of microsatellite markers under the coordination of Agrogene, SA. The proposal stimulated the cooperative effort of numerous institutions called to sequence groups of 48 cloned fragments produced by Agrogene from grape genomic libraries, enriched with 10 different types of di- and tri-nucleotide repeats. This work resulted in the production of nearly 1,000 SSR markers. Several hundreds of these markers were polymorphic and segregated in mapping populations (Doligez et al. 2002; Grando et al. 2003; Adam-Blondon et al. 2004; Fisher et al. 2004; Mandl et al. 2006; Riaz et al. 2006; Di Gaspero et al. 2007). Since 90% of the VMC SSRs were AG repeats, several institutions produced new genomic libraries enriched for different repeats, which were used to isolate a substantial number of new SSRs (Sefc et al. 2000; Lefort et al. 2002; Di Gaspero et al. 2005; Merdinoglu et al. 2005), while other groups isolated new SSRs from expressed sequence tag (EST) collections (Scott et al. 2000; Decroocq et al. 2003; Riaz et al. 2006). An added advantage of SSR markers is that several of them can be multiplexed by different dyes and by adjacent allele range sizes, and analyzed in a single run with an automatic sequencer, increasing the throughput of the analysis (Merdinoglu et al. 2005). After the completion of the grapevine genome sequencing projects, several thousand SSR markers became available and retrievable from the NCBI database with simple programs like Sputnik (*http://cbi.labri.fr/outils/ Pise/sputnik.html*). The most abundant class of SSRs in the grapevine genome are di-nucleotide repeats, representing 46% of the total 75,185 SSRs identified so far. Tri-, tetra-, and penta-nucleotide repeats are also cumulatively well represented in the genome, but the relative frequency of each individual type of repeat is low, except for AAT (19%), AAAT (10%), and AAAT (0.07%; Table 5-1).

A set of SSR markers with 3-nt to 5-nt core repeat, the type of choice in human forensics, were selected for covering all grapevine chromosomes and validated for their power of discrimination among individuals (Cipriani et al. 2008), in view of their potential use in grapevine fingerprinting (Cipriani et al. 2010).

Table 5-1 SSR markers identified in the 12X assembly of the PN40024 grape genome sequence (Jaillon et al. 2007) using a modified version of the Sputnik software (*http://cbi.labri.fr/ outils/ Pise/ sputnik.html.*

Type of repeat	Average repeat length (bp)[a]	No. of occurrences	Fraction of the total
di-nucleotide	*12.02*	*34,449*	*0.46*
-AT	14.09	23,674	0.31
-AG/CT	13.00	8,349	0.11
-AC/GT	12.01	2,423	0.03
- CG	9.00	3	0.00
tri-nucleotide	*7.47*	*18,747*	*0.25*
-AAT	9.08	14,522	0.19
-AAG	8.01	1,982	0.03
-ATC	7.09	710	0.01
-ACC	7.08	486	0.01
-AAC	7.07	363	0.00
-others	7.27	684	0.01
tetra-nucleotide	*5.68*	*11,279*	*0.15*
-AAAT	5.07	7,715	0.10
-AAAG	5.08	862	0.01
-AATT	5.04	721	0.01
-ACAT	7.03	470	0.01
-AATC	5.06	446	0.01
-others	5.73	1,065	0.01
penta-nucleotide	*4.08*	*10,710*	*0.14*
-AAAAT	4.05	5,623	0.07
-AAAAG	4.06	1,809	0.02
-AAATT	4.04	651	0.01
-AAAAC	4.06	525	0.01
-AATAT	4.05	348	0.00
-others	4.09	1754	0.02
TOTAL	*6.94*	*75,185*	*1.00*

[a]Minimum no. of repeat fixed for each search: 8 for di-nucleotide repeats, 6 for tri-nucleotide repeats, 5 for tetra-nucleotide repeats, and 4 for penta-nucleotide repeats.

All these sets of SSR markers are listed in Table 5-2, and most of them are available in public databases (see Chapter 13).

Table 5-2 Grapevine SSR markers listed according to their date of publication.

Code	No. of sequences	Type of repeat	Source (a)	Reference
VVS	5	GA and GT	EGL	Thomas and Scott 1993
VVMD	4	mainly CT	EGL	Bowers et al. 1996
VVMD	22	CT	EGL	Bowers et al. 1999
ssrVrZAG	18	GA	EGL	Sefc et al. 1999
scu[xx]vv	10	2- and 3-nt	EST	Scott et al. 2000
ssrVvUCH	7	mainly AG	EGL	Lefort et al. 2002
VVC	8	CA and CA	EST	Decroocq et al. 2003
VMC, VMCNg	357	mainly AG	EGL	Agrogene, unpublished (b)
VVI	169	various	EGL	Merdinoglu et al. 2005
UDV	108	AC	EGL	Di Gaspero et al. 2005
ctg, CF, AF, BM, VEST	unknown	unknown	EST	Riaz et al. 2006
uncoded	239,634	all	SG	Jaillon et al. 2007
VChr	38	3- to 5-nt	SG	Cipriani et al. 2008

(a) EGL, Enriched genomic libraries; EST, Expressed sequence tag libraries; SG, Shot-gun sequencing.
(b) Partially published in Di Gaspero et al. 2000; Pellerone et al. 2001; Adam-Blondon et al. 2004; Arroyo-Garcia and Martinez-Zapater 2004; Riaz et al. 2004; Goto-Yamamoto et al. 2006; deposited in NCBI STS database.

5.2.5 SNP

Single nucleotide polymorphism (SNP) markers are the most recently developed class of markers. They target a single base mutation in the DNA sequence. Small indels (insertion or deletion events) are also assimilated as SNP markers since they can be handled with many of the technologies designed to identify SNPs. Their discovery relies on resequencing the same DNA region from different haplotypes. In grapevine, two research groups sequenced 25 and 230 gene fragments in 9 and 11 *Vitis vinifera* genotypes, respectively and observed a frequency of one SNP every 47 to 64 bp (Salmaso et al. 2004; Lijavetzky et al. 2007). Sequencing the heterozygous Pinot Noir led to the discovery of one SNP every 250 bp, by comparing the two haplotypes (Velasco et al. 2007), however, a significant amount of variation in this frequency exists along the genome: from one SNP/60 bp to one SNP/250 bp or less. Authors estimated that the number of potential SNP markers in the Pinot Noir genome could reach 2 million, with many present in coding regions, covering approximately 87% of annotated genes (Velasco et al. 2007).

SNPs are easily amenable to massive parallel automatic detection (Rapley and Harbron 2004; Steemers et al. 2006), making them particularly useful for saturating maps, especially in coding regions (Pindo et al. 2008), or as an approach to gene discovery through linkage disequilibrium studies. Lijavetzky et al. (2007) estimated an 83% success rate for the conversion of a candidate variable nucleotide position into a technically scorable SNP marker. This conversion rate fell to 61% when SNPs were initially identified in the Pinot Noir sequence without resequencing any additional genotype (Pindo et al. 2008; Vezzulli et al. 2008a). Some 30% of SNP markers developed from sequencing *V. vinifera* cultivars are estimated to be heterozygous in any other *V. vinifera* cultivar (Lijavetzky et al. 2007; Vezzulli et al. 2008a). This rate could fall to 19% in feral accessions of *V. vinifera* and to 2% in wild species (Vezzulli et al. 2008b).

The transferability of SNPs across species is, therefore, much less likely than the transferability of SSRs and this is a major issue in mapping many traits, such as those for resistance to biotic and abiotic stress which usually entail genes present in wild species. However, due to their high density in the grapevine genome and the flexibility of multiplexing systems, it is possible to envisage working with complementary sets of SNP markers, some of which could be informative in *V. vinifera* and others in wild relatives, as it has been proposed for rice (*http://mlorieux.free.fr/Rice_Genomics/Research/RiceDiversity/CoreMap/index.html*).

Next-generation sequencing technologies now allow to rapidly identify customized sets of SNPs, for maximizing the informative content of the DNA chip in the gene pool of interest or in the parents of mapping populations. Accurate filtering of true SNPs from sequencing and alignment errors is still a challenging step, before embarking on a SNP chip experiment for linkage mapping. More confidence on this issue was generated by a case study of Myles et al. (2010), in which a 9K SNP array design from a set of 71K high-quality SNPs provided 97.7% concordance between genotype calls from the sequencing data and from the DNA chip.

5.3 Mapping Populations and Mapping Strategies

As mentioned above, perennial species pose specific biological problems in the development of mapping populations. The generation time is long (3–5 years from seed to seed), growing many individuals is cost and labor demanding, species are heterozygous, and several of them, grapevine included, are sensitive to inbreeding depression (Einset and Pratt 1975). Yet, large cross populations can be prepared without the limits of animal brood sizes.

The genetic structure of the mapping population and the dominant/co-dominant nature of the markers are the two factors that determine the observed genotypic classes and the information provided by each individual of the family (Allard 1956).

Traditionally, linkage studies in plants are carried out using the classical test cross, where two phenotypically or genotypically contrasting inbred lines are crossed to produce an F_1, which in turn is either backcrossed to one of the parents resulting in a 1:1 segregation of markers, or selfed to produce an F_2, resulting in a 1:2:1 segregation of co-dominant markers and a 3:1 segregation of dominant markers. Ritter et al. (1990) proposed the formulas for calculating the linkage between different classes of markers (dominant, co-dominant, with or without null allele) segregating in F_1 families obtained from heterozygous parents. In the case of dominant markers, the segregation of markers heterozygous in either parent is the same as in a classical backcross or testcross. This approach was adopted in 1994 by Grattapaglia and Sederoff to produce a linkage map of *Eucalyptus* species based on dominant RAPD markers (Grattapaglia and Sederoff 1994). They named this type of mapping approach the "pseudo-testcross". Lodhi et al. (1995) introduced such a "pseudo-testcross" in grapevine, which was widely used for the construction of RAPD- and AFLP-based maps (Table 5-3). However, the drawback of this strategy is that the correspondence between the homologous linkage groups in the two parents can not be established. It can only be established through markers that are heterozygous in both parents and segregate in a ratio of 3:1 (Ritter et al. 1990; Lodhi et al. 1995). Co-dominant markers, such as SSR markers, that often have segregating alleles from both parents, provide a better solution to bridging homologous linkage groups, and have been widely used for that reason since 2003 (Table 5-3). From this point of view, SNP markers are in most cases bi-allelic (Troggio et al. 2007; Vezzulli et al. 2008a), segregate from a single parent like in the pseudo-testcross, and do not contribute to the identification of homologous linkage groups.

In grapevine, mapping populations usually consist of full-sib families (F_1) derived from a cross between two highly heterozygous parents, but some have also been produced by selfing a single genotype (Table 5-3). This unexpectedly reduced segregation biases (Adam-Blondon et al. 2004; Duchêne et al. 2009), with the percentage of segregating markers falling to around 50% against 75% of markers segregating in a typical full-sib family. Parents of biparental populations were chosen for their high heterozygosity and loose genetic relatedness with the objective of minimizing the occurrence of common alleles, which hampers the identification of which parent transmits them to the progeny and in turn makes them less informative. Contrasting phenotypes for the trait(s) to be mapped are also necessary factors in the choice of parents, in order to follow the inheritance of the trait in the progeny.

The number of meioses, and thus the size of the progeny, necessary to reach an acceptable level of precision in the estimation of the genetic distance between two markers depends on the informative content of the markers,

Table 5-3 Grapevine (*Vitis* spp.) linkage maps listed according to their publication date. All but two mapping populations were F_1 from heterozygous parents and the others were obtained by selfing one heterozygous individual.

Mapping population (a)	No. of progeny	No. of markers (b)	Types of markers	No. of linkage groups (b)	Map lengths cM (b)	Software used	Reference
Cayuga White (*Ih*) x Aurore (*Ih*)	60	438	most RAPD	20–22	1,196–1,477	MapMaker	Lodhi et al. 1995
Horizon (*Ih*) x Illinois 547-1(*Ih*)	58	438	most RAPD	20	1,199–1,470	MapMaker	Dalbò et al. 2000
MTP2223-27 (*Vv*) x MTP2121-30 (*Vv*)	139	301	AFLP/SSR & other	22–23 (20)	767–816 (1,002)	MapMaker, JoinMap	Doligez et al. 2002
Moscato bianco (*Vv*) x V. riparia (*Vri*)	81	338–429	AFLP/SSR & other	20–19	1,639–1,518	MapMaker, JoinMap	Grando et al. 2003
Syrah (*Vv*) x Grenache (*Vv*)	96	177–178 (220)	SSR	19–18 (19)	1,172–1,361 (1,406)	CarHTaGene	Adam-Blondon et al. 2004
Riesling (*Vv*) selfed	96	110	SSR	19	1,192	CarHTaGene	Adam-Blondon et al. 2004
V. rupestris (*Vru*) x V. arizonica (*Va*)	116	475	most AFLP	17–19	756–1,082	JoinMap, PGRI	Doucleff et al. 2004
Regent (*Ih*) x Lemberger (*Vv*)	153	429	AFLP/ RAPD/SSR/ SCAR	20–26	1,277–1,158	JoinMap	Fischer et al. 2004
Riesling (*Vv*) x Cabernet Sauvignon (*Vv*)	153	152	SSR	20–20 (20)	1,431–1,450 (1,728)	MapMaker, JoinMap	Riaz et al. 2004
Dominga (*Vv*) x Autumn Seedless (*Vv*)	118	595	AFLP/ SAMPL/S-SAP, SSR	19	1,173–1,131	JoinMap	Cabezas et al. 2006
Five mapping populations (c)	46 to 153	502	most SSR	19	1,154–1,570 (1,647)	CarHTaGene	Doligez et al. 2006

Table 5-3 contd....

Table 5-3 contd....

Mapping population (a)	No. of progeny	No. of markers (b)	Types of markers	No. of linkage groups (b)	Map lengths cM (b)	Software used	Reference
MTP2687-85 (*Vv*) x Muscat de Hambourg (*Vv*)	174	139	SSR	>19	935–1,365 (1,265)	CarHTaGene	Doligez et al. 2006b
V. Champinii x *V. riparia*	188	205	SSR	19	1,245–1,095 (1,305)	JoinMap	Lowe et al. 2006
Welschriesling (*Vv*) x Sirius (*Ih*)	92	251	most SSR	20–20	not reported	MapMaker	Mandl et al. 2006
D8909-15 (*Vru* x *Va*) x F8909-17 (*Ih*)	188	159–158 (210)	most SSR	18–19 (19)	865–1,055 (1,154)	JoinMap	Riaz et al. 2006, then implemented by Xu et al. 2008
Chardonnay (*Vv*) x Bianca (*Ih*)	46 (116)*	332–353 (709)	SSR/RGA	19–19	1,210–1,425 (1,676)	CarHTaGene/*JoinMap	Di Gaspero et al. 2007, *then implemented by Bellin et al. 2009
Cabernet Sauvignon (*Vv*) x 20/3 (*Ih*)	46	379–349 (709)	SSR/RGA	19–19	1,254–1,418 (1,676)	CarHTaGene	Di Gaspero et al. 2007
Syrah (*Vv*) x Pinot Noir (*Vv*)	94	994	SNP/AFLP/SSR	19	1,245	TMAP	Troggio et al. 2007
Regent (*Ih*) x Lemberger (*Vv*)	153	398	SSR/RGA/SCAR	19	1,631	JoinMap	Welter et al. 2007
Italia (*Vv*) x Big Perlon (*Vv*) Moscato bianco (*Vv*) x V. riparia (*Vri*)	163 171	276–210 (341)	AFLP/SSR & other	19–19 (18)	1,353–1,130 (1,426)	JoinMap	Costantini et al. 2008 Battilana et al. 2009
Merzling (*Ih*) x Teroldego (*Vv*)	89	166–168 (247)	SNP/SSR	20–21 (19)	914–1,174 (1,309)	MapMaker/JoinMap	Salmaso et al. 2008
Three mapping populations (d)	87 to 94	1,134	AFLP/SSR/SNP	19	1,443	TMAP	Vezzulli et al. 2008

Muscat Ottonel (*Vv*) selfed	121	84	most SSR	19	569 (e)	CarHTaGen/ JoinMap	Duchêne et al. 2009
Cabernet S x *Vitis riparia* Gloire de Montpellier (*Vri*)	138	212	SSR/SSCP	19	1,270–1,410 (1,249)	CarHTaGene/JoinMap	Marguerit et al. 2009
V3125 (*Vv* 'Schiava grossa' x 'Riesling') x 'Börner' (*Vri* x *Vci*)	188	174–190 (235)	Most SSR	19–21 (19)	1,116–1,070 (1,155)	JoinMap	Zhang et al. 2009

(a) species: *Va* = *Vitis arizonica*, *Vc* = *Vitis champinii*, *Vci* = *Vitis cinerea*, *Vri* = *Vitis riparia*, *Vru* = *Vitis rupestris*, *Vv* = *Vitis vinifera*, *Ih* = *Interspecific hybrids.*
(b) Individual maps, and in parentheses the consensus/integrated map.
(c) Syrah *(Vv)* x Grenache *(Vv)*, Riesling selfed *(Vv)*, Chardonnay *(Vv)* x Bianca *(Ih)*, MTP2223-27 *(Vv)* x MTP2121-30 *(Vv)*, Riesling *(Vv)* x Cabernet Sauvignon *(Vv)*.
(d) Syrah *(Vv)* x Pinot Noir *(Vv)*, Syrah *(Vv)* x Grenache *(Vv)*, Cabernet Sauvignon *(Vv)* x Riesling *(Vv)*.
(e) The short length of map is due to the reduced number of markers used to produce a framework map.
(f) D8909-15 (*Vru* x *Va*) x F8909-17 (*Ih*), D8909-15 (*Vru* x *Va*) x B90-116 (*Vv*), D8909-15 (*Vru* x *Va*) x Airen (*Vv*).

which can be highly variable in an F_1 family depending on the segregation type (Ritter et al. 1990). The size of mapping populations in grape has ranged from 46 to 181 individuals, with most maps based on a number of individuals close to 100. Such a progeny size is a good compromise, in order to avoid inflation of the cost of map production while still providing enough statistical strength for the estimation of the correct marker order and distances. It also depends on the objective of the mapping project. For instance, to quickly get a genome-wide picture of marker order for as many markers as possible, it is possible to combine mapping data from two or more populations (Di Gaspero et al. 2007; Doligez et al. 2006a), sometimes very small (46 individuals in Di Gaspero et al. 2007). This trick allowed the mapping of an extraordinarily high proportion of markers in at least one parent (91%), while only 75 and 80% of total markers would have segregated in either population separately (Di Gaspero et al. 2007). Hence, although the two mapping populations were of limited size, the number of informative meioses per marker was 46 to 184, and 130 on average, depending on the heterozygosity of the parents for each marker (Di Gaspero et al. 2007). This high number of informative meioses per marker mitigated the conflicts in finding the most likely linear order of markers (Doligez et al. 2006a; Di Gaspero et al. 2007; Vezzulli et al. 2008a). The drawbacks of this strategy will be discussed later. When the purpose of mapping is to identify a trait locus, marker orders and estimation of genetic distances can be refined around the locus, once the chromosomal region controlling a trait of interest has been identified, by adding markers designed on the homologous region in the reference genome sequence and by using additional individuals of the same or related mapping populations (Riaz et al. 2008; Bellin et al. 2009; Coleman et al. 2009; Hwang et al. 2010; Di Gaspero and Cattonaro 2010).

5.4 Mapping Algorithms and Software Used

The process of map construction can be separated into three phases: identification of groups of linked markers, ordering these markers, and estimation of the genetic distance between two adjacent markers. Of these tasks, the second is the most challenging. The ways in which these issues have been addressed in the construction of genetic maps in grapevine is discussed in the following paragraphs.

5.4.1 Individual Parental Maps

Whatever the type of marker used (dominant or co-dominant), it is possible to follow the segregation of heterozygous markers in either parent of an F_1 population like in a backcross population, and to construct two independent maps (Lodhi et al. 1995; Dalbo et al. 2000). If the population is derived from a selfed individual, then the segregation behaves like a F_2 (Adam-Blondon et al. 2004; Duchêne et al. 2009).

The first very popular software adopted by plant geneticists and then by grape geneticists, was MapMaker (Lander et al. 1987). MapMaker calculates the probabilities of association between markers and the probabilities of marker order by the maximum likelihood approximation (Allard 1956; Lander et al. 1987). To handle the marker phase, some authors have suggested a strategy which consists of duplicating the whole set of data, inverting the allele code in the duplicated matrix, and merging the two mirror datasets. The program then constructs groups of linked markers and each group has a mirror group that is removed from the subsequent analyses. This strategy was first used in woody plants by Grattapaglia and Sederoff (1994) and in grape by Lodhi et al. (1995). Marker distances are finally calculated using functions that take into account the non-additive nature of mapping distances when they are large (Haldane 1919; Kosambi 1944).

5.4.2 Consensus Map from the Two Parental Segregations of a Cross

When a high number of co-dominant markers segregating in both parents are available, it is possible to construct a single consensus map from the genotypic dataset of the F_1 family, and the software must process heterogeneous types of segregation as described in Ritter et al. (1990). This kind of dataset cannot be handled by MapMaker. JoinMap is the first popular software that was developed for this purpose (Stam and Van Ooijen 1995; Van Ooijen and Voorrips 2001).

JoinMap automatically calculates the phase of the markers before conducting linkage analysis for full-sib families. Version 4.0 of the program includes either the classical maximum likelihood or a new Monte Carlo maximum likelihood algorithm (ML). The latter is very fast in computation and can easily make calculations from large sets of markers (100 or more) per linkage group (Van Ooijen 2006). Unfortunately it cannot be used for out-breeding full-sib families, which are the most commonly used populations in grapevine (Table 5-3). An original algorithm was developed for the calculation of distances between markers (Stam 1993). Fifteen papers report the use of different versions of JoinMap to produce linkage maps in grapevine (Table 5-3). Such popularity is likely due to the flexibility in handling different marker segregations and the user-friendly interface.

CarHTaGene is mentioned in six papers (Table 5-3) and was the second most popular software used in grapevine. TMAP was the last software to be adopted (documented in two papers listed in Table 5-3). TMAP but not CarHTaGene calculates the phase of the alleles, but both of them can build consensus maps with or without the assumption of conserved recombination ratios across different individuals (de Givry et al. 2005; Cartwright et al. 2007). In both, marker ordering is based on a Monte Carlo algorithm to find the maximum multipoint-likelihood order with automated heuristic methods for testing the orders and managing missing data (several in CarHTaGene and one in TMAP; Scheix and Gaspin 1997; de Givry et al. 2005; Carthwright et al. 2007). CarHTaGene is the only one among the three packages managing outbred populations that gives scientists the option of choosing between different final orders. However, this package is computationally more power-demanding. TMAP is the only package that calculates a probability of error for each marker, which is accounted for in the calculation of the maximum likelihood of the map order (Cartwright et al. 2007). The efficient removal of false double recombination has been shown to reduce map inflation. Finally, from a more practical point of view, TMAP is able to read data files compatible with MapMaker, JoinMap, and CarHTaGene, and has very useful procedures for automatic transformation of the data matrix (for parental maps, consensus maps, etc.) that does not exist in the other packages.

The simultaneous use of segregation from both parents is advantageous when one wants to compile a unique map, which consists of markers alternatively heterozygous in either parent, and also wants to increase the number of informative meioses for the estimate of marker distance. In the process of integrating maps, the way in which an analyst copes with heterogeneous recombination rates and with duplicated marker loci is crucial for the correct reconstruction of marker order. Low resolution due to genotype-specific suppression of recombination in one parent is compensated by the crossovers that could have occurred in the other parent. Recombination rates are averaged over the two individuals, thus providing a more reliable expectation for any other genotype of the species. If this is desirable at a genome wide level, it may be misleading for predicting the efficiency for marker-assisted selection of alleles linked to a trait in a certain parent. Integrated maps tend to wobble in regions with a low density of common markers, in stretches of linked markers with distorted segregation in one parent, and across segmental duplications in which duplicated markers are heterozygous in the two parents at either locus.

5.4.3 Integration of Maps developed from Different Populations

As the first SSR-based genetic maps became available, they were soon merged into single consensus maps (Doligez et al. 2006a; Vezzulli et al. 2008a). In this case, the problem is handling data sets where the segregation type can be heterogeneous not only between markers, but also within markers. JoinMap, CarHTaGene, and TMAP can all handle such data. The integrated map is constructed with different algorithms depending on the software used, which may lead to different outputs. Several papers describe the difference between the outputs of JoinMap and CarHTaGene (Scheix and Gaspin 1997; Doligez et al. 2006a).

Merging maps is very attractive because it provides a snapshot of the markers available at a given chromosomal region, but such consensus maps have some inherent weaknesses. For instance, map distances are less accurate if only a few markers are shared among individual maps, and map length is inflated by genotyping errors that add up. Also, the location of "private" markers, which are markers segregating in only one or a few parents tends to wobble, and their distance from common markers is not defined (Doligez et al. 2006a). These drawbacks are mainly due to the unbalanced number of meioses used to calculate the recombination among pairs of markers, which creates friction in the estimate of map distances. In some instances, instead of merging datasets, it would be preferable to build individual maps based on specific recombination rates, and then to align them using common markers, thus bringing markers mapped in different individuals into a unique map by a projection of distances (Arcade et al. 2004).

5.5 Different Generation of Maps in the *Vitaceae*

To date, genetic maps have been developed only in the *Vitis* genus, with several aims: QTL detection, physical map/chromosome anchoring, and map based cloning.

5.5.1 Maps Based on Anonymous Markers

The first grapevine linkage map was produced by the research group based at the Geneva Experimental Station in the United States (Lodhi et al. 1995). The mapping population was an F_1 obtained by crossing two highly heterozygous interspecific hybrids produced by the same research group, Cayuga White and Aurore, and consisted of 60 progeny. The authors were able to place 428 markers in total, mostly RAPD, and the maps resulted in 20 and 22 linkage groups, with total lengths of 1,196 and 1,477 cM, respectively (Table 5-3). The markers, being dominant, segregated from either parent and were recorded as presence/absence of amplified bands. MapMaker was adopted for the association analysis and construction of the linear series of markers along each linkage group. Lodhi's map was developed as an exercise for showing the potential of molecular markers in rapidly saturating genetic maps, without the need for hundreds of phenotypic characters. It is mentioned here for its historical value, being the first map ever produced in grape and the first application of the pseudo-testcross method in this species.

Several other maps were developed mainly from RAPD or AFLP markers, with a progressive introduction of SSR markers as they became available from the *Vitis* Microsatellite Consortium (e.g., Doligez et al. 2002; Grando et al. 2003). The main objective of these maps was to quickly cover all of the chromosomes of both parents in order to detect QTLs for traits of interest (e.g., Doligez et al. 2002; Fischer et al. 2004; Cabezas et al. 2006; Welter et al. 2007; Costantini et al. 2008). Some single locus traits were also mapped, such as the main locus for berry color and the locus controlling flower sex (Dalbò et al. 2000; Doligez et al. 2002).

5.5.2 Maps Based on SSR and SNP Markers

The publication of the first map completely developed from SSR markers was in 2004 (Riaz et al. 2004). Many genetic maps were published later, based mainly, if not exclusively, on SSR markers (Adam-Blondon et al. 2004; Fisher et al. 2004; Mandl et al. 2006; Riaz et al. 2006; Di Gaspero et al. 2007; Welter et al. 2007). SSR markers allowed the unification of linkage group numbering (Riaz et al. 2004; Adam-Blondon et al. 2004) and the definition of sets of markers located at regular intervals, suited for comparative QTL detection, and highly heterozygous across the existing maps. Some improvement is expected in the near future by defining an additional group of markers to better cover the chromosomal extremities and a few remaining gaps (Doligez et al. 2006a). This may require new marker development, as the main problem in these areas has been the homozygosity of the parents for the markers available (Doligez et al. 2006a). The genome sequence will help to provide many targeted candidates. The grape genome is very heterozygous on average, but some regions contain homozygous blocks that are not covered by informative markers (Velasco et al. 2007). It is still unclear if localized low heterozygosity occurred specifically and by chance in the few genotypes in which it was noticed, or if it is a common feature with a biological significance in certain regions across the genotypes.

The first map predominantly based on SNP markers appeared in 2007 (Troggio et al. 2007). The authors observed a non-random distribution of the SNP markers derived from non-coding regions (BAC-end sequences), whereas SNPs derived from coding regions were randomly distributed, like SSR markers. This map is also the most dense genetic map based on a single population (average distance between two markers of 1.3 cM), with 615 transferable markers (483 SNP and 132 SSR) and 379 AFLP markers (Troggio et al. 2007). It spans a 1,245-cM genome length, which means that 1 cM would correspond to 390 kb on average, although the correlation between genetic and physical distances is variable across the genome

(Lamoureux et al. 2006; Troggio et al. 2007). A comparison between the SSR-based linkage map of linkage group 2 from Doligez (Doligez et al. 2006a) and the same linkage group based on SSR and SNP markers published by Vezzulli (Vezzulli et al. 2008a) is shown in Fig. 5-1.

Finally, SSR markers and SNPs have also allowed gene mapping in grapevine, with a strong emphasis on candidate genes for disease resistance (Di Gaspero et al. 2007; Welter et al. 2007), as well as other traits (Salmaso et al. 2008), and targeting genes has allowed the development of markers that are more transferable across species (Lowe and Walker 2006).

5.5.3 *Use of Linkage Maps in Genome Sequencing Projects*

Genetic maps with dense saturation and high resolution are fundamental tools in genome sequencing projects to assist the anchorage of the genome sequence to the chromosomes, and as such, the Doligez et al. (2006a) and Vezzulli et al. (2008a) integrated maps were used to anchor respectively the 8.4X assembly of the PN40024 (Jaillon et al. 2007) and the Pinot Noir (Velasco et al. 2007) genome sequences, respectively.

Recently, an improved version of the Doligez et al. (2006a) map has been produced in order to anchor and order the sequence scaffolds of the final PN40024 12X assembly, by increasing the size of two mapping populations, and by genotyping a well chosen set of markers in each of them: 358 individuals of the Chardonnay x Bianca population were scored with 379 SSR markers, and 179 individuals of the Syrah x Grenache population were scored with 325 SSR markers (unpublished data). The two consensus maps were constructed using Tmap, and the Syrah x Grenache map was projected onto the Chardonnay x Bianca population using Biomercator (Arcade et al. 2004). The resulting integrated map has 514 SSR markers, 174 of them newly developed from BAC-end sequencing and scaffold sequences, and is presented in Fig. 5-2.

While the sequence assembly of the nearly homozygous genotype PN40024 made use of genetic map information mainly for anchoring ready-to-map sequence scaffolds (Jaillon et al. 2007), the assembly of the heterozygous Pinot Noir sequence required marker information to assist the sequence assembler in the construction of metacontigs, sorting out chimeric metacontigs that merged contigs from different linkage groups (Velasco et al. 2007). The linkage map was also used to find the metacontigs position and orientation. SNP markers were then developed on sequence contigs with unassigned map position and placed back onto the genetic map (Troggio et al. 2007; Velasco et al. 2007; Vezzulli et al. 2008a).

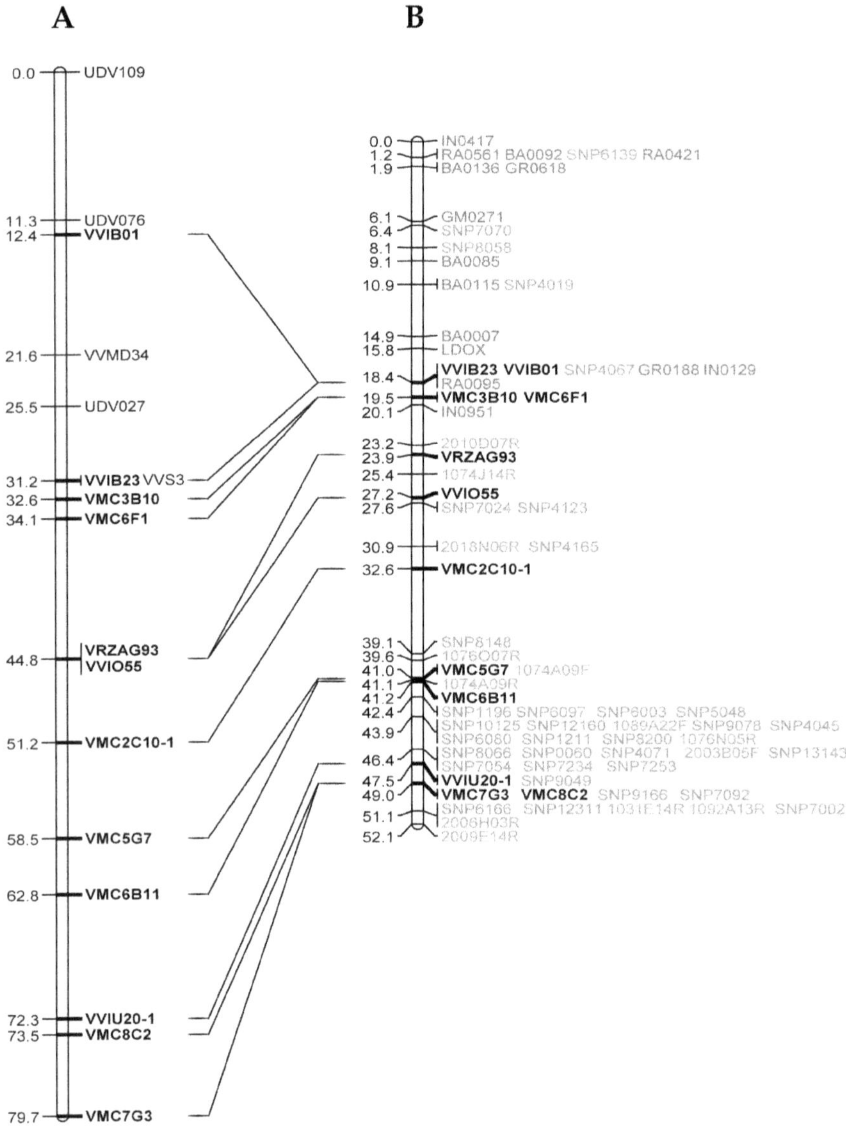

Figure 5-1 Grapevine linkage map of chromosome 2. (A) Reference map based exclusively on SSR markers (Doligez et al. 2006a); (B) the same linkage group saturated with SNP markers (Vezzulli et al. 2008a). SSR markers common to both maps are connected by lines. EST-derived SNPs are in red; BES-derived SNPs are in green (from Troggio et al. 2007); SNPs based on the heterozygous sites in the Pinot Noir sequence are in cyan (Velasco et al. 2007).

Color image of this figure appears in the color plate section at the end of the book.

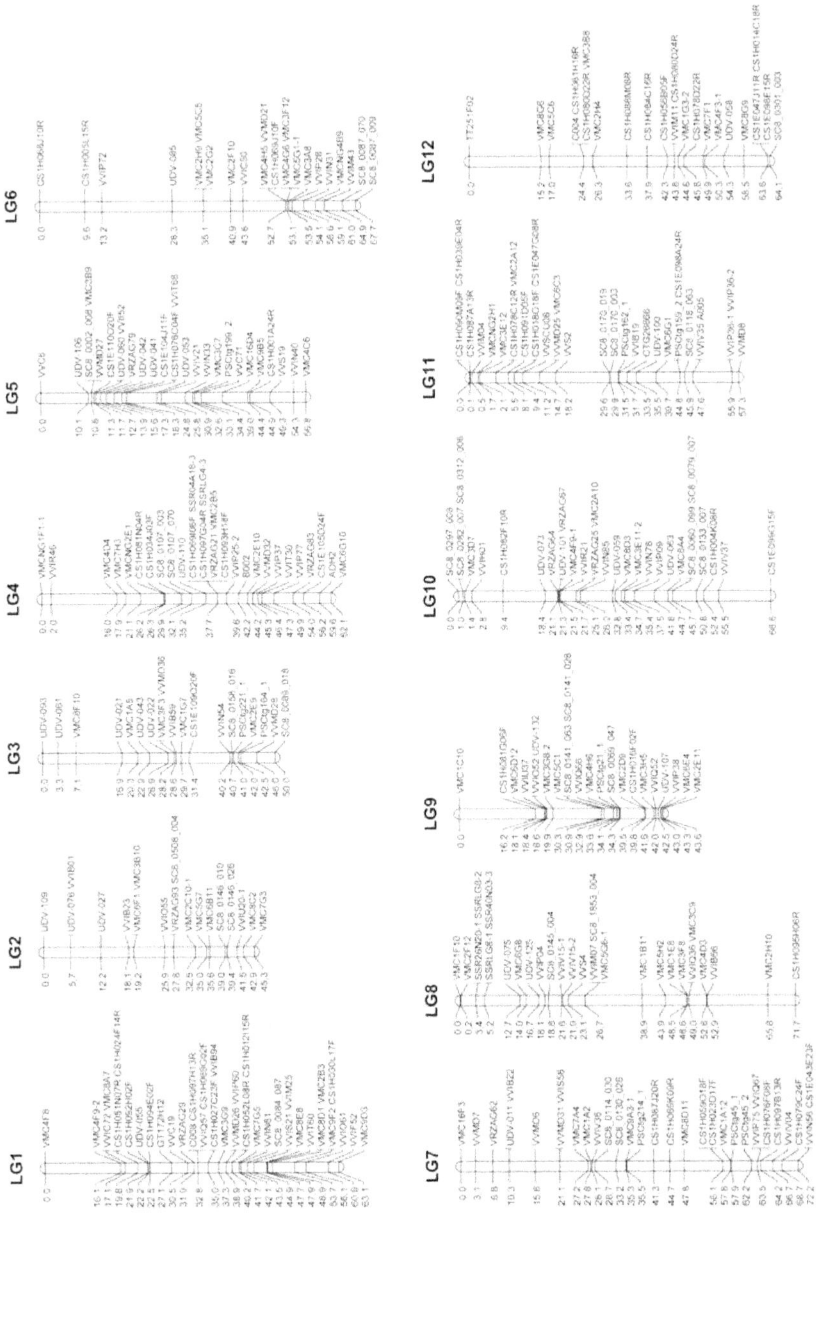
LG1
LG2
LG3
LG4
LG5
LG6
LG7
LG8
LG9
LG10
LG11
LG12

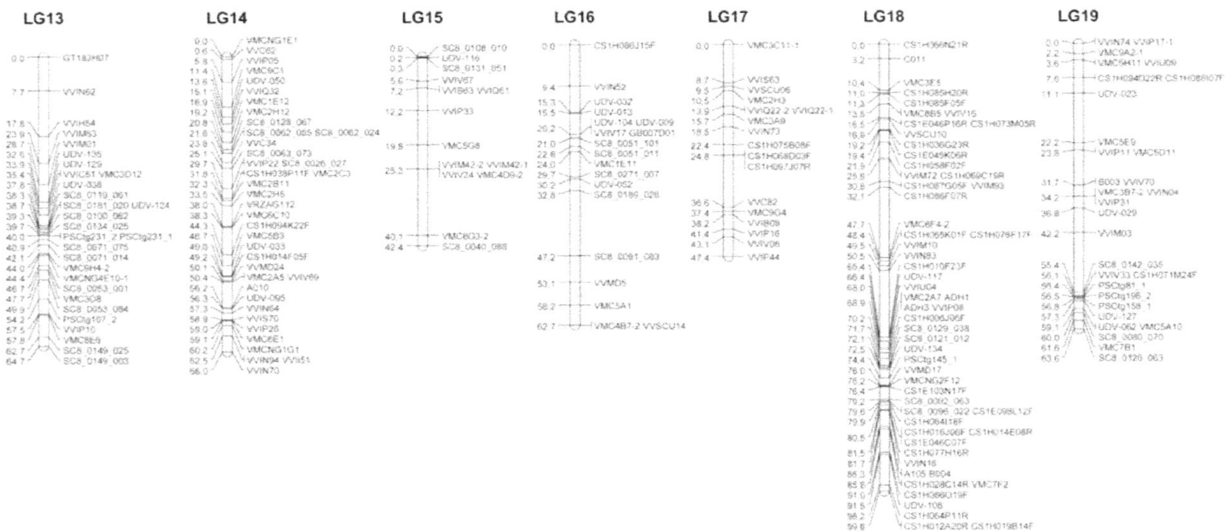

Figure 5-2 Composite SSR-based genetic map derived from the projection of a Syrah x Grenache map and a Chardonnay x Bianca map. "CS" are SSR markers derived from BAC-end sequences (Cabernet Sauvignon library; Adam-Blondon et al. 2005). "SC8"and "PSC" are SSR markers derived from PN40024 super contigs of the 8.4X genome assembly (Jaillon et al. 2007).

5.5.4 Comparative Genetic Mapping across the Vitaceae

Local marker order and estimation of their relative distances are occasionally inconsistent across different maps, but most of the time these discrepancies involve closely linked markers or markers for which the most probable order is not clear (see Doligez et al. 2006a; Di Gaspero et al. 2007; and Vezzulli et al. 2008a for discussion). The possibility of mis-ordering due to local heterogeneity in recombination rates between individuals, to genotyping errors, to intra-chromosomal segmental duplications, or to regions with skewed segregation ratio is discussed in Doligez et al. 2006a; Lowe and Walker 2006; Di Gaspero et al. 2007; and Riaz et al. 2008. Frequently, these points of attrition are associated with markers embedded into intrachromosomal segmental duplications or regions with tandemly arrayed genes (Doligez et al. 2006a; Di Gaspero et al. 2007).

Despite small local deviations from conservation of marker order, grape genetic maps are easily compared and there is so far no evidence of major disruption of synteny either between species of the genus *Vitis* (Grando et al. 2003; Doucleff et al. 2004, Lowe et al. 2006; Welter et al. 2007; Riaz et al. 2008) or within *Vitis vinifera* (Doligez et al. 2006a; Vezzulli et al. 2008a). One region with a distorted segregation on chromosome 14 was consistently found across several populations. Riaz et al. (2008) proposed that this region might contain gametophytic factors and called it *V-SDR1* for *Vitis-Segregation Disorder Region 1*.

The main use of genetic maps is to find the chromosomal region where the genetic determinants of any qualitative or quantitative trait reside. This goal may be initially achieved with a minimal set of evenly spaced markers. Some 150–200 markers covering the entire grape genome are enough for detecting the large linkage blocks, transmitted by the parents to a biparental mapping population. Once the region of interest is identified one can select sequence scaffolds anchored by the markers of the region of interest, and develop new markers in the sequence that will be used in turn to saturate the map. The methods and issues associated to map-based cloning are discussed in Chapter 8. The annotated grapevine genome can aid in the search for genetic determinants of phenotypic traits. The two-pronged process of jumping from the genetic map to the sequence scaffolds (and back) helps to integrate the positional cloning strategy with the candidate gene approach as exemplified by the linkage maps used to find the candidate gene for terpenol biosynthesis (Battilana et al. 2008; Duchêne et al. 2009) or for the powdery mildew resistance gene at the *REN1* locus (Coleman et al. 2009).

Acknowledgements

The authors are indebted to Courtney Coleman of the Missouri State University for revision of the manuscript. They thank also Isabelle Le Clainche and Audrey Weber for the technical assistance; Riccardo Velasco, Michela Troggio and Sivia Vezzulli of Fondazione Edmund Mach for the unpublished markers and the map of linkage group 2 included in Fig. 5-1 and Agnès Doligez and Thomas Scheix for discussions on mapping softwares.

References

Adam-Blondon AF, Roux C, Claux D, Butterlin G, Merdinoglu D, This P (2004) Mapping 245 SSR markers on the *Vitis vinifera* genome: a tool for grape genetics. Theor Appl Genet 109: 1017–1027.

Allard RW (1956) Formulas and tables to facilitate the calculation of recombination values in heredity. Hilgardia 24: 235–278.

Aradhya MK, Dangl GS, Prins BH, Boursiquot JM, Walker MA, Meredith CP, Simon CJ (2003) Genetic structure and differentiation in cultivated grape, *Vitis vinifera* L. Genet Res 81: 179–192.

Arcade A, Labourdette A, Falque M, Mangin B, Chardon F, Charcosset A, Joets J (2004) BioMercator: integrating genetic maps and QTL towards discovery of candidate genes. Bioinformatics 20: 2324–2326.

Arroyo-Garcia R, Martinez-Zapater JM (2004) Development and characterization of new microsatellite markers for grape. Vitis 4: 175–178.

Arulsekar S, Parfitt DE (1986) Isozyme analysis procedures for stone fruits, almond, grape, walnut, pistachio, and fig. HortScience 21: 928–933.

Battilana J, Costantini L, Emanuelli F, Sevini F, Segala C, Moser S, Velasco R, Versini G, Grando MS (2009) The 1-deoxy-D: -xylulose 5-phosphate synthase gene co-localizes with a major QTL affecting monoterpene content in grapevine. Theor Appl Genet 118: 653–669.

Beckmann JS, Soller M (1983) RFLP in genetic improvement: methodologies, mapping and costs. Theor Appl Genet 67: 35–43.

Bellin D, Peressotti E, Merdinoglu D, Wiedemann-Merdinoglu S, Adam-Blondon AF, Cipriani G, Morgante M, Testolin R, Di Gaspero G (2009) Resistance to *Plasmopara viticola* in grapevine 'Bianca' is controlled by a major dominant gene causing localised necrosis at the infection site. Theor Appl Genet 20: 163–176.

Bernatzky R, Tanksley SD (1986) Toward a saturated linkage map in tomato based on isozymes and random cDNA sequences. Genetics 112: 887–898.

Bowers JE, Dangl GS, Vignani R, Meredith CP (1996) Isolation and characterization of new polymorphic simple sequence repeat loci in grape (*Vitis vinifera* L.). Genome 39: 628–633.

Bowers JE, Dangl GS, Meredith CP (1999) Development and characterization of additional microsatellite DNA markers for grape. Am J Enol Vitic 50: 243–246.

Cabezas JA, Cervera MT, Ruiz-Garcia L, Carreno J, Martinez-Zapater JM (2006) A genetic analysis of seed and berry weight in grapevine. Genome 49: 1572–1585.

Cartwright DA, Troggio M, Velasco R, Gutin A (2007) Genetic mapping in the presence of genotyping errors. Genetics 176: 2521–2527.

Cipriani G, Marrazzo MT, Di Gaspero G, Pfeiffer A, Morgante M, Testolin R (2008) A set of microsatellite markers with long core repeat optimized for grape (*Vitis* spp.) genotyping. BMC Plant Biol 8: 127.

Cipriani G, Spadotto A, Jurman I, Di Gaspero G, Crespan M, Meneghetti S, Frare E, Vignani R,Cresti M, Morgante M, Pezzotti M, Pè E, Policriti A, Testolin R (2010) The SSR-based molecular profile of 1005 grapevine (*Vitis vinifera* L.) accessions uncovers new synonymy and parentages, and reveals a large admixture among varieties of different geographic origin. Theor Applied Genet: in press.

Coleman C, Copetti D, Cipriani G, Hoffmann S, Kozma P, Kovács L, Morgante M, Testolin R, Di Gaspero G (2009) The powdery mildew resistance gene *REN1* co-segregates with an NBS-LRR gene cluster in two Central Asian grapevines. BMC Genet 10: 89.

Costantini L, Battilana J, Lamaj F, Fanizza G, Grando MS (2008) Berry and phenology-related traits in grapevine (*Vitis vinifera* L.): from quantitative trait loci to underlying genes. BMC Plant Biol 8: 38.

Dalbò MA, Ye G-N, Weeden NF, Steinkellner H, Sefc KM, Reisch BI (2000) A gene controlling sex in grapevines place on a molecular marker-based genetic map. Genome 43: 333–340.

de Givry S, Bouchez M, Chabrier P, Milan D, Schiex T (2005) CarthaGene: multipopulation integrated genetic and radiation hybrid mapping. Bioinformatics 21: 1703–1704.

Decroocq V, Favè MG, Hagen L, Bordenave L, Decroocq S (2003) Development and transferability of apricot and grape EST microsatellite markers across taxa. Theor Appl Genet 106: 912–922.

Dib C, Fauré S, Fizames C, Samson D, Drouot D, Vignal A, Millasseau P, Marc S, Kazan J, Seboun E, Lathrop M, Gyapay G, Morissette J, Weissenbach J (1996) A comprehensive genetic map of the human genome based on 5,264 microsatellites. Nature 380: 152–154.

Di Gaspero G, Cattonaro F (2010) Application of genomics to grapevine improvement. Austr J Grape Wine Res 16: 122–130.

Di Gaspero G, Peterlunger E, Testolin R, Edwards KJ, Cipriani G (2000) Conservation of microsatellite loci within the genus Vitis. Theor Appl Genet 101: 301–308.

Di Gaspero G, Cipriani G, Marrazzo MT, Andreetta D, Prado Castro MJ, Peterlunger E, Testolin R (2005) Isolation of (AC)n-microsatellites in *Vitis vinifera* L. and analysis of genetic background in grapevines under marker assisted selection. Mol Breed 15: 11–20.

Di Gaspero G, Cipriani G, Adam-Blondon AF, Testolin R (2007) Linkage maps of grapevine displaying the chromosomal locations of 420 microsatellite markers and 82 markers from R-gene candidates. Theor Appl Genet 114: 1249–1263.

Doligez A, Bouquet A, Danglot Y, Lahogue F, Riaz S, Meredith CP, Edwards KJ, This P (2002) Genetic mapping of grapevine (*Vitis vinifera* L.) applied to the detection of QTLs for seedlessness and berry weight. Theor Appl Genet 105: 780–795.

Doligez A, Adam-Blondon AF, Cipriani G, Di Gaspero G, Laucou V, Merdinoglu D, Meredith CP, Riaz S, Roux C, This P (2006a) An integrated SSR map of grapevine based on five mapping populations. Theor Appl Genet 113: 369–382.

Doligez A, Audiot E, Baumes R, This P (2006b) QTLs for muscat flavor and monoterpenic odorant content in grapevine (*Vitis vinifera* L.). Mol Breed 18: 109–125.

Doucleff M, Jin Y, Gao F, Riaz S, Krivanek AF, Walker MA (2004) A genetic linkage map of grape, utilizing *Vitis rupestris* and *Vitis arizonica*. Theor Appl Genet 109: 1178–1187.

Duchêne E, Butterlin G, Claudel P, Dumas V, Jaegli N, Merdinoglu D (2009) A grapevine (*Vitis vinifera* L.) deoxy-D: -xylulose synthase gene colocates with a major quantitative trait loci for terpenol content. Theor Appl Genet 118: 541–552.

Einset J, Pratt C (1975) Advances in Fruit Breeding. Purdue Univ Press, USA.

Fischer BM, Salakhutdinov I, Akkurt M, Eibach R, Edwards KJ, Töpfer R, Zyprian EM (2004) Quantitative trait locus analysis of fungal disease resistance factors on a molecular map of grapevine. Theor Appl Genet 108: 501–515.

Goto-Yamamoto N, Hiroshi M, Azumi M, Edwards KJ (2006) Development of grape microsatellite markers and microsatellite analysis including oriental cultivars. Am J Enol Vitic 57: 105–108.

Grando MS, Bellin D, Edwards KJ, Pozzi C, Stefanini M, Velasco R (2003) Molecular linkage maps of *Vitis vinifera* L. and *Vitis riparia* Mchx. Theor Appl Genet 106: 1213–1224.

Grattapaglia D, Sederoff R (1994) Genetic linkage maps of *Eucalyptus grandis* and *Eucalyptus urophylla* using a pseudo-testcross: mapping strategy and RAPD markers. Genetics 137: 1121–1137.

Haldane JBS (1919) The combination of linkage values, and the calculation of distances between the loci of linked factors. J Genet 8: 299–309.

Helentjaris T, Weber D, Wright S (1988) Identification of the Genomic Locations of Duplicate Nucleotide Sequences in Maize by Analysis of Restriction Fragment Length Polymorphisms. Genetics 118: 353–363.

Hwang CF, Xu K, Hu R, Zhou R, Riaz S, Walker MA (2010). Cloning and characterization of *XiR1*, a locus responsible for dagger nematode resistance in grape. Theor Appl Genet DOI: 10.1007/s00122-010-1349-y.

Jaillon O, Aury J-M, Noel B, Policriti A, Clepet C, Casagrande A, Choisne N, Aubourg S, Vitulo N, Jubin C, Vezzi, A, Legeai F, Hugueney P, Dasilva C, Horner D, Mica E, Jublot D, Poulain J, Bruyere C, Billault A, Segurens B, Gouyvenoux M, Ugarte E, Cattonaro F, Anthouard V, Vico V, DelFabbro C, Alaux M, DiGaspero G, Dumas V, Felice N, Paillard S, Juman I, Moroldo M, Scalabrin S, Canaguier A, LeClainche I, Malacrida G, Durand E, Pesole G, Laucou V, Chatelet P, Merdinoglu D, Delledonne M, Pezotti M, Lecharny A, Scarpelli C, Artiguenave F, Pe ME, Valle G, Morgante M, Caboche M, Adam-Blondon A-F, Weissenbach J, Quetier F, Wincker O (2007) The grapevine genome 24 sequence suggests ancestral hexaploidization in major angiosperm phyla. Nature 449: 463–468.

Kosambi DD (1944) The estimation of map distances from recombination values. Ann Eugen 12: 172–175.

Labra M, Imazio S, Grassi F, Rossoni M, Sala F (2004) Vine-1 retrotransposon-based sequence-specific amplified polymorphism for *Vitis vinifera* L. genotyping. Plant Breed 123: 180–185.

Lahogue F, This P, Bouquet A (1998). Identification of a codominant scar marker linked to the seedlessness character in grapevine. Theor Appl Genet 97: 950–959.

Lamoureux D, Bernole A, Le Clainche I, Tual S, Thareau V, Paillard S, Legeai F, Dossat C, Wincker P, Oswald M, Merdinoglu D, Vignault C, Delrot S, Caboche M, Chalhoub B, Adam-Blondon AF (2006) Anchoring of a large set of markers onto a BAC library for the development of a draft physical map of the grapevine genome. Theor Appl Genet 113: 344–356.

Lander ES, Green P, Abrahamson J, Barlow A, Daly MJ, Lincoln SE, Newburg L (1987) MAPMAKER: an interactive computer package for constructing primary genetic linkage maps of experimental and natural populations. Genomics 1: 174–181.

Lefort F, Kyvelos CJ, Zervou M, Edwards KJ, Roubelakis-Angelakis KA (2002) Characterization of new microsatellite loci from *Vitis vinifera* and their conservation in some *Vitis* species and hybrids. Mol Ecol Notes 2: 20–21.

Lijavetzky D, Cabezas JA, Ibáñez A, Rodríguez V, Martínez-Zapater JM (2007) High throughput SNP discovery and genotyping in grapevine (*Vitis vinifera* L.) by combining a re-sequencing approach and SNPlex technology. BMC Genomics 8: 424.

Lodhi MA, Ye G-N, Weeden NF, Reisch BI (1994) A simple and efficient method for DNA extraction from grapevine cultivars and *Vitis* species. Plant Mol Biol Rep 12: 6–13.

Lodhi MA, Daly MJ, Ye GN, Weeden NF, Reisch BI (1995) A molecular marker based linkage map of *Vitis*. Genome 38: 786–794.

Loukas M, Stavrakakis MN, Krimbas CB (1983) Inheritance of polymorphic isoenzymes in grape cultivars. J Hered 74: 181–183.

Lowe KM, Walker MA (2006) Genetic linkage map of the interspecific grape rootstock cross Ramsey (*Vitis champinii*) x Riparia Gloire (*Vitis riparia*). Theor Appl Genet 112: 1582–1592.

Mandl K, Santiago J-L, Hack R, Fardossi A, Regner F (2006) A genetic map of Welschriesling x Sirius for the identification of magnesium-deficiency by QTL analysis. Euphytica 149: 133–144.

Marguerit E, Boury C, Manicki A, Donnart M, Butterlin G, Némorin A, Wiedemann-Merdinoglu S, Merdinoglu D, Ollat N, Decroocq S (2009) Genetic dissection of sex determinism, inflorescence morphology and downy mildew resistance in grapevine. Theor Appl Genet 118: 1261–1278.

Merdinoglu D, Butterlin G, Bevilacqua L, Chiquet V, Adam-Blondon AF, Decroocq S (2005) Development and characterization of a large set of microsatellite markers in grapevine (*Vitis vinifera* L.) suitable for multiplex PCR. Mol Breed 15: 349–366.

Moore SS, Sargeant LL, King TJ, Mattick JS, Georges M, Hetzel DJS (1991) The conservation of dinucleotide microsatellites among mammalian genomes allows the use of heterologous PCR primer pairs in closely related species. Genomics 10: 654–660.

Morgante M, Hanafey M, Powell W (2002) Microsatellites are preferentially associated with nonrepetitive DNA in plant genomes. Nat Genet 30: 194–200.

Myles S, Chia JM, Hurwitz B, Simon C, Zhong GY, Buckler E, Ware D (2010) Rapid genomic characterization of the genus *Vitis*. PLoS One 5(1): e8219.

Pauquet J, Bouquet A, This P, Adam-Blondon AF (2001) Establishment of a local map of AFLP markers around the powdery mildew resistance gene *Run1* in grapevine and assessment of their usefulness for marker assisted selection. Theor Appl Genet 103: 1201–1210.

Pellerone FI, Edwards KJ, Thomas MR (2001) Grapevine microsatellite repeats: Isolation, characterisation and use for genotyping of grape germplasm from Southern Italy. *Vitis* 40: 179–186.

Pelsy F (2007) Untranslated leader region polymorphism of *Tvv1*, a retrotransposon family, is a novel marker useful for analyzing genetic diversity and relatedness in the genus *Vitis*. Theor Appl Genet 116: 15–27.

Pelsy F, Schehrer L, Merdinoglu D 2003. Development of grapevine retrotransposon-based molecular markers (S-SAP). Acta horticulturae 603: 83–87.

Pindo M, Vezzulli S, Coppola G, Cartwright DA, Zharkikh A, Velasco R, Troggio M (2008) SNP high-throughput screening in grapevine using the SNPlex™ genotyping system. BMC Plant Biol 8: 12.

Rapley R, Harbron S (2004) Molecular analysis and genome discovery. John Wiley, Chichester, UK.

Riaz S, Dangl GS, Edwards KJ, Meredith CP (2004) A microsatellite marker based framework linkage map of *Vitis vinifera* L. Theor Appl Genet 108: 864–872.

Riaz S, Krivanek AF, Xu K, Walker MA (2006) Refined mapping of the Pierce's disease resistance locus, PdR1, and sex on the extended genetic map of *Vitis rupestris* x *V. arizonica*. Theor Appl Genet 113: 1317–1329.

Riaz S, Tenscher AC, Rubin J, Graziani R, Pao SS, Walker MA (2008) Fine-scale genetic mapping of two Pierce's disease resistance loci and a major segregation distortion region on chromosome 14 of grape. Theor Appl Genet 117: 671–681.

Ritter E, Gebhardt C, Salamini F (1990) Estimation of recombination frequencies and construction of RFLP linkage maps in plants from crosses between heterozygous parents. Genetics 125: 645–654.

Salmaso M, Faes G, Segala C, Stefanini M, Salakhutdinov L, Zyprian E, Toepfer R, Grando MS, Velasco R (2004) Genome diversity and gene haplotypes in the grapevine (*V. vinifera* L.), as revealed by single nucleotide polymorphisms. Mol Breed 14: 385–395.

Salmaso M, Malacarne G, Troggio M, Faes G, Stefanini M, Grando MS, Velasco R (2008) A grapevine (*Vitis vinifera* L) genetic map integrating the position of 139 expressed genes. Theor Appl Genet 116: 1129–1143.

Scheix T, Gaspin C (1997) Constructing and joining maximum likelihood genetic maps. Proc Int Symp Mol Biol AAI: *www.aai.org*

Scott KD, Eggler P, Seaton G, Rossetto M, Ablett EM, Lee LS, Henry RJ (2000) Analysis of SSRs derived from grape ESTs. Theor Appl Genet 100: 723–726.

Sefc KM, Regner F, Turetschek E, Glössl J, Steinkellner H (1999) Identification of microsatellite sequences in *Vitis riparia* and their applicability for genotyping of different *Vitis* species. Genome 42: 367–373.

Sefc KM, Lopes MS, Lefort F, Botta R, Roubelakis-Angelakis KA, Ibanez J, Pejic I, Wagner HW, Glossl J, Steinkellner H (2000) Microsatellite variability in grapevine cultivars from different European regions and evaluation of assignment testing to assess the geographic origin of cultivars. Theor Appl Genet 100: 498–505.

Stam P (1993) Construction of integrated genetic-linkage maps by means of a new computer package-Joinmap. Plant J 3: 739–744.

Stam P, van Ooijen JW (1995) JoinMap version 2.0: Software for the calculation of genetic linkage maps. Wageningen, The Netherlands.

Steemers FJ, Weihua Chang W, Grace Lee G, David L Barker DL, Richard Shen R, Kevin L, Gunderson KL (2006) Whole-genome genotyping with the single-base extension assay. Nat Meth 3: 31–33.

Tanksley SD, Young ND, Paterson AH, Bonierbale MW (1989) RFLP maping in plant breeding—new tools for an old science. Biotechnology 7: 257–264.

Thomas MR, Scott NS (1993) Microsatellite repeats in grapevine reveal DNA polymorphisms when analysed as sequence-tagged sites (STSs). Theor Appl Genet 86: 985–990.

Troggio M, Malacarne G, Coppola G, Segala C, Cartwright DA, Pindo M, Stefanini M, Mank R, Moroldo M, Morgante M, Grando MS, Velasco R (2007) A dense single-nucleotide polymorphism based genetic linkage map of grapevine (*Vitis vinifera* L.) anchoring Pinot noir bacterial artificial chromosome contigs. Genetics 176: 2637–2650.

Tulsieram LK, Glaubitz JC, Kiss G, Carlson JE (1992) Single tree genetic linkage mapping in conifers using haploid DNA from megagametophytes. Biotechnology 10: 686–690.

Van Ooijen JW (2006) JoinMap 4.0, Software for the calculation of genetic linkage maps in experimental populations. Kyazma BV, Wageningen, The Netherlands.

Van Ooijen JW, Voorrips RE (2001) JoinMap 3.0, Software for the calculation of genetic linkage maps. Plant Research International, Wageningen, The Netherlands.

Velasco R, Zharkikh A, Troggio M, Cartwright DA, Cestaro A, Pruss A, Pruss 1 D, Pindo M, FitzGerald LM, Vezulli S, Reid J, Malacarne G, Iliev D, Coppola G, Wardell B, Micheletti D, Macalma T, Facci M, Mitchell JT, Perrazzolli M, Eldredge G, Gatto P, Oyzerski R, Moretto M, Gutin N, Stefanini M, Chen Y, Segala C, Davenport C, Dematte L, Mraz A, Battilana J, Stormo K, Costa F, Tao Q, Si- Ammour A, Harkins T, Lackey A, Perbost C, Taillon B, Stella A, Solovyev V, Fawcett JA, Sterck L, Vandepoele K, Grando SM, Toppo S, Moser C, Lanchbury J, Bogden R, Skolnick M, Sgaramella V, Bhatnagar SK, Fontana P, Gutin A, Van de Peer Y, Salamini F, Viola R (2007) A high quality draft consensus sequence of the genome of a heterozygous grapevine variety. PloS One 2(12) e: 1326 Vezzulli S, Troggio M, Coppola G, Jermakow A, Cartwright D, Zharkikh A, Stefanini M, Grando MS, Viola R, Adam-Blondon AF, Thomas M, This P, Velasco R (2008a) A reference integrated map for cultivated grapevine (*Vitis vinifera* L.) from three crosses, based on 283 SSR and 501 SNP-based markers. Theor Appl Genet 117: 499–511.

Vezzulli S, Micheletti D, Riaz S, Pindo M, Viola R, This P, Walker MA, Troggio M, Velasco R (2008b) A SNP transferability survey within the genus *Vitis*. BMC Plant Biol 8: 128.

Vos P, Hogers R, Bleeker M, Reijans M, Vandelee T, Hornes M, Frijters A, Pot J, Peleman J, Kuiper M, Zabeau M (1995) AFLP—a new technique for DNA-fingerprinting. Nucl Acids Res 23: 4407–4414.

Welter LJ, Göktürk-Baydar N, Akkurt M, Maul E, Eibach R, Töpfer R, Zyprian EM (2007) Genetic mapping and localization of quantitative trait loci affecting fungal disease resistance and leaf morphology in grapevine (*Vitis vinifera* L.). Mol Breed 20: 359–374.

Williams JGK, Kubelik AR, Livak J, Rafalski JA, Tingey SV (1990) DNA polymorphisms amplified by arbitrary primers are useful as genetic markers. Nucl Acids Res 18: 6531–6535.

Xu K, Riaz S, Roncoroni NC, Jin Y, Hu R, Zhou R, Walker MA (2008) Genetic and QTL analysis of resistance to *Xiphinema index* in a grapevine cross. Theor Appl Genet 116: 305–311.

Zhang J, Hausmann L, Eibach R, Welter LJ, Töpfer R, Zyprian EM (2009). A framework map from grapevine V3125 (*Vitis vinifera* 'Schiava grossa' x 'Riesling') x rootstock cultivar 'Börner' (*Vitis riparia* x *Vitis cinerea*) to localize genetic determinants of phylloxera root resistance. Theor Appl Genet 119: 1039–1051.

6

Basics of Grapevine Genetic Analysis

Leocir J. Welter,[1,*] *M. Stella Grando*[2] and *Eva Zyprian*[3]

ABSTRACT

Grapevine (*Vitis vinifera* L. ssp. *vinifera*) cultivars produce hermaphrodite flowers as a domestication trait. They are able to self but also partially outcrossing and amenable to controlled crosses. Introgression of valued traits from wild *Vitis* sp. is possible due to interfertility. Agronomic traits of interest for breeding may be qualitative but are predominantly quantitative in nature. To develop robust trait-linked markers and to understand the responsible genes their genetic mapping is required.

Grapevine plants are largely heterozygous demanding application of the double pseudo testcross strategy to develop genetic maps based on linkage and recombination analysis of molecular markers. Qualitative traits can be directly integrated in such maps or analyzed with the BSA strategy. Quantitative traits have to be analyzed using marker information in combination with the statistical approach of QTL analysis. In both cases the resolution power of the methods is affected by the size of the segregating population and marker coverage of the genome. Marker density may now easily be increased in a target region of interest by exploiting information from the grapevine model genome sequences. Map-based cloning of major genes or major QTL is considerably facilitated by these new options. The current knowledge on major genes and QTL data on phenological and berry traits as well as pathogen resistances is summarized.

Keywords: Agronomic traits, double pseudo testcross, grapevine flowers, genetic maps, major genes, QTL analysis, segregating population

[1]Federal University of Pampa, Rua Luiz Joaquim de Sá Britto, 97650-000, Itaqui, Rio Grande do Sul, Brazil.
[2]IASMA Research and Innovation Centre, Fondazione Edmund Mach—Genomics and Crop Biology Area, Via Mach, 1 38010 San Michele all'Adige (Trento), Italy.
[3]Julius Kühn Institut, Federal Research Centre for Cultivated Plants, Institute for Grapevine Breeding Geilweilerhof, 76833 Siebeldingen, Germany.
*Corresponding author: *eva.zyprian@jki.bund.de*

6.1 Reproductive Biology of Grapevine

Wild grapevines (*Vitis vinifera* ssp. *sylvestris* and *Vitis* sp.) are generally dioecious plants, requiring cross-pollination for fertilization and fruit set. However, its unisexual flowers often contain rudimentary parts of the opposite sex and wild *Vitis* individuals with functionally complete hermaphrodite flowers have been occasionally found in the native habitat. This dysfunctional partial hermaphroditism may represent a relictal trait from very ancient primitive monoecious forms of *Vitis* from Tertiary times (Meneghetti et al. 2006) that was shifted to unisexual flowers promoting sexual recombination during evolution of the wild species. In contrast to the wild grapes, the cultivated European grapevine varieties (*V. vinifera* ssp. *vinifera*) generally have functionally perfect hermaphrodite flowers. Failure in one of the two organs (male or female) may occur (cf. OIV descriptor 151, *http://news.reseau-concept.net/images/oiv_uk/Client/Code_descripteurs_2ed_EN.pdf*), as described for the *V. vinifera* variety "Picolit giallo", in which problems with pollen germination were detected (Lombardo et al. 1976). Most likely hermaphroditism was secondarily re-established due to selective forces imposed during six- to eight thousand years of viticultural domestication.

The majority of cultivated forms of grapevine with hermaphrodite flowers are also self-fertile. Exception to this rule exists, like the varieties "Mahrabasi" and "Ag uzumu", described as self-incompatible (Kelen et al. 1997). Self-fertile plants are most likely selected for their facilitated berry production and better yield without the necessity of pollen donor varieties. However *V. vinifera* ssp. *vinifera* flowers still contain nectaries (Boss et al. 2003) and produce floral volatiles in their male parts (Martin et al. 2009) serving to attract pollinators. The morphology of the hermaphrodite flowers shows five fused petals forming a closed cap that comes off upon flowering. Anthesis occurs at the same time enabling the flower to pollinate in its still "closed" state. This structure together with self-compatibility facilitates self-pollination. Grapevine cultivated forms are considered as partially cleistogamic self-pollinated species, although variable degrees of cross-pollination have been observed (Vasconcelos et al. 2009). The ratio of self-versus cross-pollination seems to be genotype dependent.

The preferentially self-pollinating nature could suggest that cultivated forms of grapevine are extensively homozygous. It may come as a surprise that cultivated grapevine varieties are highly heterozygous (H_o = 0.755) (Di Vecchi-Staraz et al. 2008). The preservation of this high level of heterozygosity, a characteristic of cross-pollinated species, in the cultivars of grapevine is probably due to vegetative (clonal) propagation implemented from the earliest time of viticulture to maintain the favorite genotypes. The heterozygous genome of cultivated grapevine varieties may carry a heavy load of deleterious recessive alleles. The development of endogamous (highly homozygous) grapevine varieties is quite difficult due to strong inbreeding depression. In contrast, interspecific crosses can be performed. All known *Vitis* species can be experimentally crossed and the F_1 hybrids are fertile (Olmo 1976).

6.2 Inheritance of Agronomic Characters in Grapevine

Grapevine has been neglected in classical genetic studies on important crops due to its long juvenile phase, high level of heterozygosity, strong inbreeding depression and perennial nature. Most of the agronomically relevant traits of grapevine are quantitative or polygenic in nature. A classical example of a quantitative trait is wine quality, which is governed by multiple genes and is also strongly affected by the environment. Berry skin color is a typical qualitative trait, whose variation can roughly be classified into red (containing anthocyanins) and white, two discrete classes.

With the advent of molecular marker techniques based on the polymerase chain reaction (PCR) (around the 1990s) genetic studies in plants have been widely stimulated. For the first time it became possible to develop dense genetic maps for diverse plant species. The first grapevine genetic maps were constructed only six years after the first plant genetic map based mainly on molecular markers had been published. This delay was caused by the lack of an adequate mapping strategy for largely heterozygous plants. This bottleneck was overcome with the development of the double pseudo-testcross strategy (Weeden 1994). In this approach two highly heterozygous plants are crossed and the resulting F_1 segregating population is genotyped with a large set of molecular markers. The markers are analyzed for their parental origin and recombination from the female and/or male parent. Two separate genetic maps are constructed based on the recombination frequencies that are observed after gamete formation (meiosis) of both genitors: one map for each parent (Fig. 6-1). Through the use of co-dominant or doubly heterozygous dominant molecular markers the two parental maps can further be integrated into a single map. This strategy combined with the polymorphism detection power of molecular marker techniques has allowed the construction of the first genetic maps for grapevine (Lodhi et al. 1995). Nowadays several grapevine genetic maps based on divergent cross populations are available.

The construction of densely covered linkage and recombination-derived maps using segregating populations obtained by the cross of phenotypically contrasting parents finally permits the genomic localization of both qualitative and quantitative traits. These kinds of investigations are usually focussed on traits of agronomic interest, like disease resistances or characteristics associated with berry morphology and phenology. Soon the

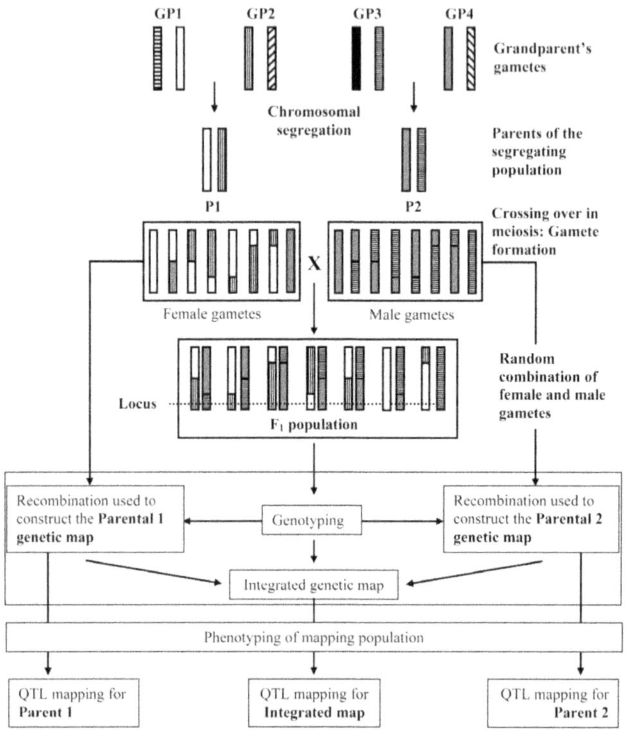

Figure 6-1 Flow-chart showing the strategy used to localize quantitative traits (QTLs) in grapevine. Firstly, two highly heterozygous parents are selected and crossed to generate a segregating F_1 population. This population must segregate for the traits under investigation. Generally, parents contrasting in the phenotypes of interest are employed. If available, phenotypic information of the grandparents may be used to support the choice of the genitors. Another strategy to improve the success of QTL detection is the selection of unrelated parents.

Secondly, the segregating population is genotyped with a large set of molecular markers. Ideally, these markers should be uniformly distributed throughout the grapevine genome. This may be optimized by selecting microsatellite markers based on the model genome sequences for even spacing (about 10 cM distances).

The genotypic information obtained from the molecular markers is then used in linkage and recombination analysis supported by statistical means. At the end, markers are aligned into linear linkage groups representing the chromosomes.

When applying the double pseudo-test cross strategy, two parental maps are generated. The parental maps are constructed based on the frequency of traceable cross-overs that occur during gamete formation of each parent independently. Therefore, the frequency of recombination, and consequently the map size, may slightly differ between the two parental maps.

If codominant markers (e.g., microsatellites) are employed to genotype the mapping population, the parental maps may be integrated into one single map. In the case of dominant markers, doubly heterozygous markers are necessary to achieve this integration.

Parallel to genotyping, the mapping population is phenotypically scored for the expression levels of the traits under investigation. Association between the phenotype and the allelic variation of the markers are searched by means of statistical methods. Up to three QTL analyses based on the three genetic maps obtained may be performed.

scope of such investigations will increase with the emerging availability of the genome sequence (Jaillon et al. 2007; Velasco et al. 2007). The genetic mapping of traits (forward genetics) is one of the most efficient strategies for the characterization of gene function. Positioning them in a genetic map is a prerequisite for the development of trait-linked markers for modern "smart" breeding and is required for physical cloning of the underlying genes. The genetic mapping of qualitative and quantitative traits will be described in more detail in the following sections.

6.3 Molecular Mapping of Qualitative Traits

Genetics started with the analysis of qualitative traits, so the first genetic localizations of agronomic traits were based on the observation of their segregation as presence or absence. In these cases the genotype of the individuals in a segregating population can be inferred from its phenotype. In this way, the recombination frequency between different traits can be calculated and their relative position is determined building a linear map. The major limitation of this method in the past was the low number of easily scorable and clearly qualitative traits, prohibiting the construction of dense genetic maps. Currently, the use of different types of molecular markers permits the construction of very dense genetic maps covering the whole genome. In case of a clear qualitative phenotype, it is possible to estimate the recombination frequencies between this one and the flanking molecular loci, thereby allowing the direct integration of major genes into genetic maps. Applying the genome-spanning genetic maps virtually all polymorphic qualitative traits detected in a segregating population can be positioned in relation to molecular markers. In grapevine, the major genes considered responsible for qualitative traits such as those governing flower sex (Dalbó et al. 2000; Lowe and Walker 2006; Riaz et al. 2006), berry color (Doligez et al. 2002; Fischer et al. 2004; Mejia et al. 2007; Salmaso et al. 2008; Costantini et al. 2008) and seedlessness (Cabezas et al. 2006; Mejia et al. 2007; Costantini et al. 2008) have been positioned in genetic maps (Table 6-1). The berry color locus was found to co-locate with the gene encoding *VvMybA1*, a transcription factor regulating the expression of the gene for the enzyme of the last biosynthetic step of anthocyanin formation (Kobayashi et al. 2004; Lijavetzky et al. 2006). A candidate gene for the control of sexual traits in grapevine linked to the sex locus has been proposed (Marguerit et al. 2009).

The construction of genome wide genetic maps is not necessarily required to localize qualitative traits. Alternatively, bulked segregant analysis (BSA) can be used to identify molecular makers linked to major genes (Michelmore et al. 1991). This strategy was successfully applied in grapevine to construct local maps, especially for major disease resistance

Table 6-1 Qualitative and quantitative traits related to flower sex, berry color, berry weight/size, phenology, seedlessness and disease resistance mapped in grapevine.

Character	Partent 1	Parent 2	LG	Progeny size	Mapping strategy	*Reference*
Flower sex	Horizon (Complex hybrid)	Illinois 547-1 (*V. rupestris* x *V. cinerea*)	2	58	WGM	Dalbó et al. 2000
	Ramsey (*V. champinii*)	Riparia Gloire (*V. riparia*)	2	188	WGM	Lowe et al. 2006
	D8909-15 (*V. rupestris* x *V. arizonica*)	F8909-17 (*V. rupestris* x *V. arizonica*)	2	181	WGM	Riaz et al. 2006
Inflorescence morpholgy	Cabernet Sauvignon (*V. vinifera*)	Gloire de Montpellier (*V. riparia*)	2 (major effect) Many minor QTLs were found	138	QTL analysis	Marguerit et al. 2009
Flower morphology	Cabernet Sauvignon (*V. vinifera*)	Gloire de Montpellier (*V. riparia*)	2 (major effetct) 6, 7, 10	138	QTL analysis	Marguerit et al. 2009
Fleshless Berry (*Flb*)	Chardonnay (*V. vinifera*)	Mutated Ugni Blanc (*V. vinifera*)	18	55	BSA	Fernandez et al. 2006
Berry color	MTP2223-27 (*V. vinifera*)	MTP2121-30 (*V. vinifera*)	2	139	WGM	Doligez et al. 2002
	Regent (with complex resistance origin)	Lemberger (*V. vinifera*)	2	153	WGM	Fischer et al. 2004
	Rubi Seedless (*V. vinifera*)	Thompson Seedless (*V. vinifera*)	2	144	WGM	Mejia et al. 2007
	Merzling (with complex resistance origin)	Teroldego (*V. vinifera*)	2	89	WGM	Salmaso et al. 2008
	Italia (*V. vinifera*)	Big Perlon (*V. vinifera*)	2	163	WGM	Costantini et al. 2008

Berry weight/size	MTP2223-27 (partially seedless; *V. vinifera*)	MTP2121-30 (partially seedless; *V. vinifera*)	18 (major effect) minor QTLs were found	139	QTL analysis	Doligez et al. 2002
	Regent (with complex resistance origin)	Lemberger (*V. vinifera*)	5, 13	153	QTL analysis	Fischer et al. 2004
	Dominga (seeded; *V. vinifera*)	Autumn Seedless (*V. vinifera*)	18 (major effect) 15 (minor effect)*	118	QTL analysis	Cabezas et al. 2006
	Rubi Seedless (*V. vinifera*)	Thompson Seedless (*V. vinifera*)	18 (major QTL) 15 (minor QTL)	144	QTL analysis	Mejia et al. 2007
	Italia (*V. vinifera*)	Big Perlon (*V. vinifera*)	18 (major effect) 1, 12 (minor effect)	163	QTL analysis	Costantini et al. 2008
Monoterpene content	Italia (*V. vinifera*)	Big Perlon (*V. vinifera*)	5 and 10 (major effect) 1, 2, 7, 10	163	QTL analysis	Battilana et al. 2009
	Moscato Bianco (*V. vinifera*)	*V. riparia*	5 (major effect) 2, 12, 10, 15	174	QTL analysis	Battilana et al. 2009
Terpenol content	Muscat Ottonel (*V. vinifera*) (selfing)		5 (major effect), 1, 10, 13, 15	121	QTL analysis	Duchêne et al. 2009
Phenology-related traits	Regent (with complex resistance origin)	Lemberger (*V. vinifera*)	7, 8, 16	153	QTL analysis	Fischer et al. 2004
	Rubi Seedless (*V. vinifera*)	Thompson Seedless (*V. vinifera*)	18	144	QTL analysis	Mejia et al. 2007
	Italia (*V. vinifera*)	Big Perlon (*V. vinifera*)	1, 2, 6, 16	163	QTL analysis	Costantini et al. 2008

Table 6-1 contd....

Table 6-1 contd....

Character	Partent 1	Parent 2	LG	Progeny size	Mapping strategy	*Reference*
Seedlessness and subtraits	MTP2223-27 (partially seedless; *V. vinifera*)	MTP2121-30 (partially seedless; *V. vinifera*)	18	139	BSA	Lahogue et al. 1998
	MTP2223-27 (partially seedless; *V. vinifera*)	MTP2121-30 (partially seedless; *V. vinifera*)	18 (major effect) minor QTLs were found	139	QTL analysis	Doligez et al. 2002
	Dominga (seeded; *V. vinifera*)	Autumn Seedless (*V. vinifera*)	18 (major effect) 1, 3, 10, 14 (minor effect)*	118	QTL analysis	Cabezas et al. 2006
	Rubi Seedless (*V. vinifera*)	Thompson Seedless (*V. vinifera*)	18	144	WGM	Mejia et al. 2007
	Rubi Seedless (*V. vinifera*)	Thompson Seedless (*V. vinifera*)	18 (major QTL) 4, 16 (minor QTL)*	144	QTL analysis	Mejia et al. 2007
	Italia (*V. vinifera*)	Big Perlon (*V. vinifera*)	18	163	WGM	Costantini et al. 2008
	Italia (*V. vinifera*)	Big Perlon (*V. vinifera*)	18 (major effect) 2, 6, 10, 13, 15 (minor effect)	163	QTL analysis	Costantini et al. 2008
Resistance to powdery mildew	VRH3082–1-42 (BC4; *Run 1* from *M. rotundifolia* introgressed into *V. vinifera*)	Cabernet Sauvignon (*V. vinifera*)	12	157	BSA	Pauquet et al. 2001 and Donald et al. 2002
	Nimrang (*V. vinifera*)	Kishmish Vatkana (*V.vinifera*); donor of *Ren1* resistance gene	13	310	BSA	Hoffman et al. 2008
	Dzhandzhal kara; donor of *Ren1* resistance gene	Lasta	13	151	-	Coleman et al.2009
	Regent (with complex resistance origin)	Lemberger (*V. vinifera*)	15	153; 144	QTL analysis	Fischer et al. 2004; Welter et al. 2007

Resistance to downy mildew	*Muscadinia rotundifolia* (donor of *Rpv1*)	BC2 (*V. vinifera*)	12	139	BSA	Merdinoglu 2003
	Regent (QTL donor)	Lemberger (*V. vinifera*)	18 (major effect) 5 (minor effect) 4 (minor effect)	153; 144	QTL analysis	Fischer et al. 2004; Welter et al. 2007
	Cabernet Sauvignon (*V. vinifera*)	Gloire de Montpellier (*V. riparia*) QTL donor	9 12	138	QTL analysis	Marguerit et al. 2009
Resistance to anthracnose	Various crosses	Various crosses	-	No local map was constructed	BSA	Kim et al. 2008
Resistance to Pierce's disease	D8909-15 (*V. rupestris* x *V. arizonica*)	F8909-17 (*V. rupestris* x *V. arizonica*/*V. candicans*) QTL donor	14 (major effect)	73	QTL analysis	Krivanek et al. 2006
Resistance to dagger nematode	D8909-15 (*V. rupestris* x *V. arizonica*) QTL donor	F8909-17 (*V. rupestris* x *V. arizonica*/*V. candicans*	19	188	QTL analysis	Xu et al. 2008
Root resistance to phylloxera	V3125 (*V. vinifera*)	Boerner (*V. riparia* x *V. cinerea*) QTL donor	13	188	QTL analysis	Zhang et al. 2009

WGM: Whole genome mapping strategy, BSA: Bulked segregant analysis.
*Only the most significant minor QTLs were considered.

genes. For these investigations, initially two contrasting bulks containing approximately 10 plants each (B1 and B2) are constructed: one containing only resistant plants and the other containing only susceptible plants. Usually the resistant and the susceptible plants are selected from a larger segregating population obtained from the cross between a resistant and a susceptible variety. The parents and the two bulks are genotyped with a large number of molecular markers to find markers whose allele(s) co-segregate with the expression of resistance. In other words, a molecular marker is linked to the corresponding resistance gene when at least one of its alleles is preferentially or exclusively present in the resistant bulk and the other or other alleles are preferentially or exclusively present in the susceptible bulk. In this case the alleles of this marker are not statistically distributed between the ~ 20 individuals represented in the two bulks, suggesting that the marker is physically linked to the corresponding gene. The putatively linked markers are further used to genotype the whole mapping population to confirm their correlation to the trait and to obtain a local genetic map. Using the BSA strategy the following resistance genes were localized in grapevine: *Run*1 (Pauquet et al. 2001 and Donald et al. 2002) and *Ren* 1 (Hoffmann et al. 2008) conferring qualitative resistance to powdery mildew, *Rpv*1 conferring resistance to downy mildew (Merdinoglu et al. 2003) and a major gene conferring resistance to anthracnose or black spot (Kim et al. 2008). Microsatellite markers linked to all these genes, except the resistance gene to anthracnose, were detected. The same markers were included in the construction of whole genome genetic maps, thus positioning these genes into chromosomes (linkage groups) and anchoring them to the grapevine genome sequence. This strategy can also be used for traits other than resistance to pathogens. In grapevine it was successfully employed to map the fleshless berry (*flb*) mutation of "Ugni Blanc" (*V. vinifera*) (Fernandez et al. 2006). A summary of the qualitative traits mapped in grapevine is shown in Table 6-1.

6.4 Cloning of Major Genes

The ultimate goal of mapping agronomical traits is the identification and isolation of the underlying genes. This approach aiming to associate a phenotype to a genotype is known as "forward genetics". In forward genetics, map-based cloning (or positional cloning) is a commonly used strategy to isolate genes governing major traits. Initially, molecular markers physically linked to the trait of interest have to be identified through whole genome or local genetic mapping. Traditionally, a local fine map is constructed with the aim to assign the trait to the smallest possible genetic region, thus reducing subsequent sequencing efforts. This means that the region of interest will be targeted and densely covered by molecular markers in order to detect markers flanking the gene of interest as closely as possible. When a marker is closely related to a gene, low recombination frequency is expected and large segregating populations are required to detect the rare recombinant individuals.

This strategy was successfully used to physically map the resistance locus *Run*1 (see Chapter 8). At first, a local map around the *Run*1 locus was constructed employing the BSA strategy (Pauquet et al. 2001). Secondly, using the same strategy, resistance gene analog (RGA)-based markers tightly linked to the *Run*1 locus were detected (Donald et al. 2002). For fine mapping, three independent populations segregating for the resistance locus *Run*1, in total 996 recombinant individuals, were employed (Barker et al. 2005). Fine mapping allowed the identification of two flanking microsatellite markers showing a very low recombination frequency (tight linkage) with the *Run*1 locus, thus defining a short genetic interval for the locus. These two flanking markers were used together with three markers that co-segregated with *Run*1, to screen a bacterial artificial chromosome (BAC)-library constructed from the genomic DNA of a plant carrying *Run*1. In this way the physical mapping of the region spanning the resistance locus was performed. Marker-carrying BAC clones were end-sequenced and assembled into extended contigs after identification of overlapping BACs. This allowed an overall coverage of the region spanning the locus *Run*1. After the sequencing of the contig spanning the genomic region associated to *Run1*, positional candidate genes were selected for functional analysis.

The recent availability of the grapevine genome sequence opens up new options to accelerate the map-based cloning of genes. Molecular markers surrounding the trait of interest may be anchored to the genome sequence. This genomic region may then be searched for positional candidate genes, based on their predicted functional role. Molecular markers tagging such candidate genes can be developed and tested for their association to the trait. Additionally, markers (e.g., single nucleotide polymorphisms, SNPs) with a progressive physical distance from the trait of interest may be developed in both directions and used to estimate their recombination frequencies to the trait. Markers closely flanking the underlying gene(s) are expected to be identified, reducing the posterior sequencing effort. However, it is important to be aware of the fact that specific genes such as resistance genes may not be present in the currently elaborated model genome sequences, as these have been derived from susceptible grapevines. It remains to be seen from further grapevine genome sequencing to what extend the two model genome sequences of "Pinot noir" are colinear to other grapevine cultivars, breeding material and *Vitis* wild species accessions on large- and fine scale levels.

6.5 Molecular Mapping of Quantitative Traits

An enormous natural variation is present within the genus *Vitis* and the species *Muscadinia rotundifolia*. The exploitation of this genetic variation has been limited because of its mostly quantitative nature. Contrary to qualitative traits, the genotype of quantitative traits cannot be inferred simply based on the absent or present phenotype, thus creating more difficulty for analysis. Several genes may be involved and some influence of environmental conditions on trait expression is generally observed. Construction of dense genetic maps with numerous recombining markers however enables to analyze the distributions of phenotypic trait expression within the individuals belonging to the allelic classes of the segregating markers. In the case of a marker linked to genetic factors governing the trait, a statistical separation of the distributions of the phenotypic classes will be evident in its different genotypic classes. A separation of phenotypic distributions will not be observable in the genotypic classes of any unlinked marker. This "statistical" approach designated quantitative trait loci (QTL) analysis is widely used in plants to localize factors governing quantitative traits in genetic maps (Tanksley 1993). QTL analysis permits the resolution of quantitative traits into their Mendelian components (indicating several factors respectively markers and different genomic positions that contribute to trait expression levels) and determines the part of the phenotypic variation explained by each component.

For QTL analysis, the first step is the development of a population segregating for the trait(s) of interest, which will be: i) characterized with a large set of molecular markers for the construction of a genetic map, and ii) phenotypically evaluated for the corresponding trait(s). The cautious selection of the parental types is fundamental for success of the analysis because only segregating polymorphic loci can be detected. Secondly, any association between the phenotypic expression of the quantitative trait and the allelic variation of the mapped markers is searched by statistical methods. There are various statistical applications available to determine if a quantitative trait is linked to molecular markers but they all share the basic principle outlined above. By these approaches, markers associated with the trait may eventually be detected, in this way delimiting the genomic regions containing genetic factors influencing the trait. The quality of QTL mapping (the number of QTL detected, the accuracy of their map position and estimates of their effects) depends largely on experimental parameters, such as size and type of the mapping population and coverage of the molecular genetic map. In addition, intrinsic factors such as the heritability of the trait, the magnitude of the QTL effect and the amount of recombination observed in the segregating population will play a role (Alonso-Blanco and Koornneef 2000).

6.5.1 Segregating populations

For annual plants, the three types of segregating populations most widely used in genetic studies are F_2, backcross and recombinant inbreed lines (RILs). However, the development of pure homozygous lines in grapevine is not possible due to its high level of inbreeding depression and the long seed to seed cycle. Therefore, most QTL analyses in grapevine were based on F_1 populations obtained by the cross of two highly heterozygous plants. In contrast to most annual crops, where pure lines are crossed and only two alleles segregate for each QTL, up to four alleles can segregate for each QTL in the mapping populations in grapevine. By the use of the double pseudo-testcross strategy two independent genetic maps are constructed, one for each parent and eventually they are used to develop an integrated map. Thus, up to three independent QTL analyses may be performed based on each of the three maps. For each parental map, the allelic substitution of all marker loci will be tested for its effect on the phenotypic variation of the corresponding trait. Selecting parents whose (grand)parents are genetically widely divergent may improve the success of QTL detection, because it may increase the effect of allelic variation detectable on the phenotype.

Because QTL analysis in F_1 populations is usually performed using the parental genetic maps (considering the two parental alleles independently), but the phenotype of the trait of interest is influenced by up to four alleles per locus in the mapping population, some kind of differential interaction between the parental alleles may interfere in QTL analysis. To avoid these possibly negative effects, a self-pollinated population could be used. Virtually, any grapevine could be selfed to generate a mapping population due to its high heterozygosity. Based on this strategy, one single map would be constructed and only two alleles per polymorphic locus would be segregating in the population. However, some deleterious loci could fall into homozygosity, leading to segregation distortion of these genomic regions, eventually affecting QTL analysis.

Nevertheless, this approach has been successfully used for the analysis of major QTL determining the content of isoprenoid berry skin aroma compounds in "Gewürtraminer" and "Muscat Ottonel" selfings identifying two major QTL on linkage groups (LGs) 5 and 10. The QTL on linkage group 5 suggests a candidate gene involved in plastidial isoprenoid biosynthesis underlying this trait (Duchêne et al. 2009).

Mapping populations for QTL studies are usually generated from a cross between parents contrasting for the quantitative trait of interest (e.g., resistant x susceptible). However, another approach could be to use parents with an intermediate phenotype of the corresponding quantitative trait, exploring transgressive segregation (i.e., appearance of individuals in segregating populations that fall beyond their parental phenotypes). The magnitude and frequency of transgressive phenotypes should increase as the proportion of segregating QTLs with opposing effects increases.

6.5.2 Size of Segregating Population and Marker Density of Genetic Maps

QTLs affecting a trait can only be detected when their phenotypic effect is sufficiently large (i.e., the phenotypic effect must be larger than a statistical threshold). QTLs whose effects fall under the statistical threshold of detection will be missed. The statistical threshold of detection is largely dependent on the size of the mapping population used in the analysis and the quality of the genetic map. The larger the population, the more likely it is to statistically detect QTLs of lesser effects (Tanksley 1993). The resolution of QTL mapping increases proportionally with the size of the mapping population (i.e., the number of recombination events observable in segregating individuals) and marker density. To fine-map QTLs in F_1 populations a large population is required. This will allow detecting rare recombination events in meioses (using 1,000 individuals it is possible to detect crossing-over with a frequency down to 0.1%) and construct refined maps. With large mapping populations it is possible to position the molecular markers more precisely and at shorter inter marker distance in cM. The marker density should be proportional to the size of the mapping population. Using a large set of molecular markers to genotype a small population does not increase the resolution of QTL analysis. Construction of genome-wide fine maps requires high costs and much work. Often, populations of up to 200 segregating individuals are used to roughly localize the QTLs. As follow-up, the size of the population is significantly increased and it is evaluated for the trait of interest. Then a fine local map is constructed targeted to the region where the QTL were previously localized. This strategy can be successfully used for the characterization of QTLs with major effect on the phenotype, however QTLs of minor effects may be missed because their influence on the phenotype will probably fall under the statistical threshold of detection in the originally used small mapping populations.

6.5.3 Bi-parental x Multi-parental Populations

In general, the QTL mapped for grapevine up to now were based on bi-parental populations (actually the segregation of the homologous chromosomes of the parents is considered). No matter how cautiously the parents are selected, the segregating population will represent only a small slice of the overall variation of the corresponding trait. Multi-parental populations were used to localize QTLs in some annual crops, such as maize

(Blanc et al. 2006). This strategy was shown to be more powerful in QTL detection, especially when connected multi-parental populations sharing one of the parental types were employed. Such experimental design could contribute to improve QTL detection in grapevine. Careful selection of parents, taking into consideration their genealogy (using unrelated parents), would allow to analyze the quantitative variation of a trait more globally.

6.5.4 QTL x Environment Interactions

Phenotypic variation is essential for QTL analysis. The phenotype of an individual is conditioned by the genotype (g), the environment (e) and the interaction between both (g x e). Therefore, QTLs with an important effect on the phenotype of a specific trait in one environment may be less important for the phenotype in another environment. All QTL analyses performed up to now in grapevine were based only on studies in one single environment. The next step will be to evaluate mapping populations under several different environmental conditions to identify the effects of genotype x environment interactions. This information can be decisive for the exploitation of traits of interest in breeding programs.

6.6 QTLs Identified in Grapevine

6.6.1 Disease Resistance

Using F_1 populations a number of QTL controlling different quantitative traits were detected in grapevine (Table 6-1). For disease resistance, QTL were found for powdery mildew (Dalbó et al. 2000; Fischer et al. 2004; Welter et al. 2007; Hoffmann et al. 2008; Coleman et al. 2009), downy mildew (Fischer et al. 2004; Welter et al. 2007; Marguerit et al. 2009) and Pierce's disease (Krivanek et al. 2006; Riaz et al. 2006). Although the phenotypic distribution of mildew resistance traits in the mapping population was continuous, major QTLs, explaining up to 72% of phenotypic variation, were found for the diseases investigated. Minor QTLs were detected only for downy mildew. Due to the limited size of the mapping populations employed in these QTL analyses, it is not possible to decide if the resistance QTLs detected comprise only single gene or many linked genes (a cluster of genes). For the moment, in the genomic region covered by the major QTL conferring resistance to downy mildew identified in the variety "Regent", a cluster of resistance gene analogs (RGA) are present in the PN40024 DNA sequence. These predicted genes are termed RGA because they share homology to resistance genes (R-genes) isolated from other plants. These genes in general confer resistance to biotrophic pathogens, such as downy mildew. This co-localization could indicate that in this region more than one resistance gene may be responsible for the expression of the trait. Further map-based cloning and functional gene analysis will provide more insight.

Besides fungal and bacterial diseases, resistance QTL affecting resistance to the dagger nematode *Xiphinema index* (Xu et al. 2008) and the damaging insect *Daktulosphaira vitifoliae* (grape phylloxera) on roots (Zhang et al. 2009) have been identified. Both are of special interest for rootstock breeding.

6.6.2 Traits Related to Phenology and Berry

Phenology is of primary importance in viticulture, both for table grapes and for quality wine production. In table grapes it is important to have grapes maturing during the widest period of time, which means the combination of early, middle and late ripening varieties. The localization of genomic regions controlling traits related to phenology can substantially help in the breeding of new table grape varieties, supporting the development of molecular tools for marker assisted selection.

Because of its importance, phenology-related traits have been the focus of investigation in the last years. Three independent investigations, considering different mapping populations, report the localization of QTLs associated with the phenology (Fischer et al. 2004; Mejía et al. 2007; Costantini et al. 2008). The parents of the crosses employed in the QTL analysis are wine grapes (Fischer et al. 2004), seeded x seedless table grapes (Costantini et al. 2008) and seedless table grapes (Mejía et al. 2007). In total, eight main ripening-related QTLs were detected, the great majority of them (seven of eight) being localized on different chromosomes. Genomic regions controlling ripening time or véraison time, for example, were detected on LG-17 and LG-18 (Mejía et al. 2007), LG-7, LG-8 and LG-16 (Fischer et al. 2004) and LG-2, LG-6 and LG-16 (Costantini et al. 2008). Unfortunately, at present it is not possible to resolve whether the two QTLs detected on LG-16 by the two independent investigations are the same. But, for the moment, this QTL showed the largest effect on the phenotype (21.1–45.5% of the total variance) (Costantini et al. 2008), thus being less responsive to environmental variation. The overall dissimilar results obtained in these three investigations suggest a strong influence of the environment on this trait, as expected for quantitative traits, but this lack of overlapping results could also be due to the divergent genetic backgrounds of the grapevine varieties.

More consistent results in different genetic backgrounds were obtained for several quantitative subtraits of seedlessness (i.e., seed number, seed fresh weight, seed dry weight, seed dry matter) and for berry size. Doligez et al. (2002), Cabezas et al. (2006), Mejía et al. (2007) and Costantini et al. (2008) all reported a major QTL affecting both seed content and berry weight on LG-18, which coincides with the *SdI* (seed development inhibitor) locus (Bouquet and Danglot 1996; Lahogue et al. 1998). The known relationship between gibberellins produced by seeds and berry growth have suggested that correlation between berry weight and seedlessness subtraits, observed at both phenotypic and genetic levels, may be due to pleiotropy rather than to tight linkage. Moreover, based on different experiments, QTLs on LG-1 and LG-12 (Mejía et al. 2007; Costantini et al. 2008), LG-11 (Doligez et al. 2002) and LG-15 (Cabezas et al. 2006) were shown to regulate berry weight without affecting seedlessness. These findings along with several QTLs specific for seed content identified in one or more progenies might allow to dissociate the unfavorable correlation between berry size and seedlessness in breeding programs.

Major effect QTLs were also reported for the content of linalool, nerol and geraniol, which are the main monoterpene compounds actually contributing to sensory perception of Muscat flavor. Monoterpenols are present in grapes and wines as free or glycosylated compounds. Muscat cultivars as well as some non-Muscat aromatic cultivars (e.g., "Riesling", "Gewürztraminer", "Silvaner") specifically contain volatile monoterpenols in much higher levels than the berries of the so-called neutral cultivars. They exhibit a rather particular aroma related to this composition, which is greatly appreciated for fresh grape consumption and plays an essential role in high-quality winemaking. Therefore, the identification of genomic regions involved in the determination of Muscat aroma represents a major contribution to understanding the genetic control of traits of large interest for grape breeding.

In accordance with the first findings reported by Doligez et al. (2006), Battilana et al. (2009) demonstrated the existence of a major QTL for the ln-transformed content of linalool, nerol and geraniol on LG-5 analyzing both a "Moscato bianco" x *Vitis riparia* and "Italia" x "Big Perlon" segregating progenies. Additional QTLs were identified for linalool content on LG-2 of the "Moscato bianco" map, as reported by Doligez et al. 2006, and on LG-10 of the "Italia" x "Big Perlon" map. These results indicated that the observed phenotypic correlations between the three monoterpenes are in a large part of genetic origin. A stronger correlation was actually found between nerol and geraniol than between linalool and nerol or geraniol, which is in agreement with the existence of common QTLs, namely loci regulating the accumulation of distinct molecules, and linalool-specific QTLs. The QTL on LG-5 had a maximum LOD at the position where an EST-derived marker related to 1-deoxy-D-xylulose 5-phosphate synthase (DXS) was placed in Battilana et al. (2009) as well as in Duchêne et al. (2009) mapping experiments. The effects of this QTL are consistent with the role of a *DXS* gene acting upstream in the non-mevalonate pathway of isopentenyl diphosphate (IPP) biosynthesis and expected to modify the availability of the geranyl diphosphate (GPP), the main substrate for monoterpenes biosynthesis. These findings represent the first indication of a putative key gene in the genetic determination of Muscat flavor in grapevine and are in complete agreement with observations made in other plant systems.

6.7 Molecular Cloning of Quantitative Trait Loci

Although the isolation of QTLs is more challenging than the isolation of major genes, the principle is the same. The majority of QTLs isolated up to now were done by map-based cloning (or positional cloning). In grapevine, segregating populations with less than 200 F_1 individuals were originally employed for QTL detection. Although this number of recombinant individuals allows the detection of QTLs governing a specific trait, the confidence interval of the QTL is relatively large, spanning 10–30 cM. This genetic distance may represent millions of base pares, challenging the sequencing process and the identification of underlying genes. To refine the confidence interval of the QTL of interest, fine local maps have to be constructed (based on large segregating population and high marker density in targeted regions). Fine mapping allows to localize the QTL more precisely. By anchoring the QTL of interest to the grapevine genome sequence, PCR-based markers may be developed, improving the marker density around the target locus. This should facilitate the identification of recombinant plants closely around the locus of interest, delimiting the target trait to the smallest genomic region possible. The QTL may then be isolated by positional cloning. This strategy should work especially for QTLs of major effect. Up to now, no QTL cloning of grapevine was reported, although projects are underway.

6.7.1 Mendelizing Quantitative Trait Loci

For a few traits, especially in disease resistance, single major QTLs were detected (Fischer et al. 2004; Krivanek et al. 2006; Welter et al. 2007). These QTLs explain the major part of the phenotypic variation. However, some other complex traits, such as those of leaf morphology, are controlled by a larger number of loci, frequently spread around the genome (Welter et al. 2007). These QTLs may interact with each other in different ways, a challenge for their characterization and isolation. Within the population individuals segregating for only one of these QTLs may be selected. Through such selection of different plants all the QTL may be individualized. Progenies of these plants should segregate as monogenic traits, facilitating their characterization enormously. This process, commonly referred to as "Mendelization of a QTL", characterizes each QTL individually. QTL Mendelization is widely used for characterization of QTLs in annual plants.

Although it may also be applied to grapevine, as exemplified previously, it is more challenging especially due to the heterozygous nature of grape. The simplest way to generate progenies in grapevine is by selfing the plants. This will give rise to a 1:2:1 segregation rate for the Mendelized QTLs. In annual crop plants for which pure lines may be generated, QTL Mendelization is frequently accomplished by the construction of near-isogenic lines (NILs). In grapevine this is not possible due to the high degree of inbreeding depression.

6.8 Opportunities Offered by the Grapevine Genome Sequence

The availability of the grapevine genome sequence will substantially reduce the work and time required for QTL isolation. Molecular markers positioned at the confidence interval of the QTLs detected may be anchored to the grapevine genome sequence. Markers explaining the largest phenotypical variance can be used to indicate the location of the underlying genes. Positional candidate genes, located in the proximity of this marker, can be selected for functional analysis. These candidate genes are also selected based on their predicted functional role, which hypothetically can play an important role on the corresponding trait.

Complementing the identification of positional candidate genes, the genome sequence may be used to support the development of molecular markers specifically tagging a region of interest. Without the genome information, screening with large numbers of markers is required to identify those that are located in the region of interest, expending work and money. With the genome sequence, markers can be specifically designed spanning the whole region of interest, at a predictable physical distance. Single nucleotide polymorphisms (SNPs) have been demonstrated to be a very efficient marker technique (Lijavetzky et al. 2007; Salmaso et al. 2008). This strategy combined with large mapping populations will greatly facilitate the fine mapping of QTLs and consequently gene isolation. The fine mapping of some traits may not be required when the underlying genes can be selected based on the positional candidate gene approach. However, some QTLs may not be present in the sequenced genomes, as probably is the case with some resistance genes that are original from other species than *V. vinifera*.

6.9 Perspectives

Current development regarding sequencing technologies will make genomic sequencing much faster and much more affordable. Therefore it is to be expected that more genome sequences from other grapevine varieties will become available in the next few years. Even distinct genotypes of *Vitis* sp. wild accessions or hybrids carrying interesting traits for breeding (e.g., pathogen resistance) will be amenable to whole genome analysis. The problem of correct sequence assembly of two partially different copies of each chromosome in the diploid, largely heterozygous grapes will be addressed by using the genome sequence of the rather homozygous "Pinot Noir" derivative (PN40024; Jaillon et al. 2007) as master template. Another option is the development of sophisticated software in integration with genetic and physical mapping data, enabling to distinguish the two haplotypes (the two different versions) of homologous chromosome pairs. Tools have to be developed or adapted from work in other well-investigated plant species to study the information of the 475 to 500 Mb of the grapevine genome.

The new sequence data will allow comparing the genome structure in different grapes, but also to decipher genetic differences in specific genomic regions where QTLs for important traits are known to be located. Sequence comparison and alignments are expected to exhibit a large number of SNPs. Some of those may be in determinant genes and linked to traits rendering them directly useful for marker-assisted breeding. Thorough testing of such SNP markers in large numbers of phenotypically diverse and well evaluated plants will be required to confirm any association between expression of a trait and a specific molecular marker. Molecular analysis of genetic diversity in specifically targeted genomic regions linked to traits of interest will be the next step. This will allow to generate molecular markers for efficient "smart breeding" to generate cultivars of wine and table grapes improved in regard to quality and resistance for future sustainable viticulture.

References

Alonso-Blanco C, Koornneef M (2000) Naturally occurring variation in *Arabidopsis*: an underexploited resource for plant genetics. Trends Plant Sci 5: 22–29.

Barker CL, Donald T, Pauquet J, Ratnaparkhe MB, Bouquet A, Adam-Blondon AF, Thomas MR, Dry I (2005) Genetic and physical mapping of the grapevine powdery mildew resistance gene, *Run1*, using a bacterial artificial chromosome library. Theor Appl Genet 111: 370–377.

Battilana J, Costantini L, Emanuelli F, Sevini F, Segala C, Moser S, Velasco R, Versini G, Grando MS (2009) The 1-deoxy-D: -xylulose 5-phosphate synthase gene co-localizes with a major QTL affecting monoterpene content in grapevine. Theor Appl Genet 118: 653–669.

Blanc G, Charcosset A, Mangin B, Gallais A, Moreau L (2006) Connected populations for detecting quantitative trait loci and testing for epistasis: an application in maize. Theor Appl Genet 113: 206–224.

Boss PK, Buckeridge EJ, Poole A, Thomas MR (2003) New insights into grapevine flowering. Funct Plant Biol 30: 593–606.

Bouquet A, Danglot Y (1996) Inheritance of seedlessness in grapevine (*Vitis vinifera* L.). Vitis 35: 35–42.

Cabezas JA, Cervera MT, Ruiz-Garcia L, Carreño J, Martínez-Zapater JM (2006) A genetic analysis of seed and berry weight in grapevine. Genome 49: 1572–1585.

Coleman C, Copetti D, Cipriani G, Hoffmann S, Kozma P, Kovacs L, Morgante M, Testolin R, Di Gaspero G (2009) The powdery mildew resistante gene *REN1* co-segregates with an NBS-LRR gene cluster in two central Asian grapevines. BMC Genet 10:89 doi: 10.1186/1471-2156-10-89.

Costantini L, Battilana J, Lamaj F, Fanizza G, Grando MS (2008) Berry and phenology-related traits in grapevine (*Vitis vinifera* L.). From quantitative trait loci to underlying genes. BMC Plant Biol 8: 38.

Dalbó MA, Ye GN, Weeden NF, Steinkellner H, Sefc KM, Reisch BI (2000) A gene controlling sex in grapevine placed on a molecular marker-based genetic map. Genome 43: 333–340.

Di Vecchi-Staraz M, Laucou V, Bruno G, Lacombe T, Gerber S, Bourse T, Boselli M, This P (2008) Low level of pollen-mediated gene flow from cultivated to wild grapevine: consequences for the evolution of the endangered subspecies *Vitis vinifera* L. subsp. *silvestris*. J Hered 100: 66–75.

Doligez A, Bouquet A, Danglot Y, Lahogue F, Riaz S, Meredith CP, Edwards KJ, This P (2002) Genetic mapping of grapevine (*Vitis vinifera* L.) applied to the detection of QTLs for seedlessness and berry weight. Theor Appl Genet 105: 780–795.

Doligez A, Audiot E, Baumes R, This P (2006). QTLs for Muscat flavor and monoterpenic odorant content in grapevine (*Vitis vinifera* L.). Mol Breed 18: 109–125.

Donald TM, Pellerone F, Adam-Blondon AF, Bouquet A, Thomas MR, Dry IB (2002) Identification of resistance gene analogs linked to a powdery mildew resistance locus in grapevine. Theor Appl Genet 104: 610–618.

Duchêne E, Butterlin G, Claudel P, Dumas V, Jaegli N, Merdinoglu D (2009) A grapevine (*Vitis vinifera* L.) deoxy-D-xylulose synthase gene colocates with a major quantitative trait loci for terpenol content. Theor Appl Genet 118: 541–552.

Fernandez L, Doligez A, Lopez G, Thomas MR, Bouquet A, Torregrosa L (2006) Somatic chimerism, genetic inheritance, and mapping of the fleshless berry (*flb*) mutation in grapevine (*Vitis vinifera* L.). Genome 49: 721–728.

Fischer B, Salakhutdinov I, Akkurt M, Eibach R, Edwards KJ, Töpfer R, Zyprian E (2004) Quantitative trait locus analysis of fungal disease resistance factors on a molecular map of grapevine. Theor Appl Genet 108: 505–515.

Hoffmann S, Di Gaspero G, Kovács L, Howard S, Kiss E, Galbács Z, Testolin R, Kozma P (2008) Resistance to *Erysiphe necator* in the grapevine 'Kishmish vatkana' is controlled by a single locus through restriction of hyphal growth. Theor Appl Genet 116: 427–438.

Jaillon O, Aury J-M, Noel B, Policriti A, Clepet C, Casagrande A, Choisne N, Aubourg S, Vitulo N, Jubin C, Vezzi A, Legeai F, Hugueney P, Dasilva C, Horner D, Mica E, Jublot D, Poulain J, Bruyere C, Billault A, Segurens B, Gouyvenoux M, Ugarte E, Cattonaro F, Anthouard V, Vico V, Del Fabbro C, Alaux M, Di Gaspero G, Dumas V, Felice N, Paillard S, Juman I, Moroldo M, Scalabrin S, Canaguier A, Le Clainche I, Malacrida G, Durand E, Pesole G, Laucou V, Chatelet P, Merdinoglu D, Delledonne M, Pezzotti M, Lecharny A, Scarpelli C, Artiguenave F, Pé E, Valle G, Morgante M, Caboche M, Adam-Blondon A-F, Weissenbach J, Quétier F, Wincker P (2007) The grapevine genome sequence suggests ancestral hexaploidization in major angiosperm phyla. Nature 449: 463–468.

Kelen M, Sutyemez M, Beyhan Ö, Yalinkiliç A (1997) A study on fertilization biology of some grape varieties. Acta Hort 441: 433–438.

Kim GH, Yun HK, Choi CS, Park JH, Jung YJ, Park KS, Dane F, Kang KK (2008) Identification of AFLP and RAPD markers linked to anthracnose resistance in grapes and their conversion to SCAR markers. Plant Breed 127: 418–423.

Kobayashi S, Yamamoto NG, Hirochika H (2004) Retrotransposon-induced mutations in grape skin color. Science 304: 982.

Krivanek AF, Riaz S, Walker MA (2006) Identification and molecular mapping of *PdR1*, a primary resistance gene to Pierce's disease in *Vitis*. Theor. Appl. Genet. 112: 1125–1131.

Lahogue F, This P, Bouquet A (1998) Identification of a codominant marker linked to the seedlessness character in grapevine. Theor Appl Genet 97: 950–959.

Lijavetzky D, Ruiz-Garcia L, Cabezas JA, De Andres MT, Bravo G, Ibanez A, Carreno J, Cabello F, Ibanez J, Martinez-Zapater JM (2006) Molecular genetics of berry colour variation in table grape. Mol Genet Genom 276: 427–435.

Lijavetzky D, Cabezas, Ibáñez A, Rodríguez V, Martínez-Zapater JM (2007) High throughput SNP discovery and genotyping in grapevine (*Vitis vinifera* L.) by combining a re-sequencing approach and SNPlex technology. BMC Genom 8: 424.

Lodhi MA, Daly MJ, Ye FN, Weeden NF, Reisch BI (1995) A molecular marker based linkage map of *Vitis*. Genome 38: 786–794.

Lombardo G, Carraro L, Cargnello G, Bassi M (1976) Ultrastructure of pollen of *Vitis vinifera* L. cv. `Picolit giallo´ and its behaviour in experiments of self- and cross-pollination. Vitis, 15: 73–81.

Lowe KM, Walker MA (2006) Genetic linkage map of the interspecific grape rootstock cross Ramsey (*Vitis champinii*) x Riparia Gloire (*Vitis riparia*). Theor Appl Genet 112: 1582–1592.

Marguerit E, Boury C, Manicki A, Donnart M, Butterlin G, Némorin A, Wiedemann-Merdinoglu S, Merdinoglu D, Ollat N, Decroocq S (2009) Genetic dissection of sex determinism, inflorescence morphology and downy mildew resistance in grapevine. Theor Appl Genet 118: 1261–1278.

Martin D, Toub O, Chiang A, Lo BC, Ohse S, Lund ST, Bohlmann J (2009) The bouquet of grapevine (*Vitis vinifera* L. cv. Cabernet Sauvignon) flowers arises from the biosynthesis of sesquiterpene volatiles in pollen grains. Proc Natl Acad Sci USA 106: 7245–7250.

Mejía N, Gebauer M, Muñoz L, Hewstone N, Muñoz C, Hinrichsen P (2007) Identification of QTLs for seedlessness, berry size, and ripening date in a seedless x seedless table grape progeny. Am J Enol Vitic 58: 499–507.

Meneghetti S, Gardiman M, Calo A (2006) Flower biology of grapevine. A review Adv Hort Sci. 20: 317–325.

Merdinoglu D, Wiedemann-Merdinoglu S, Coste P, Dumas V, Haetty S, Butterlin G, Greif C (2003) Genetic analysis of downy mildew resistance derived from *Muscadinia rotundifolia*. Acta Hort 603: 451–456.

Michelmore RW, Paran I, Kesseli RV (1991) Identification of markers linked to disease-resistance genes by bulked segregant analysis: a rapid method to detect markers in specific genomic regions by using segregating populations. PNAS, 88: 9828–9832.

Olmo HP (1976) Grapes. In: NW Simonds (ed) Evolution of Crop Plants. Longman, London, UK, pp 294–298.

Pauquet J, Bouquet A, This P, Adam-Blondon AF (2001) Establishment of a local map of AFLP markers around the powdery mildew resistance gene *Run1* in grapevine and assessment of their usefulness for marker assisted selection. Theor Appl Genet 103: 1201–1210.

Riaz S, Krivanek AF, Xu K, Walker MA (2006) Refined mapping of the pierces disease resistance locus, *PdR1*, and sex on the extended genetic map of *Vitis rupestris* x *V. arizonica*. Theor Appl Genet 113: 1317–1329.

Salmaso M, Malacarne G, Troggio M, Faes G, Stefanini M, Grando MS, Velasco R (2008) A grapevine (*Vitis vinifera* L.) genetic map integrating the position of 139 expressed genes. Theor Appl Genet 116: 1129–1143.

Tanksley SD (1993) Mapping polygenes. Annu Rev Genet 27: 205–233.

Vasconcelos MC, Greven M, Winefiled CS, Trought MCT, Raw V (2009) The flowering process of *Vitis vinifera*: A Review Am J Enol Vitic 60: 411–434.

Velasco R, Zharkikh A, Troggio M, Cartwright DA, Cestaro A, Pruss D, Pindo M, Fitzgerald LM, Vezzulli S, Reid J, Malacarne G, Iliev D, Coppola G, Wardell B, Micheletti D, Macalma T, Facci M, Mitchell JT, Perazzolli M, Eldredge G, Gatto P, Oyzerski R, Moretto M, Gutin N, Stefanini M, Chen Y, Segala C, Davenport C, Demattè L, Mraz A, Battilana J, Stormo K, Costa F, Tao Q, Si-Ammour A, Harkins T, Lackey A, Perbost C, Taillon B, Stella A, Solovyev V, Fawcett JA, Sterck L, Vandepoele K, Grando SM, Toppo S, Moser C, Lanchbury J, Bogden R, Skolnick M, Sgaramella V, Bhatnagar SK, Fontana P, Gutin A, Van de Peer Y, Salamini F, Viola R (2007) A high quality draft consensus sequence of the genome of a heterozygous grapevine variety. PLoS ONE 2: e1326.

Weeden NF (1994) Approaches to Mapping in Horticultural Crops. In: PM Gresshoff (ed) Plant genome analysis. CRC Press, Boca Raton, USA, pp 57–68.

Welter LJ, Göktürk-Baydar N, Akkurt M, Maul E, Eibach R, Toepfer R, Zyprian E (2007) Genetic mapping and localization of quantitative trait loci affecting fungal disease resistance and leaf morphology in grapevine (*Vitis vinifera* L). Mol Breed 20: 359–374.

Xu K, Riaz S, Roncoroni NC, Jin Y, Hu R, Zhou R, Walker MA (2008) Genetic and QTL analysis of resistance to *Xiphinema index* in a grapevine cross. Theor Appl Genet 116: 305–311.

Zhang J, Hausmann L, Eibach R, Welter LJ, Töpfer R, Zyprian E (2009) A framework map of grapevine V3125 (*Vitis vinifera* `Schiava grossa´ x `Riesling´) x rootstock cultivar `Börner´(*Vitis riparia* x *Vitis cinerea*) to localize genetic determinants of phylloxera root resistance. Theor Appl Genet 119: 1039–1051.

7

Molecular Breeding

*Reinhard Töpfer,** *Ludger Hausmann* and *Rudolf Eibach*

ABSTRACT

Grapevines belong to the oldest domesticated crops and as they are propagated vegetatively some of the still existing cultivars can be traced back to the Middle Ages. These cultivars are considered occasional selections. Grapevine breeding, however, was initiated due to devastating pests and diseases attacking the vine. The principals and problems underlying classical breeding for wine grapes, e.g., a long generation cycle, high heterozygosity and inbreeding depression, as well as the complexity of traits resulted in purely empirical grapevine breeding and caused slow progress for decades. In light of molecular techniques which became available at the beginning of the 21st century a paradigm shift to knowledge based breeding was possible. In particular, aspects of marker-assisted selection (MAS), pyramiding resistances, marker-assisted backcrossing (MABC) and genetically modified organisms (GMOs) in modern grapevine breeding are discussed. Marker and genome sequencing techniques will soon allow high-throughput (HT) genotyping enhancing efficiency within the breeding process. Several resistances against major diseases are being currently combined by classical breeding supported by MAS. Also the isolation of responsible genes is at the onset. Thus, viticulture will receive solutions for the urgent phytopathological problems. Whether these are plants from classical breeding or GMOs will largely be a political decision.

Keywords: Grapevine breeding, marker-assisted selection, pyramiding resistances, GMO

7.1 Introduction

During European and Middle East history thousands of varieties of cultivated grapevines emerged and have been or are used for wine, table

JKI Institute for Grapevine Breeding Geilweilerhof, D-76833 Siebeldingen, Germany.
*Corresponding author: *reinhard.toepfer@jki.bund.de*

grape or dry fruit (raisin) production (Table 7-1). These cultivars belong to only the *Vitis* species endemic to Europe and Asia Minor, *Vitis vinifera* L. It exists as the cultivated subspecies *vinifera* and its wild relative *V. vinifera* L. subsp. *sylvestris* Gmel. One of the oldest known cultivars traces back to 1283 when Philippe de Beaumanoir mentioned the cultivar Gouais Blanc (Gwäss or Heunisch), which together with Pinot forms the parentage of numerous cultivars of present importance (Bowers et al. 1999; Boursiquot et al. 2004).

Table 7-1 Distribution of the existing grapevine cultivars according to their utilization (Source: *www.vivc.de*) and world production for these sectors (Source OIV: *http://www.oiv.int/de/accueil/index.php*).

Utilization	Number of cultivars	Percentage	World production 2005
Wine grapes	9,461	56.8	44.1 Mill tons (282.3 Mill hlit wine)
Table grapes	4,497	27.0	17.8 Mill tons
Table and Wine grapes	1,222	7.3	
Raisin grapes	110	0.7	1.3 Mill tons
Rootstocks	1,367	8.2	

The old cultivars and their descent have to be considered as occasional selections rather than the outcome of planned breeding activities. The first cross-breeding was initiated in America and extended to Europe in particular France due to the devastating impact of diseases like powdery mildew (*Erysiphe necator* 1845), phylloxera (*Daktulosphaira vitifoliae* 1863), and downy mildew (*Plasmopara viticola* 1878), which were introduced from North America. The so-called American Hybrids and French Hybrids were genotypes conferring resistance but they were disappointing in terms of wine quality (Fig. 7-1).

At the beginning of the 20th century when the Mendelian genetics could first be applied, attempts were started in breeding programs to systematically elucidate the inheritance of certain traits. Hedrick and Anthony (1915) summarizing their work with *Vitis* species described some data for inheritance of self-sterility, sex, color of berry skin, berry size, berry form, quality, and season of ripening. They recognized inbreeding depression as a problem in grapevine breeding and described that certain cultivars turned out to be rather poor parents to achieve vital and resistant plants being essentially free of off-flavors and yield good wine quality. Further analyses were made with slow progress and several scientists contributed to our understanding of inheritance in the genus *Vitis* as cited by De Lattin (1957). Husfeld (1962) explained the manifold failure of early resistance breeding and genetic dissection of the traits being largely due to their complexity and to the insufficient knowledge of the plant material used. Besides the complexity of the traits, one explanation for difficulty in unraveling the genetics could be the problem of selfings, which might have

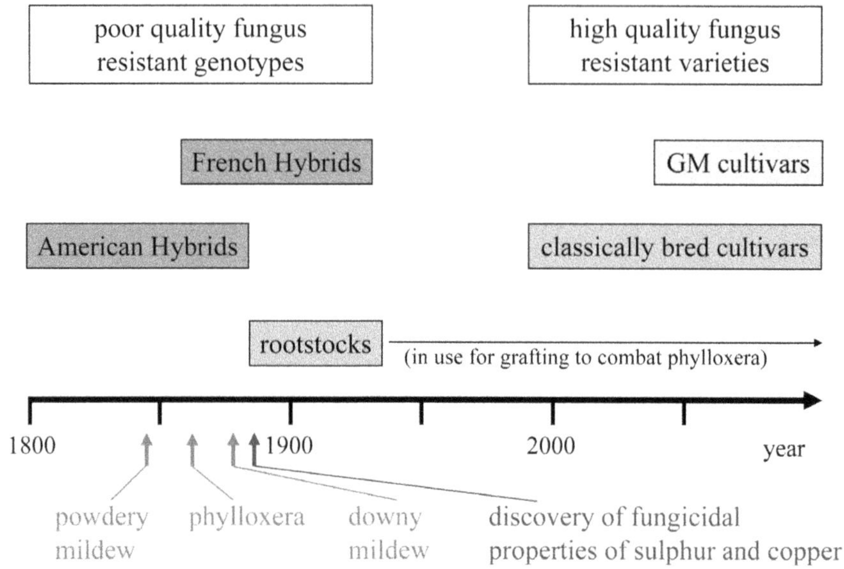

Figure 7-1 Milestones in grapevine resistance breeding on a time scale. The introduction of cultivars into the market is used as a point of reference. Red: American and French Hybrids did not succeed in the market due to poor wine quality. Even the best French Hybrids cultivated during the 20th century were almost all removed from the market. Green: Phylloxera tolerant or resistant rootstocks saved viticulture in Europe. Newly bred wine grape cultivars showing good field resistance and high wine quality entered the market around the turn of the millennium. Decoupling of resistance and quality could be proven in the 1960s but these cultivars were not accepted in the market. At the beginning of the 21st century marker-assisted selection (MAS) and marker-assisted backcrossing (MABC) became relevant for breeding programs. Yellow: Genetically modified cultivars will become available for the market in about two decades if consumer acceptance will be given. Appearance of mildew fungi and phylloxera in Europe and the discovery of sulfur and copper as fungicides are indicated.

Color image of this figure appears in the color plate section at the end of the book.

occurred instead of crosses resulting in distorted segregation patterns and misleading data. Many traits in grapevine are polygenic and are subjected to a great environmental impact, thus being difficult to be resolved by classical approaches. High inbreeding depression, slow propagation rates during vegetative propagation and first fruit setting three years after planting were additional retarding factors for resolving the genetics of traits in grapevine. Therefore, for grape breeders methods permitting an early selection are desired since a long time.

In the 1990s entirely new tools emerged opening up novel means to depict grapevine genetics. Several marker techniques showed up. While other crops already received some benefits within the breeding process from the use of marker techniques like isoenzyme analysis (Shiraishi et al. 1994;

Dzheneev et al. 1998) or DNA-based markers as restriction fragment length polymorphism (RFLP; Zyprian 1998), the breakthrough for grapevine came when PCR-based DNA marker analyses became available. The first genetic mapping studies for grapevine using DNA-markers (randomly amplified polymorphic DNA, RAPD) were described by Weeden et al. (1994) and the first genetic map followed soon (Lodhi et al. 1995). Based on genetic fingerprints and new technological achievements the highly demanded methods for early selection are becoming available (Adam-Blondon et al. 2001; Eibach et al. 2007; Molnar et al. 2007). A major achievement for further development is the recent success in sequencing the grapevine genome (Jaillon et al. 2007; Velasco et al. 2007). Consequently, the beginning of the 21st century represents a paradigm shift in grapevine breeding: from empiric to knowledge based breeding. In future the molecular breeding (or SMART breeding = selection with markers and advanced reproductive technologies) will result in a more efficient and much more targeted grapevine selection.

7.2 Germplasm Characterization

During domestication and subsequent millennia yield increase was most likely the motivation to search for new vines. During the last few centuries quality aspects became more and more important, too. However, during the last 200 years until today a major impetus for grapevine selection and breeding was the development of cultivars resistant to major diseases and stress factors. Old cultivars, the American and French Hybrids, numerous breeding strains and wild vines represent today's genetic resources for grapevine breeding. They are found in about 130 *ex situ* grapevine collections all over the world (*www.vivc.de, www.eu-vitis.de*). Major *in situ* genetic resources are also found in their native habitats reflecting the gene centers of (1) Europe and Asia Minor, (2) North America, and (3) Asia. Access to the *in situ* genetic resources is difficult and time consuming and in all gene centers vines are threatened with extinction due to the changing environment with human activities.

Thus, in 1983 Alleweldt and co-workers started to make an inventory of cultivars, and *Vitis* genotypes existing in the grapevine collections worldwide resulting in the *Vitis* International Variety Catalog (VIVC database). In 1996, VIVC became publicly available at *www.vivc.de* providing genotype descriptions for breeders, researchers, wine growers, and the public. A second database has been established giving accession related information in the EU-Vitis database (*www.eu-vitis.de*). It was established within the EU-project GenRes CT96 No81 and is continued in a second project AGRI GenRes#008.

In order to make use of the genetic resources these databases play an important role to identify certain genotypes conferring interesting traits for breeding. As an intermediate step, collections representing a large set of genetic diversity could be of great help to find association of genetic fingerprints and traits. Some so called core-collections have been developed (Le Cunff et al. 2008; J.-M. Martinez-Zapater et al. unpublished; E. Zyprian et al. unpublished). Le Cunff et al. (2008) analyzed 2.262 unique genotypes to construct genetic core collections based on the diversity identified by using 20 simple sequence repeat (SSR) markers. Subsets of 12, 24, 48 and 92 cultivars represent 58, 73, 83 and 100% of total SSR diversity, respectively. On 233 individuals the authors verified the observation by sequencing three genes showing most single nucleotide polymorphism (SNP) diversity in the core set of 92 cultivars compared with a set selected on the basis of 50 morphological traits. Thus, the approach shows clearly that core collections can effectively be used for SNP discovery (Le Cunff et al. 2008).

As methods for high-throughput genotyping (HT-genotyping) are becoming available based on marker technologies like SNPs and appropriate technology platforms (e.g., Lijavetzky et al. 2007; Pindo et al. 2008; Vezzulli et al. 2008a) an important stopover is being achieved. One of the missing links for further improvement of breeding is a thorough evaluation of the genetic resources. Efficient screening tools for evaluation of large numbers of vines *ex situ* or in a greenhouse are to be developed. Such high-throughput phenotyping solutions (HT-phenotyping) are a prerequisite for further acceleration of grapevine breeding. The existing grapevine collections could be very helpful and provide the possibility of evaluation at various environmental conditions. A harmonization of evaluation protocols can be envisaged and particularly achieved based on OIV descriptors (*http://www.oiv.int/de/accueil/index.php*).

Characterization of grapevine genetic resources on the genotype level became most successful when DNA microsatellite analysis using sequence tagged microsatellite sites (STMS, Beckmann and Soller 1990), also called simple sequence repeats (SSR) became available. Microsatellites were used first for genotyping studies and proved extremely helpful to identify misnaming and to resolve the parentage of cultivars (e.g., Botta et al. 1995; Bowers and Meredith 1997; Sefc et al. 1998; Bowers et al. 1999; Dettweiler et al. 2000; Boursiquot et al. 2004; This et al. 2004). Furthermore, mutations which have accumulated during centuries in the clonally propagated vines could be traced back to different cell layers identifying certain plants as periclinal chimeras in L1/L2 layers (Franks et al. 2002). Using SSR even triallelic variants have been detected like in "Pinot meunier" a cultivar, which shows hairy leaves and is discussed as a mutant from "Pinot gris" (Franks et al. 2002; Hocquigny et al. 2004).

7.3 Marker-assisted Gene Introgression and Pyramiding Resistance Loci

In order to introduce resistances into the gene pool of *V. vinifera*, breeders generated F_1-plants by interspecific crosses. But results soon showed that the wine quality of these plants was poor and unacceptable. Consequently, Baur (1922) suggested using interspecific hybrids between *V. vinifera* cultivars and a wild species as a resistance donor to generate a moderatly selected (resistance, vitality) F_1 generation, allow sister fertilisation, and to perform the main selection at the F_2 generation level. The strategy was the consequent application of the Mendelian rules rediscovered in 1900. However, it became clear that it requires more than two generations from the wild to select cultivable genotypes and even more crosses to obtain really elite lines and new cultivars. During the 20th century some insights were gained in the genetics of the traits of interest but in most cases breeders did not have a clear understanding.

In the 1990s the picture started to change dramatically when DNA marker techniques in particular SSR were introduced into grapevine genetic mapping (e.g., Riaz et al. 2004; Doligez et al. 2006a). Several genetic maps have been developed thus far using SSR or other marker types (Table 7-2) providing the basis for quantitative trait loci (QTL) mapping and thus combining genotype and phenotype information. This sophisticated analysis permits the dissection of complex traits and the integration of monogenic traits or QTLs into a genetic map. Furthermore, due to the overall genome synteny within grapevine it was possible to integrate the maps of several different genotypes to reference maps (Doligez et al. 2006a; Di Gaspero et al. 2007; Vezzulli et al. 2008b). Genetic maps from other *Vitis* species were then linked to these reference maps using common markers so that for the first time an orientation in the grape genome became possible (Fig. 7-2) as well as a dissection of complex traits into single loci. Therefore, molecular marker techniques and the overall genetic map of the *Vitis* genome enable the marker-assisted selection (MAS), introgression and pyramiding of resistance loci throughout the breeding process.

The principles of MAS are extensively described (for review see e.g., Mohan et al. 1997; Collard et al. 2005; Peleman et al. 2005; Collard and Mackill 2008). MAS is applied to (i) optimize parental selection, (ii) introgress a single locus into an elite line, (iii) pyramiding different loci in a single cross in one genotype, and (iv) select for progeny with the highest proportion of the recurrent parent genome. The major prerequisite is the availability of co-segregating markers closely linked to the trait loci. By using two markers flanking the trait the distance should be generally less than 5 cM on either side or 1 cM if just a single marker exists (Collard and Mackill 2008). Further requirements are appropriate technical equipment and a reliable and cost-effective high-throughput screening method. There are two major advantages of MAS accelerated grapevine breeding. First, MAS can substitute phenotyping. This is especially powerful in screenings of traits which are difficult, laborious or expensive to measure (e.g., phylloxera or *Xiphinema index* resistance). Second, MAS can be carried out early at the seedling stage. Time can be saved significantly in particular if the traits are developmentally late expressed (e.g., berry color, flavor).

In the last years the first genetic maps with tightly linked markers for important trait loci, mostly resistance genes, were published being the starting point of application of MAS in grapevine breeding. Analyzing the genetics of cv. "Regent", Fischer et al. (2004) and Welter et al. (2007) identified one major QTL for powdery mildew (*Ren*3, chromosome 15) and two QTLs for downy mildew (*Rpv*3 and *Rpv*4 on chromosomes 18 and 4, respectively). Further two loci for powdery mildew resistance (*Run*1, *Ren*1) are available in different resources. Bouquet et al. (2000) and Pauquet et al. (2001) characterized the *Run*1 locus located on chromosome 12 and tightly linked markers were identified as later being suitable for MAS (Donald et al. 2002; Barker et al. 2005). Closely associated with the *Run*1 locus, a resistance locus against *P. viticola* assigned as *Rpv*1 was found, which is partially lost in line VRH3082-1-42 (see Chapter 8 and below). A third locus for resistance against powdery mildew, *Ren*1, could be identified on chromosome 13 in cv. "Kishmish vatkana" (Hoffmann et al. 2008) and cv. "Dzhandzhal kara" (Coleman et al. 2009). An overview of the chromosomal localization of the resistance loci is shown in Fig. 7-2. Some markers, which are useful for MAS, are given in Table 7-2.

Since most traits described thus far are resistance loci, they were used to gain first experiences with validation of markers to be used for MAS in the run-up of resistance breeding. After proof of concept further aspects have to be considered like practicability, cost and economy of MAS. Molnar et al. (2007) for instance analyzed three markers linked to the *Run*1 locus in a population derived from a ([*M. rotundifolia* x *V. vinifera*] BC_4) x *V. vinifera* cross. Their study focused on a simple method for marker detection and used agarose gels for allele scoring. Agarose gels are robust, easy to handle and do not need very expensive equipments. This method can be established in a short time and also by personnel not so well trained in molecular biology. It can, therefore, be introduced easily in classical breeding institutions. However, if the population sizes are large or a lot of markers have to be analyzed an automated allele detection system is recommended.

An example of pyramiding four resistance loci, two for resistance against *E. necator* (*Run*1, *Ren*3) and two for resistance against *P. viticola* (*Rpv*1, *Rpv*3), is given by Eibach et al. (2007). The vines selected show all major resistance loci from "Regent" (*Ren*3, *Rpv*3) and VRH3082-1-42 (*Run*1, *Rpv*1). For *P. viticola* the resistances were found to be additive showing no symptoms in test infections if all known loci were combined. Powdery

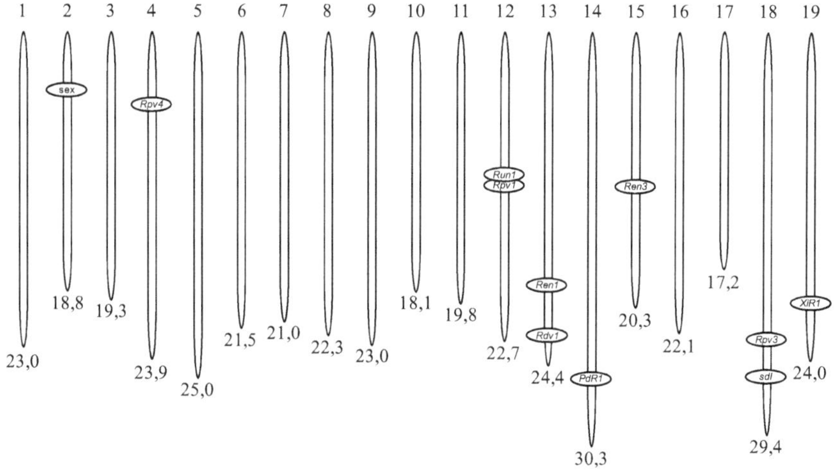

Figure 7-2 Schematic representation of the chromosome set of *Vitis*. Numbering of the 19 chromosomes (top numbering) is according to the International Grape Genome Program (IGGP). The figures below the chromosomes indicate the chromosomal length in megabases according to the 12x genome sequence of PN40024 (http://www.genoscope.cns.fr/vitis). The *sex* locus on chromosome 2 is found in several *Vitis* species. Various resistance loci from different sources are indicated. All resistances are from different and freely cross fertile *Vitis* species. The only exception is *Run*1/*Rpv*1 on chromosome 12 which is introgressed from *Muscadinia rotundifolia*. A single genotype carrying all the resistance loci as depicted does not currently exist. For further details see text and Table 7.2.

mildew resistance was found to be tight even under extreme pathogen pressure. Also from the trait of wine quality the genotypes selected proved to be very interesting breeding lines. For this particular case a combined scheme of phenotypic and genetic selection was applied (Fig. 7-3). In the first step at the seedling stage, the entire population was evaluated for downy mildew infection after artificial inoculation. Subsequently, the downy mildew resistant sub-population was screened for resistance against powdery mildew. This conventional phenotyping was highly effective and economic and resulted in a decrease in population size of about one-sixth. Up to that point there were no compelling reasons to use MAS. The remaining individuals were then subjected to MAS for the presence of the four resistance genes and a further reduction of about one-fifth was achieved. This clearly shows that without MAS the breeder is not able to select for those seedlings carrying all pyramided resistance loci. Accordingly, in a traditional breeding program without using MAS the chance to select for the right genotype would be 1:5.

The examples show that for breeding programs, which even these days are largely performed empirically, MAS is at the onset of utilization. Several other markers have been assigned to traits such as seedlessness and can already be used for selection of particular traits. For instance, seedlessness

Table 7-2 Traits relevant for breeding. Associated markers, their chromosomal localization, and the donor genotype are given. Genome position [Chr/Mb] = chromosome number and position in megabases according to the 12x genome sequence of PN40024 (*http://www.genoscope.cns.fr/vitis*).

Symbol	Resistance Trait	Associated marker	Genome Position [Chr/Mb]	authors	Mapping population (population size)	Source (origin)
*PdR*1	Pierce's disease	VMCNg3h8 VVIn64 UDV095	14/25.3 14/26.6 14/26.1	Riaz et al. 2006 ; Riaz et al. 2008	*V. rupestris* x *V. arizonica* (181)	*V. arizonica*
*Rdv*1	*Daktulosphaira vitifoliae*	Gf139 VMC8e6	13/21.9 13/22.5	Zhang et al. 2009	Gf.V3125 x "Börner"	"Börner" (*V. riparia* x *V. cinerea*)
*Rpv*1	*Plasmopara viticola*	VMC72 VVIb32	12/- 12/10.3	Merdinoglu et al. 2003	"Syrah" x 22-8-78	*Muscadinia rotundifolia*
*Rpv*3	*Plasmopara viticola*	UDV112 VVIn16 (*1)	18/- 18/23.4	Welter et al. 2007	"Regent" x "Lemberger"	"Regent", major QTL
*Rpv*4	*Plasmopara viticola*	VMC7h3 VMCNg2e2.1	4/4.7 4/5.2	Welter et al. 2007	"Regent" x "Lemberger"	"Regent"
*Ren*1	*Eriysiphe necator*	UDV-020 VMC9h4-2 VMCNg4e10-1	13/- 13/18.4 13/18.4	Hoffmann et al. 2008	"Nimrang" x "Kishmish vatkana" (310)	"Kishmish vatkana" (*V. vinifera*)
*Ren*3	*Erysiphe necator*	UDV15b VVIv67	15/7.1 15/10.9	Welter et al. 2007	"Regent" x "Lemberger"'	"Regent"
*Run*1	*Eriysiphe* (*Uncinula*) *necator*	VMC8g9 VMC4f3.1	12/20.4 12/13.1	Barker et al. 2005	VRH3082-1-42 x "Cabernet Sauvignon" (161)	VRH3082-1- 42 (*Muscadinia rotundifolia*)
sdI	seed development inhibitor	SCAR: SCC8	18/25.9	Doligez et al. 2002	MTP2223-27 x MTP2121-30 (139)	
	seedlessness	VMC7f2 VMC6f11	18/26.9 18/23.2	Cabezas et al. 2006	"Dominga" x "Autumn Seedless"	Autumn Seedless

Symbol	Resistance Trait	Associated marker	Genome Position [Chr/Mb]	authors	Mapping population (population size)	Source (origin)
sex	sex	VVS3	2/4.2	Dalbó et al. 2000	"Horizion" x Illinois 547-1	
sex	sex	VVIb23	2/4.9	Lowe and Walker 2006	"Ramsey" (*Vitis champinii*) x "Riparia Gloire" (*Vitis riparia*) (188)	
sex	sex	VVMD34 VVIb23	2/3.7 2/4.9	Riaz et al. 2006	*V. rupestris* x *V. arizonica* (181)	
*XiR*1	*Xiphinema index*	VMC5a10	19/20.9	Xu et al. 2008	*V. rupestris* x *V. arizonica* (185)	*V. arizonica*

*1 VVIn16 according to Merdinoglu et al. 2005.

could be scored easily by sequence-characterized amplified region (SCAR) markers developed by Adam-Blondon et al. (2001), or by SSR markers identified by Cabezas et al. (2006).

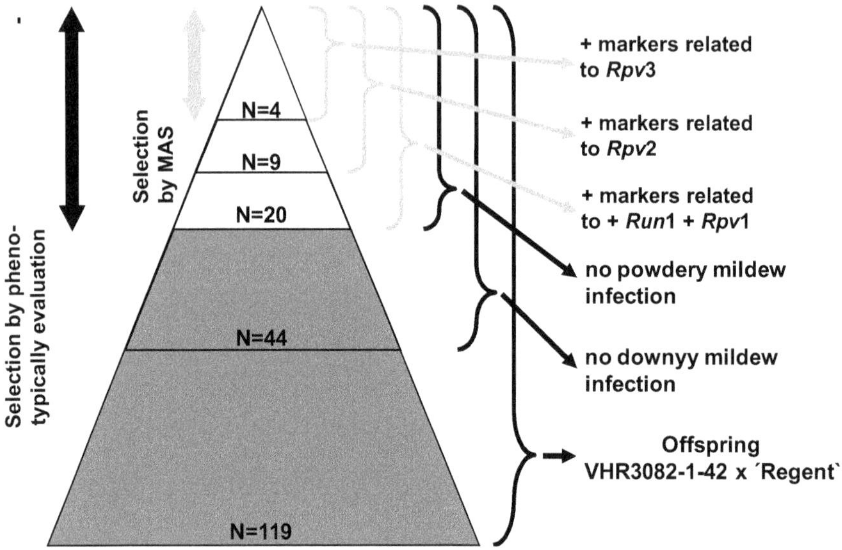

Figure 7-3 Pyramiding different mildew resistance loci (*Run*1-*Rpv*1, *Ren*3, *Rpv*3) combining conventional phenotyping and marker-assisted selection (MAS). With MAS those individuals among the phenotypically resistant sub-population were identified that carry all resistance genes.

A totally different aspect of molecular breeding is the application of MAS for the so-called "background selection" (Collard and Mackill 2008). In grapevine breeding the resistance genes to be introgressed into the genomic background of a *V. vinifera* cultivar or an elite line originate generally from non-*vinifera Vitis* species. Thus, after the first cross a series of pseudo-backcrosses has to follow to get rid of undesired traits linked to the donor genome fraction. With each further pseudo-backcross an increase in the *V. vinifera* genome proportion is achieved resulting after six rounds and 25 to 30 years at more than 99% (Table 7-3). Using unlinked background markers (markers that are scattered all over the genome) the number of pseudo-backcrosses can be reduced selecting in each backcross population for those genotypes with the highest percentage of *V. vinifera*-specific alleles. Di Gaspero et al. (2005) used informative background SSR markers to select for highly heterozygous genotypes within their breeding populations. Since grapevine is severely suffering from inbreeding depression they wanted to specifically look for "highly outbred" genotypes supposed to have a higher overall fitness.

Table 7-3 Reduction of the donor genome (P2) during backcross breeding is indicated aiming at about 1% of donor genome upon six backcrosses. Due to high inbreeding depression the recurrent parent can not be a single genotype. P1 and recurrent parents are different *V. vinifera* cultivars/genotypes resulting in pseudo-backcrosses (pBC). Marker-assisted backcrossing (MABC) is expected to save one to two generations accelerating introgression of desired genes. As outlined in the text using greenhouse steps a further reduction is anticipated (not shown) permitting MABC in a much shorter period.

Year	1	2	3	4	5	6	7	8	9	10	11	12	13	14	15	16	17	18	19	20	21	22	23	24	25
	P1				F_1				pBC_1				pBC_2				pBC_3				pBC_4				pBC_5
Crossing	x				x				x				x				x				x				x
	P2				V.v.				V.v.				V.v.				V.v.				V.v.				V.v.
Generation	F_1				pBC_1				pBC_2				pBC_3				pBC_4				pBC_5				pBC_6
Seedling																									
Flowering																									
[%] of P2		50				25				12.5				6.3				3.1				1.6			0.8

7.4 Limitations and Prospects of Marker-assisted Selection and Marker-assisted Backcrossing

Since long resistance resources are well known to be found within the American genepool of *Vitis* and *Muscadinia* (Alleweldt and Possingham 1988), but some reports provide valuable information about the Asian *Vitis* genepool (He and Wang 1986; Wan et al. 2007), too. A systematic utilization of resistance resources has rarely been done. An excellent but long lasting example of the development of introgression lines is given by Bouquet et al. (2000) for the *Run*1 locus of *Muscadinia rotundifolia*. The first crosses were made in 1974 when a breeding program was initiated at Montpellier (Bouquet 1980). Female seedlings of various *V. vinifera* cultivars were chosen and fertilized with pollen of different *M. rotundifolia* plants. Since chromosome numbers differ between *V. vinifera* (38) and *M. rotundifolia* (40) hybrids showing 39 chromosomes were identified and used for backcrossing. Despite problems with sterility in the offspring successful backcrosses could be established. Consecutive pseudo-backcrosses were made getting genotypes up to BC_6, which show a high resistance against powdery mildew and good viticultural performance (Bouquet et al. 2000). However, a second resistance (*Rpv*1), that is directed against *P. viticola*, was attenuated (Bouquet et al. 2000; Wiedemann-Merdinoglu et al. 2006b) obviously due to a recombination, which occurred close to the powdery mildew resistance locus. Using genotypes of the BC_2 this loss of resistance could be demonstrated by marker application (Wiedemann-Merdinoglu et al. 2006a). Pyramiding the resistance with other resistances was possible as mentioned above (Eibach et al. 2007).

Currently no example exists for grapevine for marker-assisted backcrossing (MABC). However, introgression of phylloxera resistance into the *V. vinifera* background could become such an example in the near future. The most important breeding objective for rootstock breeding was and is a complete resistance against the root phylloxera. Therefore, in the past great emphasis was given to the genetics of phylloxera resistance, however, without a final conclusion (Börner 1943; Breider 1969; Manty 2006). This might be due to the material analyzed which originates from a small number of genotypes representing a limited genetic basis (Schmid et al. 2007). Almost all of this material shows tolerance rather than resistance. Since rootstocks became available at the beginning of the 20th century (compare Fig. 7-1) and brought the solution of the phylloxera disaster, rootstock breeding activities declined. Small rootstock breeding programs have continued and research is directed to elucidate the genetics of certain traits. For example, Roush et al. (2007) analyzed phylloxera resistance in a F_2 progeny from a remake of AXR1 (*V. vinifera* x *V. rupestris*) for inheritance of nodosities and tuberosities. The genetic analysis revealed two loci involved in formation of nodosities and one or two loci for tuberosity formation, being recessive in each case.

A different picture was obtained for rootstock cv. "Börner" (*V. riparia* 183 Gm x *V. cinerea* Arnold), which is a completely phylloxera root resistant genotype showing a hypersensitive response (Schmid et al. 1998; Schmid et al. 2003). Genetic analysis of the progeny of Gf.V3125 ("Schiava Grossa" x "Riesling") x "Börner" indicated one major locus located on Chromosome 13 (see Table 7-2; Fig. 7-2) (Zhang et al. 2009). Since "Börner" is an interspecific hybrid between wild species, a MABC strategy would be very interesting to introduce the resistance locus in a *V. vinifera* background. Use as rootstock for just a few generations might be sufficient to get a good performance on calcareous soils and the affinity to the scion as well as the rooting ability, which are major traits in rootstock breeding. On the long term phylloxera root resistance might again become an objective in wine grape or table grape breeding. Using MABC one to two generations could be saved if the offspring is selected for the resistance locus and against the donor ("Börner") genome. If a reduction of the generation time to two years in the greenhouse providing optimal growth conditions becomes applicable, a MABC program can be performed within 10 to 15 years. An even further acceleration might become possible by using rapid cycling grapevines described as "microvines" making use of Vv*gai*1 mutant allele causing dwarfism (Chaib et al. 2010). Another example could be the introgression of Pierce´s disease resistance or nematode resistance. Riaz et al. (2004, 2008) identified a resistance locus on chromosome 14 for Pierce´s disease originating from *V. arizonica*. With this knowledge a MACB strategy might be applied. Similarly nematode resistance which was recently localized (Xu et al. 2008) becomes accessible to MABC.

For certain applications, it might be of interest to eliminate male plants from a progeny or identify females. The inheritance of sex in grapevine has been investigated by Anthony (1914) as well as Hendrick and Anthony (1915). Valleau (1915) interpreting these data gave a good hypothesis of sex determination of male, hermaphroditic, and female flowers assuming partial suppression of the determiner for male flowers. Negi and Olmo (1971) summarized the inheritance of male sex as a simple Mendelian factor which is modulated by a number of modifying genes. According to Levadoux (1946) and confirmed by Antcliff (1980) while analyzing 15,000 seedlings, a single set of three alleles is responsible for sex determination: male (MM, MH, of MF) is dominant over hermaphroditic (HH, HF) which is dominant over female (FF). This picture fits with the molecular genetic analysis by Dalbó et al. (2000), Riaz et al. (2006), and Lowe and Walker (2006) who found a major single locus which could be assigned to chromosome 2 (Table 7-2; Fig. 7-2). The markers could be very useful in early steps of developing introgression lines to reduce undesired genotypes, though removal of male plants from an early pseudo backcross progeny is not urgently required.

Most useful in the selection process would be the MAS for quality traits. High wine quality combined with good climatic adaptation and high disease resistances summarize the major objectives in grapevine breeding since the initial breeding initiatives. Berry quality is by far the most complex trait. It depends on complex sensory perceptions including taste, olfaction, and mouthfeel. Selection of good quality genotypes depends on the organoleptic perception of a testing panel and is thus rather subjective. Berry quality is difficult to evaluate for table grapes and even more difficult for wine grapes since fermentation increases complexity of the trait. Sugar, acid, fermentable nitrogen (amino acids), minerals (e.g., potassium), color, balanced (positive) aroma compounds, and lack of off-flavors are major attributes to estimate berry quality. In particular the concentration, the balance, and interference of up to 800 aroma compounds (Rapp 1998) (not all are relevant for sensory perception) is crucial for appraisal of quality. In a wine, which is free of sugar after fermentation, inharmonious taste can easily be recognized and off-flavors can emerge quickly. Storage and aging of wine need to be evaluated to uncover sensory deficits, which are attributed to the breeding line. Within a breeding program quality can be recorded at the earliest four to five years after a cross and is highly influenced by environmental factors. Furthermore, for assessment of wine quality the amount of grapes available for microvinification is limited thus the number of vines on-hand impairs fermentation and hence a quality evaluation. As indicated the assessment of berry quality is direfully complex, most time consuming, but finally the most important trait to be evaluated.

The first attempt to elucidate berry quality genetically was reported by Hedrick and Anthony (1915). The authors summarized the results of various parental combinations. Most noticeable is the very low percentage of seedlings whose quality is good or above average even when parents of the highest quality were used. The authors observed a tendency for the proportion of seedlings giving good quality to decrease with the use of parents showing poorer quality. They conclude that for breeding only high quality parents should be used. Thousands of years of selection of grapevine during domestication have raised the quality in *V. vinifera* subsp. *vinifera* to a point that it has become a powerful factor in transmitting high quality (Hedrick and Anthony 1915).

A considerable lack of knowledge and useful methods has to be ascertained to describe wine quality. However, insights into the complex trait of wine quality will be gained during the forthcoming years. A method of choice could be the use of molecular markers substituting diverse and expensive analytical methods like gas chromatography, gas chromatography-mass spectrometry, liquid chromatography and liquid chromatography-mass spectrometry. Concerning positive aroma compounds (e.g., monoterpenes) the first QTLs have been found (Eibach et al. 2003; Grando et al. 2004; Doligez et al. 2006b). But the data still need to form a clearer picture to become useful for MAS of berry quality. In contrast, the development of markers to monitor off-flavors could be much easier. They would be very useful to eliminate alleles resulting in undesirable flavor compounds (e.g., furaneol (Rapp 1998) or methylanthranilate) very rapidly from the genepool while introducing new resistance genes into *V. vinifera*.

Recently the biosynthesis of tartaric acid contributing to taste, mouthfeel, and aging potential received some interest since low acidity in hot climate is a major quality issue. DeBolt et al. (2004, 2006) gained major insight in the biosynthetic pathway of tartaric acid synthesis and the underlying enzymes. This knowledge will be used to unravel the regulation of the pathway opening the possibility to build up new selection schemes for cultivars showing an appropriate acid balance.

A typical breeding program consists of several consecutive steps decreasing the number of individuals in each selection step. The most important traits in grapevine breeding are quality and resistances. Fig. 7-4 shows the various breeding steps and the existing number of individuals of a particular breeding strain. Assuming a current breeding program for wine grapes starts with 50,000 seedlings a year, greenhouse testing and screening for mildew resistance results in about 5,000 plants to be planted in a seedling plot (requiring about one hectare). After the seedling stage all breeding steps require five to eight years for growth: year one to three to get the vine established and year four to eight for a full crop. By far the most time consuming is the evaluation for wine quality. Grapes from breeding lines showing good viticultural performance including sufficiently good resistance will be used for wine making. The first wine is already made from a single vine yielding frequently not more than one litre. This so-called microvinification is crucial in wine grape breeding since wines need to be made in a comparable manner. Cutting the time required for evaluation of wine quality could be the major step reducing the breeding time. This can be achieved by developing and application of markers monitoring distinct aspects of wine quality like sugars, acids, flavors, off-flavors, etc. or which have an important impact on quality like berry size, berry number, cluster size, cluster architecture, ripening time, ripening duration, etc. First markers for resistance loci showed the power of MAS and are used already in breeding programs and it is only a question of time when the first suitable markers for quality traits will be available.

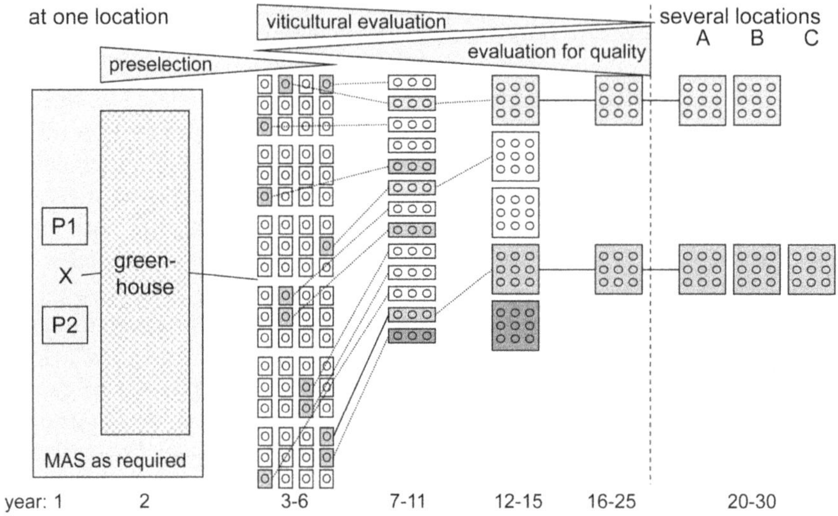

Figure 7-4 Steps in wine grape breeding: (from left to right) first selection for fungus resistance (*Plasmopara viticola* and *Erysiphe necator*) in a greenhouse, seedlings field (1 individual), pretesting (ca. 10 vines), intermediate testing (ca. 50 vines), main testing (ca. 500-1,000 vines), test planting with wine growers at different locations. A rough time scale in years is shown for each selection step.

Color image of this figure appears in the color plate section at the end of the book.

7.5 Transgenic Breeding

In 1983 first transgenic plants were established (Herrera-Estrella et al. 1983; Fraley et al. 1983) and since that point of time a tremendous development of this sector occurred. Cultivation of genetically modified (GM) plants began in 1996 and resulted in 2009 in worldwide 134 million hectare of commercially grown transgenic crops, predominantly soybean, maize, cotton, and rapeseed (Clive 2009). Herbicide resistance is the most important trait represented on 83.6 million hectare. Plants with combinations of traits, so-called stacked trait products, were by far the fastest growing trait group between 2007 and 2009 indicating the future development. Double and triple stacked traits grew at 23%, compared with 9% for herbicide and 6% for insect resistance (Clive 2008) and in 2010 a maize genotype (SmartStax™) will be released in USA carrying eight different genes coding for three traits (two pest resistances and a herbicide resistance) (Clive 2009). Even if no commercial GM vine exists currently, stacked trait vines with different transgenes conferring resistance to one pathogen will be the ultimate goal to achieve durable resistance.

In contrast to the fast development worldwide, in some countries, particularly European countries, introduction of GMOs into the market is delayed. Non acceptance is causal for the slow progress. Most disappointing in the ongoing debate is the loss of scientific basics making the discussion purely political. Facing these difficulties it remains unclear whether or not consumer acceptance will be achieved. A rapid change of the public opinion is not to be foreseen. Assuming this the debate for or against GM-grapevines will be a debate for years.

A legal solution of the GMO dispute could lead to a pertinent solution at least for vegetatively propagated plant species, in particular for grapevine that would help both humans and the environment. This solution could come from a close look and scientific assessment of transgenesis versus cisgenesis. A transgenic event is considered as a gene transfer overlapping sexual borders creating plants which naturally can not be evolved. It therefore might provide a different risk potential compared with traditionally bred cultivars. Lawmakers and regulatory authorities have paid attention to the biosafety of such a plant controlling potential risks to a maximum. In contrast, cisgenesis is referred to as the transfer of genes including their regulatory elements within the sexual barriers. Hence, plants created in such a way are considered to be as safe as plants selected by classical breeding approaches. Essentially no gene is added to the genepool. According to Rommens et al. (2007), this approach mimics classical breeding like introgression of genes but circumvents several rounds of backcrossing and long lasting screening of large progenies for desired phenotypes (compare Fig. 7-5). As outlined above even after six backcrosses about 1% of the genes of the wild genotype remains in the introgression lines (see Table 7-3), which means a lot of genetic information for potentially undesirable characters. Cisgenes minimize this linkage drag. Schouten et al. (2006) suggest updating annex 1B of GMO Directive 2002/18/EC dealing with techniques to be excluded from the Directive by adding as item *"(3) cisgenesis"*. As a consequence grapevines carrying cisgenes would not be considered as GMO.

Irrespective of the political controversy, from the current viticultural perspective several open questions with respect to GMOs need to be discussed seriously. It is anticipated that in some countries GMOs will approach the market and raise the question of product labeling. Most interesting from the point of view of viticulture and grapevine breeding is the improvement of existing cultivars for disease and pest resistance (e.g., Le Gall et al. 1994; Viss and Driver 1996; Martinelli et al. 2002; Bornhoff et al. 2005). As far as rootstocks are concerned the resulting wine in most cases will be essentially free of GMO material, however, its labeling needs to be clarified. As far as wine grapes engineered for resistance are used, the resulting wine typically does not contain a GMO or any compound related to the gene transfer. Despite this, because of legislation in Europe labeling of the product wine will be mandatory. Finally, labeling for table grapes

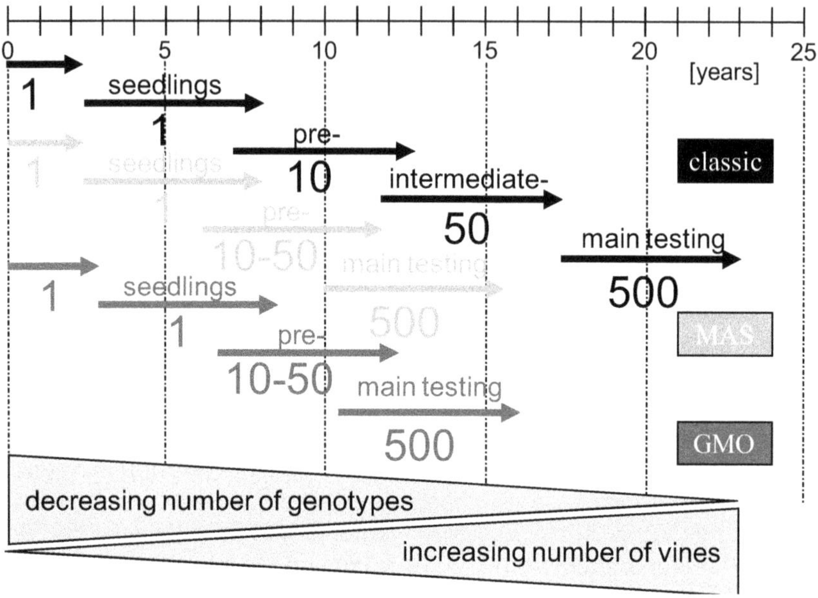

Figure 7-5 Comparison of the schedule of selection steps in a breeding program using different approaches: classical breeding, MAS, and GMO. A greenhouse test will be done for removal of susceptible genotypes in the year after the crossing/gene transfer. Subsequent steps are planting in the vineyard (restrictions for GMOs are not considered). The number of plants of a breeding strain is given below the arrows indicating that propagation of the plant material is a limiting factor to provide the market with grafted vines. Developing a new cultivar by classical breeding techniques requires usually 25 to 30 years. Applying MAS could shorten the breeding process considerably. An acceleration of up to 10 years is expected. Also transgenic breeding is expected to require less time if it is an established technology. black = classical breeding, light gray = breeding using MAS; dark gray transgenic breeding.

will be obligatory, too, since the product is a GMO. From a consumer point of view labeling will bring a maximum of transparency and opening the possibility for choosing or rejecting a product.

A second question concerns the naming of transgenic cultivars . It is irrelevant for rootstocks and for table grape cultivars as naming them is of minor importance. A new table grape cultivar can be introduced into the market without seriously affecting the consumer's perception. The cultivar name is not closely linked to the grape as a product. This is completely different for wine grapes where the consumer is very much aware of the cultivar name since it is very often used as the product name for selling the wine. Therefore, a transgenic approach maintaining the cultivars name and thus the name of the sold product would be highly appreciated for marketing the product (Töpfer 2000). The legal rules need to be elaborated.

Aside from the debate regarding labeling , the enormous advantage of a GM-vine carrying stacked resistance genes (equivalent to pyramided resistances) and being resistant to mildew diseases is the reduction in pesticide use. As outlined earlier grapevine breeding was initiated around the world due to a substantial deficit of the vines for pathogen and pest resistance. Hence, maintaining wine quality but increasing resistance will be the greatest benefit of GM-vines. It will be an advantage for the environment, the producer, and finally for the consumer.

The principles of genetic transformation of grapevine were shown more than 15 years ago (e.g., Scorza et al. 1995; Kikkert et al. 1996; Franks et al. 1998). Several protocols are available and have been updated (e.g., Perl and Eshdat 1998; Wang et al. 2005; Bouquet et al. 2006). Despite that progress transformation of grapevine is still difficult and often restricted to certain well established cultivars. Furthermore, safety issues for grapevine have to be elaborated prior to creating products for the market. Currently, the only biosafety report for grapevine is a pilot study by Harst et al. (2009) providing the first arguments for a debate. They observed pollen flow up to 150 m and out-crossing around 2 to 3% in up to 20 m distance of an experimental planting using a small number of transgenic vines.

Another important issue is the availability of suitable genes for gene transfer. First genes encoding relevant traits have been isolated and are being tested (e.g., *Run*1). The novel possibilities of genome sequencing technologies (Jaillon et al. 2007; Velasco et al. 2007) facilitate the dissection of the gene—trait relation and a molecular understanding of gene function. The crucial genes from grapevine will become available for improving characters of existing cultivars. It can be expected that for several important traits (in particular resistance traits) genes will be identified within the next few years opening a tremendous potential for application. However, the cost for introducing GM-vines into the market from the current perspective will be enormous ranging between US $ 7 and 15 million which just takes into account the figures for *Bt*-maize (Kalaitzandonakes et al. 2007). However, it is anticipated that in the EU considerably more money would be required. The question arises whether this will be affordable or not for grapevine and if new funding regimes need to be developed.

7.6 Outlook from a Breeder's Perspective

Since DNA marker technologies became available a paradigm shift was initiated for grapevine breeding: the switch from pure empirical efforts to more rational and knowledge-based endeavors. The basis is provided by the increasing knowledge of the genetics of distinct traits, which already is being used in practical grapevine breeding programs.

Since thousands of seedlings are to be tested in a breeding program both cost-effectiveness and reliability of the screening method is fundamental. The successful sequencing of the grapevine genome (Jaillon et al. 2007; Velasco et al. 2007) and future genome sequencings provide the basis to develop HT marker systems based on SNP and will dramatically alter the toolbox for grapevine breeding. However, in addition to HT genotyping technologies also HT phenotyping platforms are required. The recent achievements and the prospects will have a considerable impact on breeding programs. A substantial acceleration of the breeding process can be anticipated. For example, for wine grapes a reduction of the selection period of up to 10 years is conceivable as indicated in Fig. 7-5. MAS will be used to make decisions earlier and to omit eventually one to two screening blocks. Targeted backcross breeding programs by MABC breeding become a possibility. However, availability and propagation of the plant material will soon become a limiting factor. Once the appropriate methods are established a cost-benefit calculation will show in the end what will be accomplishable in the breeder's hands and what will remain a dream.

GM-vines are a completely different issue. Consumer acceptance is the prerequisite for their commercialization. GM-vines could be accepted, if they contribute to considerably reduce the use of pesticides in viticulture. Apart from the point of acceptance, it is expected that it will take about two decades to bring GM-cultivars to the market.

In light of this time span two options will become available: new cultivars developed by means of MAS/MABC or GM-cultivars. Both will show considerably improved traits, which will significantly contribute to a more sustainable and more environmentally friendly cultivation of grapes. From the breeding point of view, cultivars conferring resistances to pests such as powdery and downy mildew, phylloxera and Pierce's disease will be ready for the market. At that time grapevine breeding will reach its peak.

Acknowledgements

The authors gratefully acknowledge Anne-Francoise Adam-Blondon and Genoscope for access to 12x genome data of the French–Italian Public Consortium prior to publication for Grapevine Genome Characterization used for Table 7-2 and Fig. 7-2.

References

Adam-Blondon AF, Lahogue-Esnault F, Bouquet A, Boursiquot JM, This P (2001) Usefulness of two SCAR markers for marker-assisted selection of seedless grapevine cultivars. Vitis 40: 147–155.

Alleweldt G, Possingham JV (1988) Progress in grapevine breeding. Theor Appl Genet 75: 669–673.

Antcliff AJ (1980) Inheritance of sex in *Vitis*. Ann Amelior Plantes 30: 113–122.

Anthony RD (1914) Methods and Results in Grape Breeding. Proc Soc Hort Sci 11: 81–86.

Barker CL, Donald T, Pauquet J, Ratnaparkhe MB, Bouquet A, Adam-Blondon AF, Thomas MR, Dry I (2005) Genetic and physical mapping of the grapevine powdery mildew resistance gene, *Run*1, using a bacterial artificial chromosome library. Theor Appl Genet 111: 370–377.

Baur E (1922) Einige Aufgaben der Rebenzüchtung im Lichte der Vererbungswissenschaft. Beiträge zur Pflanzenzucht 5: 104–117 (oral presentation in 1914).

Beckmann JS, Soller M (1990) Toward a unified approach to genetic mapping of eukaryotes based on sequence tagged microsatellite sites. Bio/Technology 8: 930–932.

Börner C (1943) Dreißig Jahre Deutsche Rebenzüchtung. Bremer Beiträge zur Naturwissenschaft, 7. Bd., 3. Heft. Arthur Geist Verlag, Bremen. (Hrsg.: Die Wittheit zu Bremen).

Bornhoff BA, Harst M, Zyprian E, Töpfer R (2005) Transgenic plants of *Vitis vinifera* cv. Seyval blanc. Plant Cell Rep 24: 433–438.

Botta R, Scott NS, Eynard I, Thomas MR (1995) Evaluation of microsatellite sequence-tagged site markers for characterizing *Vitis vinifera* cultivars. Vitis 34: 99–102.

Bouquet A (1980) *Vitis muscadinia* hybridization: A new way in grape breeding for disease resistance in France. 3rd Int Symp Grape Breeding Davis CA USA, pp 42–61.

Bouquet A, Pauquet J, Adam-Blondon AF, Torregrosa L, Merdinoglu D, Wiedemann-Merdinoglu S (2000) Vers l'obtention de variétés de vigne résistantes à l'oidium et au mildiou par les méthodes conventionelles et biotechnologiques. Progr Agric Vitic 117: 383–389.

Bouquet A, Torregrosa L, Iocco P, Thomas MR (2006) Grapevine (*Vitis vinifera* L.). Meth Mol Biol 344: 273–85.

Boursiquot JM, Lacombe T, Bowers J, Meredith C (2004) Le Gouais, un cépage clé du patrimoine viticole européen. Bull OIV 77: 875–876.

Bowers J, Boursiquot JM, This P, Chu K, Johansson H, Meredith C (1999) Historical Genetics: The Parentage of Chardonnay, Gamay, and Other Wine Grapes of Northeastern France. Science 285: 1562–1565.

Bowers JE, Meredith CP (1997) The parentage of a classic wine grape, Cabernet Sauvignon. Nat Genet 16: 84–87.

Breider H (1969) Unterlagenzüchtung im deutschen Weinbau in historischer Sicht. Weinberg Keller 16: 169–200.

Cabezas JA, Cervera MT, Ruiz-Garcia L, Carreno J, Martinez-Zapater JM (2006) A genetic analysis of seed and berry weight in grapevine. Genome Sour 49: 1572–1585.

Chaib J, Torregrosa L, Mackenzie D, Corena P, Bouquet A, Thomas MR (2010) The grape microvine—a model system for rapid forward and reverse genetics of grapevines. Plant J 62: 1083–1092.

Clive J (2008) Global Status of Commercialized Biotech/GM Crops: 2008. ISAAA Brief No. 39. ISAAA: Ithaca, NY.

Clive J (2009) Global Status of Commercialized Biotech/GM Crops: 2009. ISAAA Brief No. 41. ISAAA: Ithaca, NY, USA.

Coleman C, Copetti D, Cipriani G, Hoffmann S, Kozman P, Kovacs L, Morgante M, Testolin R, Di Gaspero G (2009). The powdery mildew resistance gene REN1 co-segregates with an NBS-LRR gene cluster in two Central Asian grapevines. BMC Genet 10: 89.

Collard BCY, Mackill DJ (2008) Marker-assisted selection: an approach for precision plant breeding in the twenty-first century. Phil Trans Roy Soc B 363: 557–572.

Collard BCY, Jahufer MZZ, Brouwer JB, Pang ECK (2005) An introduction to markers, quantitative trait loci (QTL) Mapping and marker-assisted selection for crop improvement: The basic concepts. Euphytica 142: 169–196.

Dalbó MA, Ye GN, Weeden NF, Steinkellner H, Sefc KM, Reisch BI (2000) A gene controlling sex in grapevines placed on a molecular marker-based genetic map. Genome 43: 333–340.

DeBolt S, Hardie J, Tyerman S, Ford CM (2004) Composition and synthesis of raphide crystals and druse crystals in berries of *Vitis vinifera* L. cv. Cabernet Sauvignon: Ascorbic acid as

precursor for both oxalic and tartaric acids as revealed by radiolabelling studies. Aust J Grape Wine Res 10: 134–142.
DeBolt S, Cook DR, Ford CM (2006) L-tartaric acid synthesis from vitamin C in higher plants. Proc Natl Acad Sci USA 103: 5608–5613.
DeLattin G (1957) On the genetics of grapevines. Present results of the factor analysis of Vitis. Vitis 1: 1–8
Dettweiler E, Jung A, Zyprian E, Töpfer R (2000) Grapevine cultivar Müller-Thurgau and its true to type descent. Vitis 39: 63–65.
Di Gaspero G, Cipriani G, Marrazzo MT, Andreetta D, Prado Castro JM, Peterlunger E, Testolin E (2005) Isolation of (AC)*n*-microsatellites in *Vitis vinifera* L. and analysis of genetic background in grapevines under marker-assisted selection. Mol Breed 15: 11–20.
Di Gaspero G, Cipriani G, Adam-Blondon AF, Testolin R (2007) Linkage maps of grapevine displaying the chromosomal locations of 420 microsatellite markers and 82 markers for *R*-gene candidates. Theor Appl Genet 114: 1249–1263.
Doligez A, Bouquet A, Danglot Y, Lahogue F, Riaz S, Meredith CP, Edwards KJ, This P (2002) Theor Appl Genet 105: 780–795.
Doligez A, Adam-Blondon AF, Cipriani G, Di Gaspero G, Laucou V, Merdinoglu D, Meredith CP, Riaz S, Roux C, This P (2006a) An integrated SSR map of grapevine based on five mapping populations. Theor Appl Genet 113: 369–382.
Doligez A, Audiot E, Baumes R, This P (2006b) QTLs for muscat flavor and monoterpenic odorant content in grapevine (*Vitis vinifera* L.) Mol Breed 18: 109–125.
Donald TM, Pellerone F, Adam-Blondon AF, Bouquet A, Thomas MR, Dry IB (2002) Identification of resistance gene analogs linked to a powdery mildew resistance locus in grapevine Theor Appl Genet 104: 610–618.
Dzheneev SY, Melkonyan MV, Risovannaya VI, Polulyakh AA (1998) Analysis of interspecies hybrids of grapes by isoenzyme spectra. Tsitol Genet 32: 64–68.
Eibach R, Hastrich H, Töpfer R (2003) Inheritance of aroma compounds. In: E Hajdu, A Borbas (eds) Proc 8th Int Conf Grape Genet Breeding, Kecskemét, Hungary. Acta Hort 603: 337–344.
Eibach R, Zyprian E, Welter L, Töpfer R (2007) The use of molecular markers for pyramiding resistance genes in grapevine breeding. Vitis 46: 120–124.
Fischer BM, Salakhutdinov I, Akkurt M, Eibach R, Edwards KJ, Töpfer R, Zyprian EM (2004) Quantitative trait locus analysis of fungal disease resistance factors on a molecular map of grapevine. Theor Appl Genet 108: 501–515.
Franks T, Botta R, Thomas MR (2002) Chimerism in grapevines: Implications for cultivar identity, ancestry and genetic improvement. Theor. Appl Genet 104: 192–199.
Fraley RT, Rogers SG, Horsch RB, Sanders PR, Flick JS, Adams SP, Bittner ML, Brand LA, Fink CL, Fry JS, Galluppi GR, Goldberg SB, Hoffmann NL, Woo SC (1983) Expression of bacterial genes in plant cells. Proc Natl Acad Sci USA 80: 4803–4807.
Franks T, He DG, Thomas M. (1998) Regeneration of transgenic *Vitis vinifera* L. Sultana plants: Genotypic and phenotypic analysis. Mol Breed 4: 321–333.
Grando MS, Sevini F, Moser S, Marino R, Dalla Serra A, Versini G (2004) Genetic mapping of aroma compounds in grape. Bull OIV 77: 503–514.
Harst M, Cobanov A, Hausmann L, Eibach E, Töpfer R (2009) Evaluation of pollen dispersal and cross pollination using transgenic grapevines. Environ Biosaf Res 8: 87–99.
He P, Wang G (1986) Studies on the resistance of wild *Vitis* species native to China to downy mildew, *Plasmopara viticola*. Acta Hort Sin 13: 17–24.
Hedrick UP, Anthony RD (1915) Inheritance of certain characters of grapes. J. Agri. Res. 4: 315–330.
Herrera-Estrella L, Block D, Messens E, Hernalsteens JP, Montagu M van, Schell J (1983) Chimeric genes as dominant selectable markers in plant cells. EMBO J 2: 987–995.
Hocquigny S, Pelsy F, Dumas V, Kindt S, Heloir MC, Merdinoglu D (2004) Diversification within grapevine cultivars goes through chimeric states. Genome 47: 579–589.

Hoffmann S, Di Gaspero G, Kovács L, Howard S, Kiss E, Galbács Z, Testolin R, Kozma P (2008) Resistance to *Erysiphe necator* in the grapevine 'Kishmish vatkana' is controlled by a single locus through restriction of hyphal growth. Theor Appl Genet 116: 427–438.

Husfeld B (1962) Reben. In: H Kappert, W Rudorf (eds) Handbuch der Pflanzenzüchtung VI. Paul Parey, Berlin und Hamburg, Germany, pp 723–773.

Jaillon O, Aury JM, Noel B, Policriti A, Clepet C, Casagrande A, Choisne N, Aubourg S, Vitulo N, Jubin C, Vezzi A, Legeai F, Hugueney P, Dasilva C, Horner D, Mica E, Jublot J, Poulain J, Bruyere C, Billault A, B. Segurens B, Gouyvenoux M, Ugarte E, Cattonaro F, Anthouard V, Vico V, Del Fabbro C, Alaux M, Di Gaspero G, Dumas V, Felice N, Paillard S, Juman I, Moroldo M, Scalabrin S, Canaguier A, Le Clainche I, Malacrida G, Durand E, Pesole G, Laucou V, Chatelet P, Merdinoglu D, Delledonne M, Pezzotti M, LecharnyA, Scarpelli C, Artiguenave F, Pè ME, Valle G, Morgante M, Caboche M, Adam-Blondon AF, Weissenbach J, Quétier F, Wincker P (2007) The grapevine genome sequence suggests ancestral hexaploidization in major angiosperm phyla. Nature 449: 463–467.

Kalaitzandonakes N, Alston JM, Bradford KJ (2007) Compliance costs for regulatory approval of new biotech crops. Nat Biotechnol. 25: 509–511.

Kikkert JR, Hebert-Soule D, Wallace PG, Striem MJ, Reisch BI (1996) Transgenic plantlets of Chancellor grapevine, *Vitis* sp., from biolistic transformation of embryogenic cell suspensions. Plant Cell Rep 15: 311–316.

Le Cunff L, Fournier-Level A, Laucou V, Vezzulli S, Lacombe T, Adam-Blondon AF, Boursiquot JM, This P (2008) Construction of nested genetic core collections to optimize the exploitation of natural diversity in *Vitis vinifera* L. subsp. *sativa*. BMC Plant Biol 8: 1471–2229.

Le Gall O, Torregrosa L, Danglot Y, Candresse T, Bouquet A (1994) *Agrobacterium*-mediated genetic transformation of grapevine somatic embryos and regeneration of transgenic plants expressing the coat protein of grapevine chrome mosaic nepovirus, GCMV. Plant Sci 102: 161–170.

Levadoux L (1946) Etude de la fleur et de la sexualité chez la vigne. Ann Ec natl Agric de Montpellier 27: 1–89.

Lijavetzky D, Cabezas JA, Ibáñez A, Rodríguez V, Martínez-Zapater JM (2007) High throughput SNP discovery and genotyping in grapevine (*Vitis vinifera* L.) by combining a re-sequencing approach and SNPlex technology. BMC Genom 8 (424): 11.

Lodhi MA, Daly MJ, Ye GN, Weeden NF, Reisch BI (1995) A molecular marker based linkage map of *Vitis*. Genome 38: 786–794.

Lowe KM, Walker MA (2006) Genetic linkage map of the interspecific grape rootstock cross Ramsey (*Vitis champinii*) x Riparia Gloire (*Vitis riparia*). Theor Appl Genet 112: 1582–1592.

Manty F (2006) Hintergründe zur Entstehung der Bezeichnung der Unterlagenselektionen von Sigmund Teleki und Franz Kober. Dt Weinbau-Jahrbuch 57: 159–164.

Martinelli L, Candioli E, Costa D, Minafra A (2002) Stable insertion and expression of the movement protein gene of Grapevine Virus A (GVA) in grape (*Vitis rupestris* S.) Vitis 41: 189–193.

Mohan M, Nair S, Bhagwat A, Krishna TG, Yano M, Bhatia CR, Sasaki T (1997) Genome mapping, molecular markers and marker-assisted selection in crop plants. Mol Breed 3: 87–103.

Molnar S, Galbacs Z, Hasasz G, Hoffmann S, Kiss E, Kozma P, Veres A, Galli Z, Szöke A, Heszky L (2007) Marker assisted selection (MAS) for powdery mildew resistance in a grapevine hybrid family. Vitis 46: 212–213.

Negi SS, Olmo HP (1971) Conversion and determination of sex in *Vitis vinifera* L. (*sylvestris*). Vitis 9: 265–279.

Pauquet J, Bouquet A, This P, Adam-Blondon AF (2001) Establishment of a local map of AFLP markers around the powdery mildew resistance gene *Run*1 in grapevine and assessment of their usefulness for marker assisted selection. Theor Appl Genet 103: 1201–1210.

Peleman JD, Sorensen AP, van der Voort. JR (2005) Breeding by design: Exploiting genetic maps and molecular markers through marker-assisted selection. In: K Meksem, G Kahl (eds) The Handbook of Plant Genome Mapping. Wiley, Weinheim, Germany, pp 109–129.

Perl A, Eshdat Y (1998) DNA transfer and gene expression in transgenic grapes. Biotechnol Genet Eng Rev 15: 365–386.

Pindo M, Vezzulli S, Coppola G, Cartwright DA, ZharkikhA, Velasco R, Troggio M (2008) SNP high-throughput screening in grapevine using the SNPlex™ genotyping system. BMC Plant Biol 8 (12): 6.

Rapp A (1988) Volatile flavour of wine: Correlation between instrumental analysis and sensory perception. Nahrung 42: 351–363.

Riaz S, Dangl GS, Edwards KJ, Meredith CP (2004) A microsatellite marker based framework linkage map of *Vitis vinifera* L. Theor Appl Genet 108: 864–872.

Riaz S, Krivanek AF, Xu K, Walker MA (2006) Refined mapping of the Pierce's disease resistance locus, *PdR*1, and *Sex* on an extended genetic map of *Vitis rupestris* x *V. arizonica*. Theor Appl Genet 113: 1317–1329.

Riaz S, Tencher AC, Rubin J, Graziani R, Pao SS, Walker MA (2008) Fine-scale genetic mapping of two Pierce's disease resistance loci and a major segregation distortion region on chromosome 14 of grape. Theor Appl Genet 117: 671–681.

Rommens CM, Haring MA, Swords K, Davies HV, Belknap WR (2007) The intragenic approach as a new extension to traditional plant breeding. Trends Plant Sci 12: 397–403.

Roush TL, Granett J, Walker MA (2007) Inheritance of gall formation relative to Phylloxera resistance levels in hybrid grapevines. Am J Enol Vitic 58: 234–241.

Schmid J, Sopp E, Rühl EH (1998) Breeding rootstock varieties with complete Phylloxera resistance. Acta Hort 473: 131–135.

Schmid J, Manty F, Rühl EH (2003) Utilizing the complete Phylloxera resistance of *Vitis cinerea* Arnold in rootstock breeding. In: E Hajdu, A Borbas (eds) Proc 8th Int Conf Grape Genet Breeding, Kecskemét, Hungary. Acta Hort 603: 393–400.

Schmid J, Manty F, Cousins P (2007) Auf der Suche nach *Vitis berlandieri* in ihrem Ursprungsgebiet in Texas. Dt. Weinbau-Jahrbuch 58: 146–152.

Schouten HJ, Krens FA, Jacobsen E (2006) Cisgenic plants are similar to traditional bred plants. EMBO Rep 7: 750–753.

Scorza R, Cordts JM, Ramming DW, Emershad RL (1995)Transformation of grape (*Vitis vinifera* L.) zygotic-derived somatic embryos and regeneration of transgenic plants. Plant Cell Rep 14: 589–592.

Sefc KM, Steinkellner H, Glössl J, Kampfer S, Regner F (1998) Reconstruction of a grapevine pedigree by microsatellite analysis. Theor Appl Genet 97: 227–231.

Shiraishi S, Ohmi C, Wakana A, Hiramatsu M (1994) Variation of glucosephosphate isomerase and phosphoglucomutase isozymes in *Vitis* and their use in grape breeding. J Fac Agri Kyushu Univ 38: 255–272.

This P, Jung A, Boccacci P, Borrego J, Botta R, Costantini L, Crespan M, Dangl GS, Eisenheld C, Ferreira-Monteiro F, Grando S, Ibanez J, Lacombe T, Laucou V, Magalhaes R, Meredith CP, Milani N, Peterlunger E, Regner F, Zulini L, Maul E (2004) Development of a standard set of microsatellite reference alleles for identification of grape cultivars. Theor Appl Genet 109: 1448–1458.

Töpfer R (2000) Genetic engineering—new horizons for grapevine breeding. AIDV/IWLA Bull 22: 36–40.

Valleau WD (1915) Inheritance of sex in the grape. The American Naturalist, vol L No 597: 554–564.

Velasco R, Zharkikh A, Troggio M, Cartwright DA, Cestaro A, Pruss D, Pindo M, FitzGerald LM, Vezzulli S, Reid J, Malacarne G, Iliev D, Coppola G, Wardell B, Micheletti D, Macalma T, Facci M, Mitchell JT, Perazzolli M, Eldredge G, Gatto P, Oyzerski R, MorettoM, Gutin N, Stefanini M, Chen Y, Segala C, Davenport C, Demattè L, Mraz A, Battilana J, Stormo K, Costa F, Tao Q, Si-Ammour A, Harkins T, Lackey A, Perbost C, Taillon B, Stella A, Solovyev V, Fawcett JA, Sterck L, Vandepoele K, Grando SM, Toppo S, Moser C, Lanchbury

J, Bogden R, Skolnick M, Sgaramella V, Bhatnagar SK, Fontana P, Gutin A, Van de Peer Y, Salamini F, Viola R (2007) A high quality draft consensus sequence of the genome of a heterozygous grapevine variety. PLoS One 2: e1326.

Vezzulli S, Micheletti D, Riaz S, Pindo M, Viola R, This P, Walker MA, Troggio M, Velasco R (2008a) A SNP transferability survey within the genus *Vitis*. BMC Plant Biol 8 (128): 10.

Vezzulli S, Troggio M, Coppola G, Jermakow A, Cartwright D, Zharkikh A, Stefanini M, Grando MS, Viola R, Adam-Blondon AF, Thomas M, This P, Velasco R (2008b) A reference integrated map for cultivated grapevine (*Vitis vinifera* L.) from three crosses, based on 283 SSR and 501 SNP-based markers. Theor Appl Genet 117: 499–511.

Viss WJ, Driver JA (1996) Key grapevine pests and diseases in North America and resistance through genetic engineering. British Crop Protection Council Monograph 1–3: 125–130.

Wan Y, Schwaninger H, He P, Wang Y (2007) Comparison of resistance to powdery mildew and downy mildew in Chinese wild grapes. Vitis 46: 132–136.

Wang Q, Li P, Hanania U, Sahar N, Mawassi M, Gafny R, Sela I, Tanne E, Perl A (2005) Improvement of *Agrobacterium*-mediated transformation efficiency and transgenic plant regeneration of *Vitis vinifera* L. by optimizing selection regimes and utilizing cryopreserved cell suspensions. Plant Sci 168: 565–571.

Weeden NF, Hemmat M, Lawson DM, Lodhi M, Bell RL, Manganaris AG, Reisch BI, Brown SK, Ye GN (1994) Development and application of molecular marker linkage maps in woody fruit crops. Euphytica 77: 71–75.

Welter L, Göktürk-Baydar N, Akkurt M, Maul E, Eibach R, Töpfer R, Zyprian E (2007) Genetic mapping and localization of quantitative trait loci affecting fungal disease resistance and leaf morphology in grapevine (*Vitis vinifera* L.). Mol Breed 20: 359–374.

Wiedemann-Merdinoglu S, Prado E, Coste P, Dumas V, Butterlin G, Bouquet A, Merdinoglu D (2006a) Genetic analysis of resistance to downy mildew from *Muscadinia rotundifolia*. 9th Int Conf Grape Genet Breeding, 2–6 July 2006, Udine, Italy.

Wiedemann-Merdinoglu S, Prado E, Schneider C, Coste P, Onimus C, Dumas V, Butterlin G, Bouquet A, Merdinoglu D (2006b) Resistance to downy mildew derived form *Muscadinia rotundifolia*: genetic analysis and use of molecular markers for breeding, 28. In: I Pertot, C Gessler D, Gadoury W Gubler, H-H Kassemeyer, P Magarey (eds) Proc 5th Int Workshop on Grapevine Downy and Powdery Mildew, 18–23 June 2006, San Michele all'Adige, Italy.

Xu K, Riaz S, Roncoroni NC, Jin Y, Hu R, Zhou R, Walker MA (2008) Genetic and QTL analysis of resistance to *Xiphinema index* in a grapevine cross. Theor Appl Genet 116: 305–311.

Zhang J, Hausmann L, Eibach R, Welter L, Töpfer R, Zyprian E (2009) A framework map from grapevine Gf.V3125 (*Vitis vinifera* 'Schiava grossa' x 'Riesling') x rootstock cultivar 'Börner' (*Vitis riparia* x *Vitis cinerea*) to localize genetic determinants to Phylloxera root resistance. Theor Appl Genet 119: 1039–51.

Zyprian E (1998) Plant breeding: Genetic mapping in woody crops. Progr Bot 60: 167–189.

8

Positional Cloning of Disease Resistance Genes in Grapevine

Claire Anderson,[1,#] *Nathalie Choisne,*[2] *Anne-Françoise Adam-Blondon*[3] and *Ian B. Dry*[1,*]

ABSTRACT

The introduction of genes that confer resistance to major grapevine pests and pathogens into the highly susceptible *V. vinifera* cultivars used for wine production worldwide would be of significant economic and environmental benefit. The availability of genetic resources for grapevine has led to a dramatic increase in research on the mapping of resistance loci. Often the outcome of this work is the development of markers for marker-assisted selection of new disease-resistant inter-specific hybrids. However, positional cloning of these resistance genes offers the possibility to introduce these genes into existing elite wine grape cultivars by grapevine transformation without affecting wine quality which has been a major obstacle in the adoption of hybrid vines generated by classical breeding. Furthermore positional cloning offers the possibility to introduce resistance genes into *V. vinifera* cultivars from species which cannot be achieved through hybridization. Finally, cloning of resistance genes provides sequence information that can be used to design perfect genetic markers which will maximize the efficiency of marker assisted selection approaches. This chapter describes the strategy used, and technical challenges encountered, during the mapping and positional cloning of powdery mildew and

[1]CSIRO Plant Industry, Waite Campus, Hartley Grove, Urrbrae, SA 5060, Australia.
[2]Genoscope IG-CEA, UMR CNRS-INRA-Univ Evry Research in Plant Genomics, 2 rue Gaston Crémieux, BP 5708, 91 057 Evry cedex, France.
[3] UMR INRA UEVE ERL CNRS Génomique Végétale, 2 rue Gaston Crémieux, BP 5708, 91 057, Evry cedex, France.
*Corresponding author: *ian.dry@csiro.au*
[#]*Present address*: School of Forest Resources and Conservation, University of Florida, Gainesville, FL 2611, USA.

downy mildew resistance genes from the wild North American grape species *Muscadinia rotundifolia*.

Keywords: powdery mildew, downy mildew, resistance gene, positional cloning, *Muscadinia rotundifolia*

8.1 Introduction

The identification of plant genes by positional cloning is a lengthy process that requires a number of steps to be performed. Firstly, a mapping population must be constructed in which the gene of interest segregates in a genetically defined ratio. The mapping population can then be used to identify genetic markers linked to the locus of interest, enabling construction of a local genetic map. Conversion of the genetic map into a physical map requires a large-insert genomic library that can be used for chromosome walking or, as has become possible in recent years, the availability of a complete genome sequence. Finally, candidate genes identified in the interval between the genetic markers most closely flanking the locus of interest, must be tested by complementation to identify the functional gene.

Due to the considerable effort required, the positional cloning projects initiated in the late 20th century concentrated on largely important agricultural traits, such as those conferring disease resistance to agriculturally important plant pathogens. The first plant gene identified by positional cloning was the *Pto* gene of tomato (Martin et al. 1993) and its identification required the time-consuming development of introgressed genetic stocks and extensive marker development (Bortiri et al. 2006). *Pto* is a single dominant gene that confers resistance to the bacterial speck pathogen, *Pseudomonas syringae* pv. *tomato* and is an example of a gene that confers disease resistance in a "gene-for-gene" manner. Gene-for-gene resistance (Flor 1971) is a highly effective mechanism that relies on the direct, or indirect interaction, of the products encoded by a resistance gene in the plant and an avirulence gene in the pathogen. The interaction is highly specific and results in the activation of a suite of plant defence responses that culminate in localized cell death, called a hypersensitive response, at the site of attempted infection (Mur et al. 2008). This mechanism is most effective against biotrophic pathogens that require living tissue upon which to survive and gene-for-gene interactions have been described between plants and a wide variety of pathogen types, including bacteria, viruses, fungi, oomycetes, nematodes and aphids. Due to their strong effect and single gene basis, the isolation of plant resistance genes was a major goal of early positional cloning efforts.

Since the identification of *Pto*, over 50 plant resistance genes have been mapped and cloned from a range of plant species (Martin et al. 2003; van Ooijen et al. 2007) including woody perennials such as apple (Belfanti et al. 2004; Malnoy et al. 2008). However, we now recognize that gene-for-gene resistance is the second of two distinct mechanisms that plants use to defend themselves against pathogens. The first layer of defence is termed basal immunity, or non-host resistance, and comprises of both the preformed chemical and physical barriers that prevent pathogen penetration and the inducible immune response that can be triggered upon plant recognition of microbe-associated molecular patterns (MAMPs; reviewed in Bent and Mackey 2007). The genetic basis of resistance mediated by preformed chemical and physical barriers is likely to be complex, involving multiple loci. However, MAMP recognition is achieved by plant receptor-like kinase proteins (Chinchilla et al. 2006; Zipfel et al. 2006) that have a simple, single gene basis. Unlike avirulence (effector) proteins, which are only produced by certain races of the pathogen population, MAMPs are generic molecules forming a core component of the microorganism that cannot be sacrificed or altered significantly without seriously impairing viability such as the chitin which makes up fungal cell walls (Bittel and Robatzek 2007). Receptor activation results in cytoskeletal rearrangements, callose deposition at the site of infection and the production of antimicrobial compounds, which can be an effective form of defence against both biotrophic pathogens that require living tissue upon which to survive and necrotrophic pathogens that feed off dead plant cells. Identification of the genes controlling these two important plant defence mechanisms could prove useful in strategies to improve the disease resistance of crop plants.

The threat of disease is of great concern to grapevine growers throughout the world. The European grape species, *Vitis vinifera*, is the most widely cultivated grapevine and produces berries of excellent quality. However, this species is highly susceptible to all major grapevine pathogens, including powdery mildew, caused by the fungus *Erysiphe necator* (syn. *Uncinula necator*), downy mildew, caused by the oomycete *Plasmopara viticola*, Pierce's disease, caused by the bacterium *Xyllela fastidiosa*, and phylloxera, caused by the insect *Dactylosphaeria vitifolia* (reviewed in Riaz et al. 2007) with only one known exception (Hoffmann et al. 2008). These, and other grapevine pathogens, have caused the near disappearance of viticulture in some areas of the world (Galet 1996) and are a major barrier to the growth of European grapevine species in areas such as the south-eastern USA, and Central and South America. Where grapevine production has persisted, the threat of disease is so great that extensive use of chemical fungicides is required, particularly to control the powdery and downy mildews. For example, a study of fungicide use in European Union member states (Phytowelt

GmbH 2003) concluded that while grapevines accounted for only ~ 8% of total crop production area, they accounted for ~ 70% of all fungicides applied. This equates to costs of 150 and 75 million euros in France alone, for the chemicals used to control downy and powdery mildew, respectively, and does not take into account the increasing fuel costs associated with the application of these fungicides. Furthermore, fungal strains have been identified that have evolved resistance to a number of the commonly used fungicides (Erickson and Wilcox 1997; Savocchia et al. 2004; Baudoin et al. 2008) and it is now considered that the widespread use of agrochemicals is both environmentally detrimental and may pose a risk to human health. Consequently, the identification of genes that confer resistance to these pathogens and their incorporation into *V. vinifera* would be of significant economic and environmental benefit.

The susceptible nature of *V. vinifera* might be explained by the fact that this species arose in Europe and Asia, whereas many of the major grapevine pathogens such as powdery mildew, downy mildew and phylloxera are endemic to North America and were only introduced into Europe in the mid-19th century (Boubals 1958, 1961, 1966). Therefore, *V. vinifera* has only been exposed to them for an extremely short period of time in evolutionary terms. In contrast, many grapevine species that are endemic to North America, display varying levels of resistance to these diseases. Although the potential exists to transfer resistance from these American relatives into *V. vinifera* cultivars, this has proven difficult to achieve by conventional breeding. One major constraint is that the North American grape species often have poor berry characteristics and, when hybrids have been generated, they lack the high berry quality of the original *V. vinifera* cultivar. The second constraint is that it is difficult to produce hybrids between *V. vinifera* and some of the most resistant American grapevines. For example, *Muscadinia rotundifolia*, a wild grapevine native to the south-eastern USA, is highly resistant to a number of pathogens known to affect cultivated grapevines, including powdery mildew, downy mildew, phylloxera and nematodes (Olmo 1986) and, as such, is a rich source of potential disease resistance genes. However, *M. rotundifolia* is taxonomically separated from Euvitis species by anatomical and morphological characteristics (Planchon 1887) and a difference in chromosome number (*Vitis* $2n = 38$, *Muscadinia* $2n = 40$; Branas 1932) and when the first attempts to produce interspecific crosses between *M. rotundifolia* and *V. vinifera* were performed in the mid-1800s they achieved only limited success. For these two reasons, the transfer of resistance genes into cultivated varieties by conventional breeding is still a major challenge for grapevine geneticists (Bisson et al. 2002). Consequently, a number of groups around the world are now interested in studying the North American grapevine species to determine the genetic basis of their disease resistance (Pauquet et al. 2001; Akkurt et al. 2006; Wiedemann-Merdinoglu et al. 2006; Welter et al. 2007; Chapter 1 Section 1.5.6) as this could lead to novel control methods by generating genetically modified grapevine cultivars.

8.2 Positional Cloning of the *Run1* and *Rpv1* Genes

8.2.1 Development of Mapping Populations

In 1998, we initiated a project to isolate two important resistance genes from *M. rotundifolia* by positional cloning. Despite the problems inherent in using *M. rotundifolia* as a source of resistance in grapevine breeding programs, Detjen (1919) was successful in producing authentic hybrids using *V. vinifera* cv. Malaga as the female parent. This paved the way for construction of modern mapping populations, which is a crucial first step in any positional cloning effort. One of the hybrids produced, NC6-15, was used as the resistant parent in a series of pseudo-backcrosses (Bouquet 1986; Fig. 8-1). At each backcross step, a resistant individual was crossed with a different susceptible *V. vinifera* genotype to prevent inbreeding depression. This resulted in a number of populations from the fifth and sixth backcross generations in which resistance to powdery and downy mildew segregated in a 1:1 ratio. Complete resistance to powdery mildew was controlled by a single, dominant locus named *Run1* (for resistance to *Uncinula necator* 1) and partial resistance to downy mildew was controlled by a major quantitative trait locus (QTL) named *Rpv1* (for resistance to *Plasmopora viticola* 1; Bouquet 1986; Pauquet et al. 2001; Merdinoglu et al. 2003; Fig. 8-2). Interestingly, these two important resistance loci are co-located in the *Muscadinia* genome making them highly valuable targets that could be identified in a single positional cloning effort.

8.2.2 Isolation of Genetic Markers Linked to the Run1 and Rpv1 Loci

At the time the *Run1/Rpv1* positional cloning project was conceived, limited genetic resources were available for grapevine and very few genetic markers existed. Therefore, we used three of our mapping populations, Mtp3294, Mtp3322 and Mtp3328 (Fig. 8-1), to identify novel markers that were linked to the resistance loci (Pauquet et al. 2001; Donald et al. 2002; Barker et al. 2005). A bulked segregant analysis approach (Michelmore et al. 1991) was initially used to identify amplified fragment length polymorphism (AFLP) markers that were linked to *Run1*. Pauquet et al. (2001) constructed bulks using the DNA of 10 individuals from the Mtp3294 population known to be resistant to powdery mildew (heterozygous for *Run1*) and 10 individuals

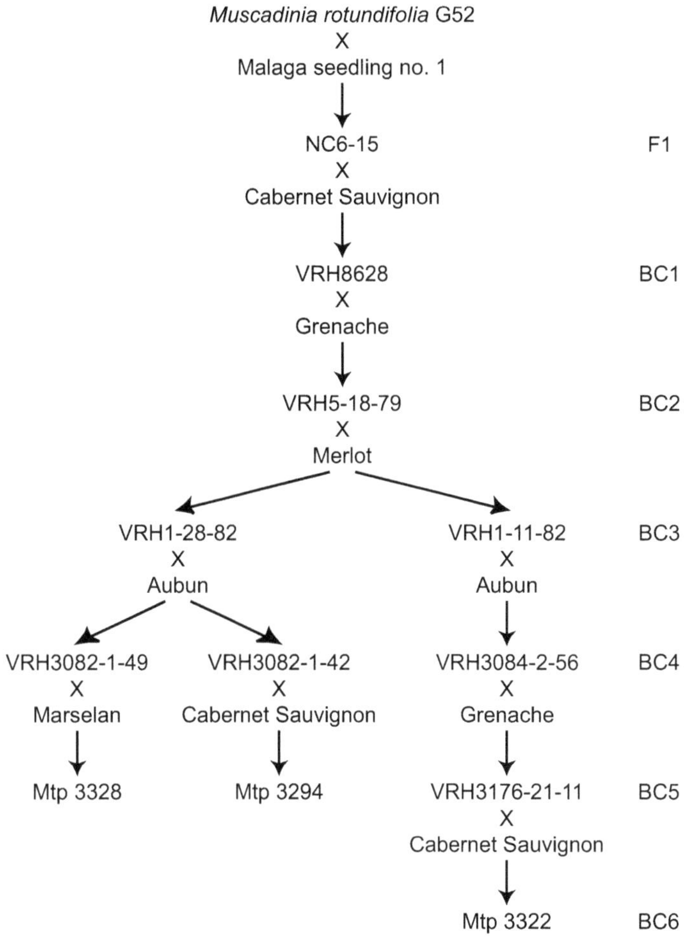

Figure 8-1 Parentage of mapping populations used to map the powdery and downy mildew resistance genes introgressed from *M. rotundifolia* cv. G52. BC_n represents the pseudo-backcross step at which the individual was obtained.

known to be susceptible. Sixty four combinations of selective primers were then used to amplify fragments from the DNA pools and 19 polymorphic fragments were identified. To confirm linkage and determine marker order, each primer pair was used to amplify fragments from the genomic DNA of 157 individuals from the Mtp3294 population that had previously been challenged with powdery mildew. All 157 individuals from the Mtp3294 population were later also screened for resistance to downy mildew (*P. viticola*). *Rpv1* was found to co-segregate with *Run1* in all resistant progeny. This approach successfully identified 14 AFLP markers that were closely linked to the *Run1*/*Rpv1* locus in VRH3082-1-42, of which 11 co-segregated with the resistance locus (Fig. 8-3).

Figure 8-2 Comparison of powdery mildew (A, B) and downy mildew (C, D) susceptibility of leaves from *V. vinifera* cv. Cabernet sauvignon (A, C) and from *M. rotundifolia* x *V. vinifera* 4th generation backcross (B, D). Powdery mildew growth is completely suppressed on the BC_4 plant but a low level of downy mildew sporulation is supported on BC_4 leaves. Patches of dead cells which have undergone programmed cell death in response to downy mildew infection are visible in panel D.

Color image of this figure appears in the color plate section at the end of the book.

To supplement the genetic map produced using AFLPs, a second class of markers was sought that are unique to the study of plant disease resistance genes. Despite the fact that they confer resistance to a wide range of pathogen types, comparison of the plant resistance genes that contribute to gene-for-gene resistance has shown that the majority encode proteins with a central nucleotide binding site (NBS) linked to a C-terminal leucine-rich repeat (LRR) domain (reviewed in Takken et al. 2006). Motifs within the NBS are highly conserved and have been used to design degenerate primers capable of amplifying novel resistance gene analogs (RGAs) from many plant species (Aarts et al. 1998; Collins et al. 1998; Shen et al. 1998). Nested primers designed to four conserved amino acid motifs present in the NBS (GVGKTT (P-loop), L(I/V/L)VLDDV (kinase-2), GLPL and MHD) were used to amplify RGAs from the genome of a BC_4 (VRH3082-1-42) resistant individual (Donald et al. 2002). Twenty eight unique grapevine RGA sequences were identified and were subdivided into 22 groups of

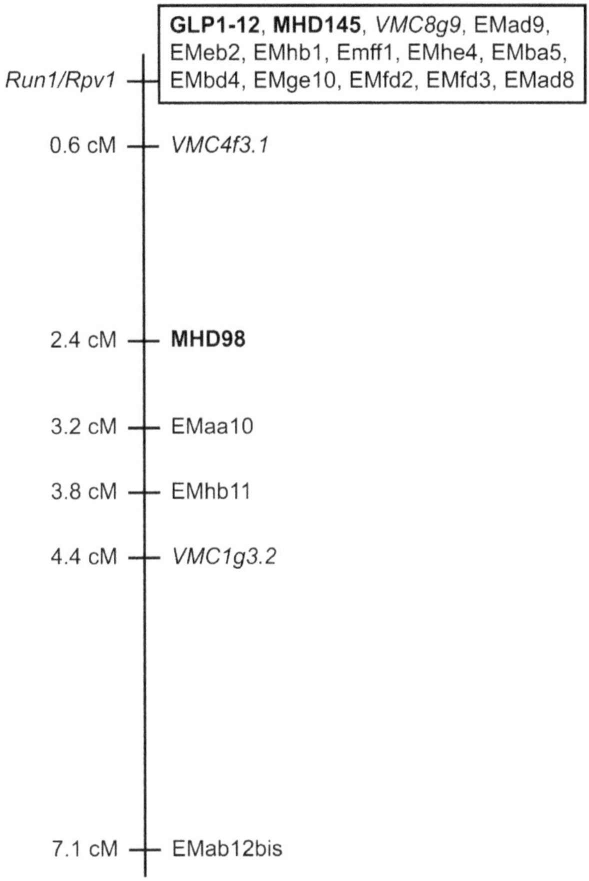

Figure 8-3 Genetic map of VRH3082-1-42 based on screening 160 individuals of Mtp3294 showing the location of SSR, AFLP and RGA markers linked to *Run1/Rpv1* locus. RGA markers are shown in bold, SSR markers in italics.

sequences that showed nucleic acid sequence identity of 70% or greater. Representatives from each group were used in a bulked segregant analysis strategy to identify restriction fragment length polymorphisms (RFLPs) linked to the powdery mildew resistance locus. This led to the identification of three markers, MHD98, MHD145 and GLP1-12, for which linkage was subsequently confirmed by analyzing the RFLP patterns produced by a selection of powdery mildew-resistant and powdery mildew-susceptible individuals from the Mtp3294 population. To facilitate screening of the entire Mtp3294 mapping population, GLP1-12 was subsequently converted into a cleaved amplified polymorphic sequence (CAPS) marker and MHD98 was converted into a dominant, PCR-based marker (Donald et al. 2002). The three markers were then examined in 167 individuals in the Mtp 3294 population, revealing that both GLP1-12 and MHD145 co-segregated with the *Run1*/*Rpv1* locus (Fig. 8-3). A similar approach was used by Di Gaspero and Cipriani (2002) to identify RGA sequences from the grape species *Vitis amurensis* and *Vitis riparia*. These species are known to display resistance to downy mildew and one of the RGA clones identified, *rgVrip064*, was linked to a downy mildew resistance locus segregating in a hybrid population.

Three further markers were added to the local genetic map surrounding *Run1* by searching for simple sequence repeat (SSR) markers that showed linkage to the resistance locus in the Mtp 3294 population (Barker et al. 2005) from those generated by the Vitis Microsatellite Consortium coordinated by Agrogène, France. Together, the data obtained by analysis of ~ 160 individuals from the Mtp3294 population generated a genetic map for the VRH3082-1-42 parent in which 14 markers co-segregated with *Run1* (Fig. 8-3).

To further refine the genetic map surrounding the *Run1*/*Rpv1* locus, an additional 419 individuals from the Mtp3322 population and 416 individuals from the Mtp3328 population were then examined (Barker et al. 2005). Mtp3322 was derived from a cross between the resistant parent, VRH3176-21-11 and the susceptible parent, *V. vinifera* cv. Cabernet Sauvignon. Mtp3328 was derived from a cross between the resistant parent, VRH3082-1-49 and the susceptible parent, *V. vinifera* cv. Marselan (Fig. 8-1). To facilitate scoring of AFLP markers, EMhb1 and EMfd3 were cloned and converted into the dominant, PCR-based marker Shb1 or the RFLP marker Sfd3, respectively. Once recombinant plants had been identified, they were phenotyped for powdery and downy mildew resistance and additional markers were scored. Two plants were identified that were recombinant between *Run1*/*Rpv1* and VMC8g9 and four plants were recombinant between *Run1*/*Rpv1* and Shb1 (Table 8-1). This allowed a marker order to be assigned in which *Run1*/*Rpv1* was located between the SSR markers, VMC4f3.1 and VMC8g9, an interval that was considered sufficiently small to enable map-based cloning of the *Run1* and *Rpv1* loci.

8.3 Physical Mapping of the *Run1/Rpv1* Locus

8.3.1 Generation of a Bacterial Artificial Chromosome library

As no genome sequence currently exists for *M. rotundifolia*, we chose to isolate the *Run1* and *Rpv1* genes using a chromosome walking approach. This required construction of a large-insert genomic library containing the allele of interest. In recent years, bacterial artificial chromosome (BAC) vectors have become the vehicle of choice for the generation of such libraries due to the large size of fragments that can be cloned and their increased stability and ease of manipulation compared to other vectors

Table 8-1 Genotype of informative recombinant plants isolated from populations Mtp3294, Mtp3322 and Mtp3328. The presence of either the resistant (+) or susceptible (–) allele of each marker is shown.

		Marker								
Population	Plant	VMC1g3.2	MHD98	VMC4f3.1	GLP1-12	*Run1/Rpv1*	Sfd3	MHD145	VMC8G9	Shb1
3294	R11	+	–	–	–	–	–	–	–	–
3294	R102	+	–	–	–	–	–	–	–	–
3294	S100	+	–	–	–	–	–	–	–	–
3294	R149	+	+	–	–	–	–	–	–	–
3294	R5	–	–	+	+	+	+	+	+	+
3294	R104	–	–	+	+	+	+	+	+	+
3294	R37	–	–	–	+	+	+	+	+	+
3328	245	–	–	–	–	–	–	–	+	+
3322	42	–	–	–	–	–	–	–	+	+
3328	174	–	–	–	–	–	–	–	–	+
3322	422	–	–	–	–	–	–	–	–	+

such as cosmids or yeast artificial chromosomes (Shizuya et al. 1992; Woo et al. 1994; Chapter 9 Section 9.2). A BAC library using the genomic DNA of a single powdery mildew-resistant plant from the Mtp3294 mapping population was generated (Barker et al. 2005). This individual was clonally propagated to generate sufficient leaf material for construction of the BAC library and megabase-sized DNA was extracted, digested with *Hind*III or *Bam*HI and size-fractionated essentially as described by Peterson et al. (2000). The library was constructed in the pINDIGOBAC-5 vector (Epicentre, Madison, WI) and individual clones were stored in 144 x 384-well plates (Barker et al. 2005). The library contained a total of 55,295 clones with an estimated average insert size of 93 kb. Less than 5% of clones did not contain an insert and approximately 1.7% of clones contained chloroplast-derived sequence as determined by hybridization of clones arrayed on a filter with the grapevine RNA polymerase B gene. Given that the grapevine genome has been estimated at 475 Mb by flow cytometry (Lodhi and Reisch 1995) and 487 Mb by genome sequencing (Jaillon et al. 2007), the library contains around 10-fold haploid genome coverage.

8.3.2 Assembly of BAC Contigs Containing Linked Genetic Markers

To initiate a physical map of the region containing the *Run1/Rpv1* locus, the BAC library was screened to identify clones containing three genetic markers that co-segregated with resistance (GLP1-12, MHD145 and Sfd3) and two flanking markers (VMC8g9 and VMC4f3.1). For VMC4f3.1, VMC8g9 and Sfd3, between seven and nine clones were isolated from the library, close to the expected representation of a single copy sequence in a library of predicted 10-fold genome coverage (Barker et al. 2005). In contrast, over 60 clones were identified that contained MHD145-hybridizing sequences and over 50 hybridized the GLP1-12 probe. This indicated that GLP1-12 and MHD145 might represent repeated elements, or gene families, present within the grapevine genome.

The individual plant chosen to generate the BAC library was heterozygous for *Run1* and *Rpv1*. Therefore, once clones containing genetic markers had been identified, they were grouped into "resistant" and "susceptible" contigs representing the chromosome containing *Run1/Rpv1* derived from the resistant parent, VRH3082-1-42, or the chromosome lacking *Run1/Rpv1* derived from the susceptible parent, *V. vinifera* cv. Cabernet Sauvignon, respectively. Comparison of restriction digest patterns was used to determine overlap between BAC clones and alignments were confirmed by PCR using primers designed from BAC end sequences. These primers were also used to amplify PCR products from a selection of resistant and susceptible genotypes in the Mtp3294 population, enabling a large number of dominant and co-dominant markers to be identified that were linked to

Run1 and *Rpv1*. Primers that amplified products of identical size from both resistant and susceptible genotypes were used to directly sequence BAC clones that contained the region of interest. Alignment of sequences enabled polymorphic nucleotides to be identified that could be utilized to generate CAPS markers. Primers were then used to isolate overlapping clones that could be used to extend BAC contigs (Fig. 8-4).

8.3.3 Assembly of BAC Contigs Containing Resistance Gene Candidates

GLP1-12 and MHD145 were originally identified as RFLP markers (Donald et al. 2002). Therefore, BACs containing these sequences were examined by RFLP analysis, enabling them to be grouped into contigs representing the resistant or susceptible chromosomes (Barker et al. 2005). Out of 49 clones examined, only two overlapping BACs were identified that contained the 2.2 kb MHD145-hybridizing *Eco*RI fragment characteristic of resistant plants. These BACs contained additional MHD145-hybridizing fragments, suggesting that each clone might contain multiple copies of a related sequence. By comparing *Eco*RI digests and RFLP patterns, other MHD145-hybridizing BACs could be aligned with those containing the fragments characteristic of the co-dominant marker and overlap between clones was confirmed by PCR amplification using primers designed from BAC end sequences. Three BACs containing MHD145-hybridizing sequences were assembled into each of the resistant and susceptible contigs. Additional markers designed using BAC end sequence information confirmed that the two contigs were haplotypes and allowed overlapping BAC clones to be identified to extend the contigs.

Out of 43 GLP1-12-hybridizing BACs examined, six contained the 1.6 kb *Eco*RI fragment characteristic of resistant plants. However, in this case the BAC clones did not directly overlap, suggesting that the hybridization signals observed upon RFLP analysis of genomic DNA might be contributed by multiple members of a gene family (Barker et al. 2005). Five out of six BACs containing the 1.6 kb *Eco*RI fragment (4M9, 26C5, 61L18, 69D12 and 101P19; Fig. 8-4) were assembled into one contig using restriction fragment digest patterns and PCR products amplified using BAC end-derived primers (Barker et al. 2005). However, the sixth BAC containing the characteristic RFLP fragment (97A19; Fig. 8-4) did not overlap the main GLP1-12-containing contig, despite the fact that primers designed using BAC end sequence from this clone (CB33.34) could be used as a PCR-based marker that co-segregated with *Run1* and *Rpv1*. Therefore, both the main GLP1-12-containing contig and 97A19 were used to initiate chromosome walks.

Extension of the main GLP1-12-containing contig allowed identification of five overlapping BAC clones (7D20, 19G13, 87P7, 130N7 and 134D9) that

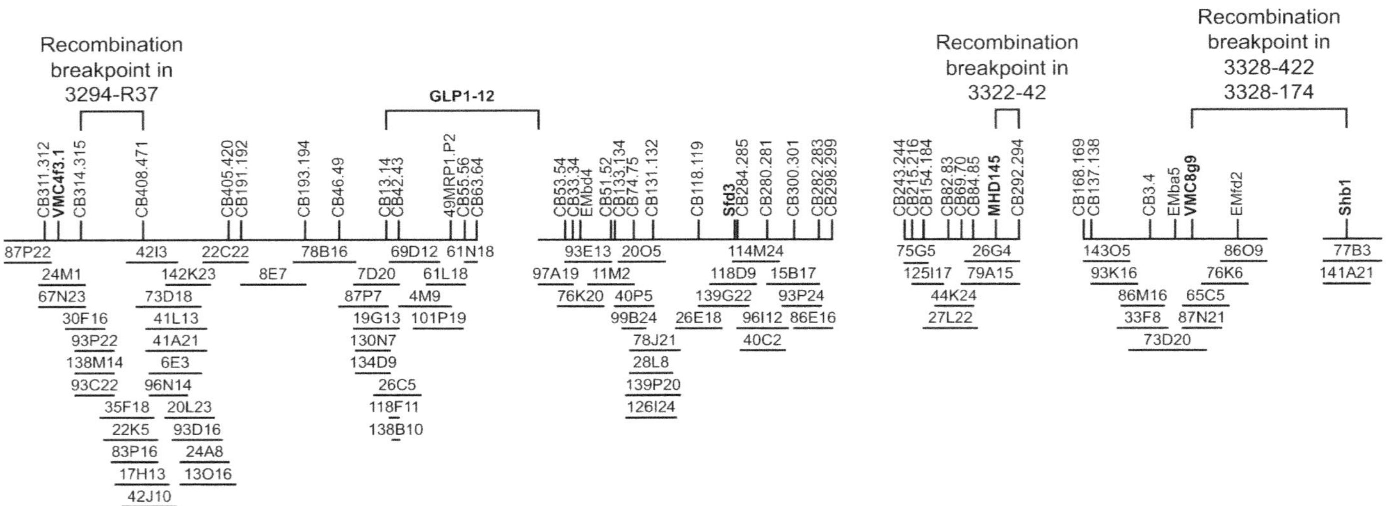

Figure 8-4 A physical map of the region surrounding the *Run1/Rpv1* locus. BAC clones were identified that contained genetic markers linked to *Run1*. Clones were assembled into contigs representing the resistant chromosome and extended by identification of overlapping BACs. Genetic markers used to initiate contigs are shown in bold and the location of multiple GLP1-12-hybridizing sequences is bracketed. Unique genetic markers that facilitated contig assembly are shown and the approximate location of the recombination breakpoints found in four plants from the Mtp3294, Mtp3322 and Mtp3328 mapping populations are indicated.

contained additional GLP1-12-hybridizing sequences. By RFLP analysis of PCR products amplified from all GLP1-12-hybridizing BACs in the contig, Barker et al. (2005) showed that this region contained at least four members of a gene family. PCR-based markers designed from BAC end sequences were used to identify additional overlapping BACs. Although none of the BACs added to the contig beyond 87P7 contained GLP1-12-hybridizing sequences, the contig eventually spanned the region between GLP1-12 and the SSR marker, VMC4f3.1, providing evidence that these two markers were adjacent in the genome.

No clones were present in our BAC library that could extend the main GLP1-12 contig beyond 61N18. Similarly, no clones were identified that overlapped the right-hand BAC end of 97A19. However, two overlapping BACs were identified using the PCR-based marker CB33.34 that had been designed from the left-hand BAC end sequence of 97A19. A further 16 clones were added to this contig, resulting in an assembly of 19 BAC clones that formed a contiguous region from 97A19 to 86E16. This region contained both the AFLP marker EMbd4, as determined using primers designed from the cloned AFLP sequence, and the RFLP marker Sfd3.

8.3.4 Genetic Mapping using BAC-derived Markers

A large number of new genetic markers were identified during the physical mapping process. Therefore, all Mtp3294, Mtp3322 and Mtp3328 progeny that had been identified as recombinant between VMC4f3.1 and Shb1 were retested using the panel of new BAC end sequence-derived markers. This identified recombination breakpoints in 3328-422 and 3328-174 between VMC8g9 and Shb1, which allowed the orientation of the VMC8g9 contig to be determined. The recombination breakpoint in 3322-42 was located between markers MHD145 and CB292.294, which allowed orientation of the MHD145 contig and confirmed its placement between Sfd3 and VMC8g9 (Barker et al. 2005). The recombination breakpoint in 3294-R37 was located between CB314.315 and CB407.471 which, when taken together with the information obtained from 3322-42, suggests that *Run1* and *Rpv1* are located in the interval between CB314.315 and CB292.294 (Fig. 8-4).

8.4 Technical Issues Associated with the Chromosome Walking Approach

8.4.1 Gaps in the BAC Library

Despite the fact that our BAC library was equipped to contain 10-fold haploid genome coverage, it proved impossible to join the three major contigs containing VMC4f3.1, GLP1-12, Sfd3 and MHD145 (Fig. 8-4).

The BAC library contained 49,920 clones generated using *Hind*III and 5,378 clones generated using *Bam*HI. Therefore, gaps in the library could have been caused by bias in restriction site distribution as might occur in centromeres, telomeres and highly repetitive regions. For this reason it has been suggested that BAC libraries should be generated from genomic DNA that has been digested with a number of different restriction enzymes (Wu et al. 2004; Suetsugu et al. 2007), or randomly sheared (Osoegawa 2007), to improve representation of problematic and repeat-rich regions at telomeres and in centromeric heterochromatin.

To supplement our BAC library we made two additional libraries using genomic DNA from the same powdery mildew-resistant plant that had been used for BAC library construction. For both libraries, nuclei were prepared according to Peterson et al. (2000). The genomic DNA was then either mechanically sheared and cloned into a pWEB::TNC™ cosmid vector (Epicentre, Madison, USA) or partially digested with *Sau*3AI and cloned into a λBlue STAR lambda vector (Novagen, Madison, USA) according to the individual manufacturer's protocols. The cosmid library contained 148,400 clones with an average insert size of 35.3 kb and the lambda library contained 395,600 clones with an average insert size of 12 kb. However, despite the fact that these two libraries contained approximately 11-fold and 10-fold genome coverage respectively, when the libraries were screened with BAC end sequences for which no overlapping BAC clones could be found, these sequences were also missing from the cosmid and lambda libraries. This suggested that the lack of representation of clones in this region might be caused by clone instability or lethality, rather than restriction enzyme bias.

Evidence of clone instability was obtained when the gap in the GLP1-12 contig was examined in more detail. When the BAC library was screened with the BAC end marker CB55.56 (derived from clone 61L18) to find overlapping BACs, only one additional clone was identified in the library. This BAC clone, 82O20, contained an insert of 31.5 kb which was significantly smaller than the average BAC library insert size of 93 kb. Upon sequencing 82O20, it became apparent that while it contained large regions of homologous sequence to 61L18, these two clones did not overlap (Fig. 8-5). The BAC library was rescreened with a new marker designed to the end of 82O20 (CB63.64) and this identified BAC clone 61N18 which was also unusually short (31.4 kb) and which again did not overlap. A further marker (CB90.488) was designed to the end of 61N18. This marker did not identify any new clones in the BAC library, but a 40 kb clone (Cos106A) was identified in the cosmid library. Sequence comparison of BAC clones 61L18, 61N18, 82O20 and cosmid clone Cos106A indicated the presence of large regions of DNA sequence that is locally repeated in the *M. rotundifolia*-derived introgression (Fig. 8-5).

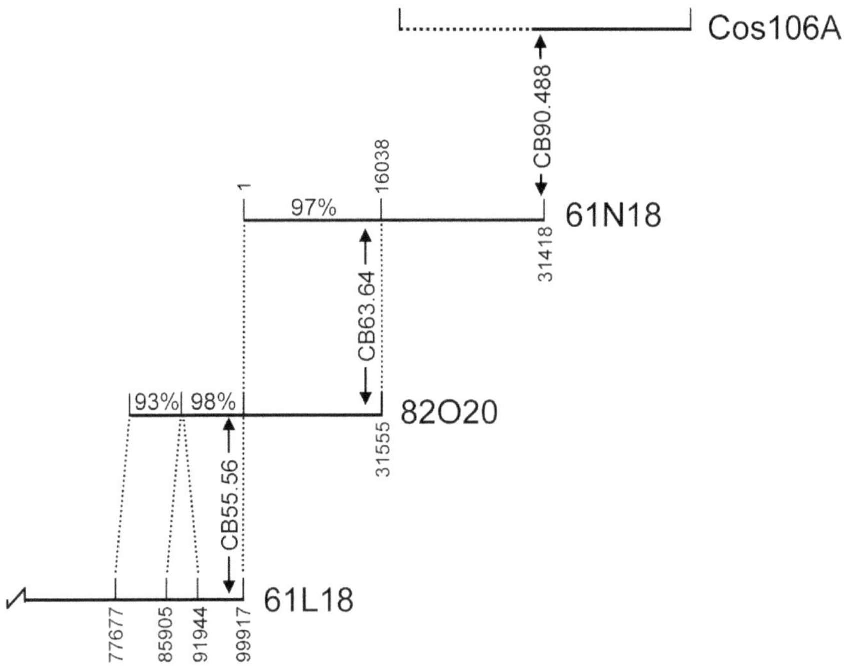

Figure 8-5 Homology of BAC and cosmid clones that contain repeated DNA. The identity and position of primers used to screen for overlapping clones (as outlined in the text) are shown by arrows. The percentage nucleotide identity and position of homologous regions within the clones are indicated. Only approximately half of Cos106A (solid line) could be successfully sequenced. Primers CB90.488 amplified a 1kb fragment from Cos106A that was 99% identical to the sequence from 61N18. BAC clones 61L18 and 61N18 were subsequently demonstrated to be adjacent within the genome as shown in Fig. 8-4.

The small insert size of BAC clones 82O20 and 61N18 compared to the average insert size of our library also suggested that these clones contained sequences that have caused rearrangement and deletion during culture. Cosmid clone Cos106A also proved to be highly unstable and the estimated 40 kb insert in this clone would spontaneously and unpredictably revert to an approximately 15 kb insert in culture. Sequencing of this clone proved to be extremely difficult and assembly was hampered by the presence of numerous repeated sequences, including a partial tandem repeat 45S rDNA unit (Fig. 8-6). Eukaryotes generally have many copies of rDNA organized in tandem repeats (Lafontaine and Tollervey 2006) and it has been shown that BAC clones containing tandemly repeated DNA elements are not stable and can undergo drastic deletions during routine library maintenance and DNA preparation (Song et al. 2001). The presence of these tandem repeat units within the genome may therefore explain at least some of the gaps observed in the contig assembly.

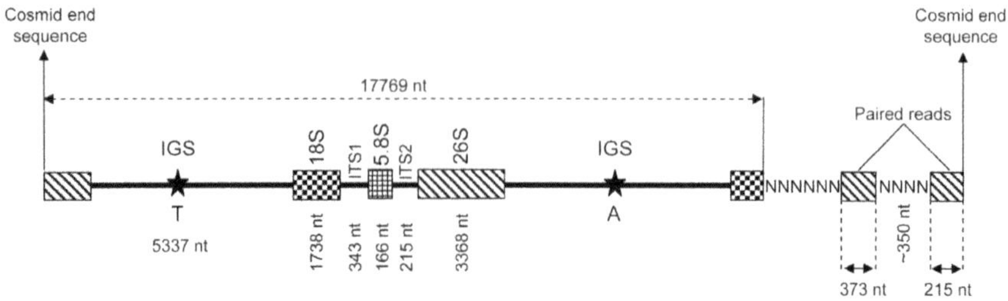

Figure 8-6 Sequence assembly of Cos106A using PHRED, PHRAP and in-house Genoscope programs taking into account paired reads and a manual curation of the repeated sequences. The sequence obtained corresponded to a partial tandem repeat of a 45S rDNA unit. N's indicate where joining sequence could not be resolved. Abbreviations: IGS: InterGenic Spacer; ITS: Internally transcribed spacer.

8.4.2 Lack of Correlation between Physical and Genetic Distance

Our genetic mapping studies indicated that the *Run1/Rpv1 locus* is located in the interval between the SSR markers, VMC4f3.1 and VMC8g9. These two markers were placed just 0.6 cM apart in the local genetic map of VRH3082-1-42, the resistant parent of the Mtp3294 mapping population (Fig. 3; Barker et al. 2005), suggesting that either the markers were located physically close to one another in the grapevine genome, or that recombination rates were unusually low in this region. Riaz et al. (2004) and Adam-Blondon et al. (2004) have produced genome linkage maps for *V. vinifera* that span 1,728 cM and 2,200 cM, respectively. Given that the grapevine genome has been estimated at 475–487 Mb (Lodhi and Reisch 1995; Jaillon et al. 2007), this predicts that on average 1 cM is equivalent to 216–282 kb. Thus, the interval between VMC4f3.1 and VMC8g9 should be approximately 130–169 kb. However, as contigs spanning well over 1 Mb were constructed, this is clearly not the case. Indeed, the actual estimated physical distance between VMC4f3.1 and VMC8g9, based on alignments of the sequenced BAC contigs with the current version of the *Vitis* genome sequence (improved from Jaillon et al. 2007) is approximately 5.4 Mbp (N. Choisne, unpublished results).

Reduced recombination frequencies have been observed around resistance loci in wheat, barley and poplar where resistance has been introduced via an introgressed region from a related species (Wei et al. 1999; Stirling et al. 2001; Neu et al. 2002). Therefore, the dramatically reduced recombination rate in the vicinity of *Run1* might be due to the presence of divergent DNA contributed by the introgression from *M. rotundifolia*. The implication of low recombination frequencies in this region for future isolation of the *Run1* gene is that large populations may be required in order to obtain sufficient recombinant individuals to delineate the resistance locus with suitable precision.

8.5 Fine Mapping of the *Run1/Rpv1* Locus and Functional Evaluation of Candidate Resistance Genes

As the physical distance between VMC4f3.1 and VMC8g9 was clearly much greater than the genetic distance had suggested, extended mapping populations were examined to identify rare recombinants that would allow the *Run1* and *Rpv1* loci to be delineated with greater precision. To achieve this, a further 3,300 progeny of the Mtp3294 and Mtp3322 populations (Fig. 8-1) were generated and screened for recombination between the markers 49MRP1.P2, CB69.70 and CB137.138 (Fig. 8-4) using a high-throughput multiplex PCR system. This identified 21 plants in which a recombination event was observed between the markers of interest. The reaction of these recombinant plants to powdery and downy mildew was determined, allowing the location of *Run1* and *Rpv1* to be localized to a ~ 1 Mb region of introgressed DNA. Sequencing of this region has revealed the presence of a cluster of 11 resistance gene analogs (RGAs) encoding TIR-NBS-LRR type resistance proteins which are very similar in structure to the *Mla* powdery mildew resistance genes isolated from barley (Zhou et al. 2001). Seven of the RGAs encode full-length proteins which are highly homologous apart from a variable number of leucine-rich repeat domains and real-time expression analysis indicates that all of the RGA candidates are expressed in resistant progeny. Full-length genomic clones including the encompassing, the endogenous promoter and terminator regions of all seven RGA candidates are currently being transformed individually into powdery and downy mildew susceptible *V. vinifera* cultivars to determine which of the RGA candidates function as powdery or downy mildew resistance genes.

8.6 The Future of Positional Cloning in Grapevine

Manipulation of disease resistance is a major challenge to be met by grapevine geneticists (Bisson et al. 2002) and, as such, has been the first goal of fine mapping and map-based cloning approaches in grapevine (Table 8-2). In the future, we expect to see an increase in the application of positional cloning approaches for the identification of key genes involved in grapevine development (Fernandez et al. 2006) and adaptation to abiotic stress. Indeed, even in model species like Arabidopsis, where large collections of mutants are available which can be used to determine gene function, there are still many projects utilizing positional cloning techniques to gain an understanding of the genetic basis of trait variation (see as an example Loudet et al. 2003; Mouille et al. 2006).

Since the commencement of the *Run1/Rpv1* mapping project, the availability of genetic resources for grapevine has improved dramatically.

Table 8-2 Fine mapping projects published so far in grapevine.

Locus	Trait	Mapping population	n° individuals	Marker types	Interval between the closest markers	Reference
Run1/Rpv1	Resistance to powdery & downy mildew	See Fig. 1	996	AFLP CAPS RFLP PCR SSR	0.6 cM	Barker et al. 2005
Pdr1a	Resistance to Pierce disease	V. vinifera x (V. rupestris x V. arizonica)	425	SSR	0.6 cM	Riaz et al. 2008
Pdr1b	Resistance to Pierce disease	V. vinifera x (V. rupestris x V. arizonica)	361	SSR	0.4 cM	Riaz et al. 2008
REN1	Resistance to powdery mildew	V. vinifera x V. vinifera	310	SSR	7.1 cM; 0.9 cM to the closest marker	Hoffmann et al. 2008

Comprehensive genome linkage maps have been constructed (Adam-Blondon et al. 2004; Riaz et al. 2004) and a large number of SSR loci have been identified that are now the markers of choice for positional cloning projects due to their transferability between species. For example, Hoffmann et al. (2008) quickly localized a *V. vinifera* resistance gene to powdery mildew, *REN1*, in a 7.1 cM interval by screening 195 well scattered SSR markers by bulk segregant analysis. The release of the complete *V. vinifera* genome sequence (Velasco et al. 2007; Jaillon et al. 2007) will also significantly increase the speed with which future genes of all types can be identified by positional cloning. However, as *V. vinifera* appears to be lacking resistance to many of the major pathogens, the published genome sequence can only act as a guide to the types of genes that may be present within the region of interest. Analysis of the *V. vinifera* genome sequence has revealed that many of the quantitative trait loci for resistance which have been mapped to date to disease co-localize with clusters of NBS-LRR genes (Di Gaspero et al. 2007; Moroldo et al. 2008). However, identification of the alleles conferring resistance will still require sequencing of the genome in question.

This then leads to the question of the most cost-efficient way to obtain genomic sequence from the species containing the gene of interest, i.e., library construction/chromosome walking or whole genome sequencing? With the continued development of new, cost-efficient sequencing technologies, it may ultimately be quicker and cheaper to sequence an entire genome rather than construct a BAC library, assemble a contig and sequence individual clones, as we have done. Furthermore, as the new sequencing technologies do not require cloned DNA, this will reduce the risk of gaps occurring due to clone instability/lethality. Therefore, it is possible that *Run1* and *Rpv1* may be the first and last grapevine genes isolated by chromosome walking. However, these new sequencing techniques are still not well suited to dealing with regions of highly repetitive DNA, which may still require construction of large insert libraries to resolve these regions.

Despite the tremendous improvements in genomic technologies over the last decade, the identification of unknown genes will still require significant resources and challenges in terms of the management of large populations of perennial plants, accurate phenotyping of traits, the generation of recombinant genotypes within the region of interest and the issue of non-homogeneous recombination rates within the genome. Most projects aiming at the identification of disease resistance genes will involve the generation of integrated genetic and physical maps in non-*vinifera* species, which will also generate interesting information about genome evolution in the *Vitaceae* family.

Unlike most other genes isolated from grapevine which can be functionally evaluated in model plants such as *Arabidopsis*, tobacco or tomato which are readily transformable, NBS-LRR type resistance genes which make up the majority of resistance genes within a plant species, show "restricted taxonomic functionality" (Michelmore 2003) which means they are only function within species of the same taxonomic family. Furthermore, many biotrophic pathogens such as *E. necator* and *P. viticola* are specifically adapted to the grapevine host and will not infect model plant species. This means that functional evaluation of most cloned candidate resistance genes will require this evaluation to be undertaken in transgenic grapevines, the generation of which remains a time-consuming and inefficient process. The development of highly efficient transient assay systems involving agroinfiltration (Santos-Rosa et al. 2008) or particle bombardment will be crucial for the rapid evaluation of resistance gene candidates in grapevine.

Finally, an important decision for researchers to make in the future will be whether to devote time and energy into cloning and functionally characterizing individual resistance genes, which can be used for genetic transformation of existing cultivars or alternatively whether to concentrate on the development of tightly-linked markers for use in classical breeding approaches such as marker-assisted selection. However, such a decision is more likely to be influenced by political and social issues associated with the adoption of transgenic vines for wine-making (discussed in Riaz et al. 2007; DeFrancesco 2008) rather than by the scientific merits of each approach.

Acknowledgements

The Australian component of this research is supported by the Grape and Wine Research and Development Corporation. The excellent technical assistance of Ange Jermakow, Karina Swann and Pat Corena in the generation of these results is gratefully acknowledged. We also wish to acknowledge the tremendous contribution of Alain Bouquet who initiated this project, produced all of the progeny populations used in the mapping studies and continued to contribute to the project through the generation of transgenic grapevines for functional analysis of the resistance gene candidates.

References

Aarts MGM, Lintel TE, Hekkert B, Holub EB, Beynon JL, Stiekema WJ, Pereira A (1998) Identification of R-gene homologous DNA fragments genetically linked to disease resistance loci in *Arabidopsis thaliana*. Mol Plant-Microbe Interact 11: 251–258.

Adam-Blondon A-F, Roux C, Claux D, Butterlin G, Merdinoglu D, This P (2004) Mapping 245 SSR markers on the *Vitis vinifera* genome: a tool for grape genetics. Theor Appl Genet 109: 1017–1027.

Akkurt M, Welter L, Maul E, Töpfer R, Zyprian E (2006) Development of SCAR markers linked to powdery mildew (*Uncinula necator*) resistance in grapevine (*Vitis vinifera* L. and *Vitis* sp.). Mol Breed 19: 103–111.

Barker CL, Donald T, Pauquet J, Ratnaparkhe MB, Bouquet A, Adam-Blondon AF, Thomas MR, Dry IB (2005) Genetic and physical mapping of the grapevine powdery mildew resistance gene, *Run1*, using a bacterial artifcial chromosome library. Theor Appl Genet 111: 370–377.

Baudoin A, Olaya G, Delmotte F, Colcol JF, Sierotzki H (2008) QoI resistance of *Plasmopara viticola* and *Erysiphe necator* in the mid-Atlantic United States. Online. Plant Health Progress doi: 10.1094/PHP-2008-0211-02-RS.

Belfanti E, Silfverberg-Dilworth E, Tartarini S, Patocchi A, Barbieri M, Zhu J, Vinatzer BA, Gianfranceschi L, Gessler C, Sansavini S (2004) The *HcrVf2* gene from a wild apple confers scab resistance to a transgenic cultivated variety. Proc Natl Acad Sci USA 101: 886–890.

Bent AF, Mackey D (2007) Elicitors, effectors, and R genes: the new paradigm and a lifetime supply of questions. Annu Rev Phytopathol 45: 399–436.

Bisson LF, Waterhouse AL, Ebeler SE, Walker MA, Lapsley JT (2002) The present and future of the international wine industry. Nature 418: 696–699.

Bittel P, Robatzek S (2007) Microbe-associated molecular patterns (MAMPs) probe plant immunity. Curr Opin Plant Biol 10: 335–341.

Bortiri E, Jackson D, Hake S (2006) Advances in maize genomics: the emergence of positional cloning. Curr Opin Plant Biol 9: 164–171.

Boubals D (1958) Contribution à l'étude des causes de la résistance des Vitacées au mildiou de la vigne (*Plasmopara viticola* (B and C) *Berl.* Et de *T.*) et de leur mode de transmission héréditaire. Ann Amélior Plantes 1: 1–236.

Boubals D (1961) Etude des causes de la résistance des Vitacées à l'oïdium de la vigne (*Uncinula necator* (*Schw.*) *Burr.*) et de leur mode de transmission héréditaire. Ann Amélior Plantes 11: 401–500.

Boubals D (1966) Etude de la distribution et des causes de la résistance au phylloxera redicicole chez les Vitacées. Ann Amélior Plantes 16: 145–184.

Bouquet A (1986) Introduction dans l'espèce *Vitis vinifera* L. d'un caractère de résistance à l'oidium (*Uncinula necator* Schw. Burr.) issu de l'espèce *Muscadinia rotundifolia* (Michx.) Small. Vignevini 12(suppl): 141–146.

Branas MM (1932) Sur la caryologie des Ampélidées. CR Acad Sci Paris 194: 121–123.

Chinchilla D, Bauer Z, Regenass M, Boller T, Felix G (2006) The *Arabidopsis* receptor kinase FLS2 binds flg22 and determines the specificity of flagellin perception. Plant Cell 18: 465–476.

Collins NC, Webb CA, Seah S, Ellis JG, Hulbert SH, Pryor A (1998) The isolation and mapping of disease resistance gene analogues in maize. Mol Plant-Microbe Interact 11: 968–978.

Detjen LR (1919) Some F1 hybrids of *Vitis rotundifolia* with related species and genera. North Carolina Agri Exp Stat Bull 18: 1–50.

DeFrancesco L (2008) Vintage genetic engineering. Nature Biotechnol 26: 261–263.

Di Gaspero G, Cipriani G (2002) Resistance gene analogs are candidate markers for disease-resistance genes in grape (*Vitis* spp.). Theor Appl Genet 106: 163–172.

Di Gaspero G, Cipriani G, Adam-Blondon AF, Testolin R (2007) Linkage maps of grapevine displaying the chromosomal locations of 420 microsatellite markers and 82 markers from *R*-gene candidates. Theor Appl Genet 114: 1249–1263.

Donald TM, Pellerone F, Adam-Blondon A-F, Bouquet A, Thomas MR, Dry IB (2002) Identification of resistance gene analogs linked to a powdery mildew resistance locus in grapevine. Theor Appl Genet 104: 610–618.

Erickson EO, Wilcox WF (1997) Distributions of sensitivities to three sterol demethylation inhibitor fungicides among populations of *Uncinula necator* sensitive and resistant to triadimefon. Phytopathology 87: 784–791.

Fernandez L, Doligez A, Lopez G, Thomas MR, Bouquet A, Torregrosa L (2006) Somatic chimerism, genetic inheritance, and mapping of the fleshless berry (flb) mutation in grapevine (*Vitis vinifera* L.). Genome 49: 721–728.

Flor HH (*1971)* Current status of the gene-for-gene concept. Annu Rev Phytopathol 9: 275–276.

Galet P (1996) Grape Diseases. Oeno Plurimedia Chaintre FranceHoffmann S, Di Gaspero G, Kovács L, Howard S, Kiss E, Galbács Z, Testolin R, Kozma P (2008) Resistance to *Erysiphe necator* in the grapevine 'Kishmish vatkana' is controlled by a single locus through restriction of hyphal growth. Theor Appl Genet 116: 427–438.

Hoffmann S, Di Gaspero G, Kovács L, Howard S, Kiss E, Galbács Z, Testolin R, Kozma P (2008) Resistance to Erysiphe necator in the grapevine 'Kishmish vatkana' is controlled by a single locus through restriction of hyphal growth. Theor Appl Genet 116: 427–438.

Jaillon O, Aury J-M, Noel B, Policriti A, Clepet C, Casagrande A, Choisne N, Aubourg S, Vitulo N, Jubin C, Vezzi A, Legeai F, Hugueney P, Dasilva C, Horner D, Mica E, Jublot D, Poulain J, Bruyere C, Billault A, Segurens B, Gouyvenoux M, Ugarte E, Cattonaro F, Anthouard V, Vico V, Del Fabbro C, Alaux M, Di Gaspero G, Dumas V, Felice N, Paillard S, Juman I, Moroldo M, Scalabrin S, Canaguier A, Le Clainche I, Malacrida G, Durand E, Pesole G, Laucou V, Chatelet P, Merdinoglu D, Delledonne M, Pezzotti M, Lecharny A, Scarpelli C, Artiguenave F, Pé E, Valle G, Morgante M, Caboche M, Adam-Blondon A-F, Weissenbach J, Quétier F, Wincker P (2007) The grapevine genome sequence suggests ancestral hexaploidization in major angiosperm phyla. Nature 449: 463–468.

Lafontaine DLJ, Tollervey D (2006) Ribosomal RNA. In: Encyclopedia of Life Sciences. John Wiley, Chichester, UK: *http: //www.els.net/* [10.1038/npg.els.0003832].

Lodhi MA, Reisch BI (1995) Nuclear DNA content of *Vitis* species, cultivars, and other genera of the Viticeae. Theor Appl Genet 90: 11–16.

Loudet O, Chaillou S, Krapp A, Daniel-Vedele F (2003) Quantitative trait loci analysis of water and anion contents in interaction with nitrogen availability in *Arabidopsis thaliana.* Genetics 163: 711–722.

Malnoy M, Xu M, Borejsza-Wysocka E, Korban SS, Aldwinckle HS (2008) Two Receptor-like genes, *Vfa1* and *Vfa2*, confer resistance to the fungal pathogen V*enturia inaequalis* inciting Apple Scab disease. Mol Plant-Microbe Interact 21: 448–458.

Martin GB, Brommonschenkel SH, Chunwongse J, Frary A, Ganal MW, Spivey R, Wu T, Earle ED, Tanksley SD (1993) Map-based cloning of a protein kinase gene conferring disease resistance in tomato. Science 262: 1432–1436.

Martin GB, Bogdanove AJ, Sessa G (2003) Understanding the functions of plant disease resistance proteins. Annu Rev Plant Biol 54: 23–61.

Merdinoglu D, Wiedeman-Merdinoglu S, Coste P, Dumas V, Haetty S, Butterlin G, Greif C (2003) Genetic analysis of downy mildew resistance derived from *Muscadinia rotundifolia.* Acta Hort 603: 451–456.

Michelmore RW (2003) The impact zone: genomics and breeding for durable resistance. Curr Opin Plant Biol 6: 397–404.

Michelmore RW, Paran I, Kesseli RV (1991) Identification of markers linked to disease-resistance genes by Bulked Segregant Analysis: a rapid method to detect markers in specific genomic regions by using segregating populations. Proc Natl Acad Sci USA 88: 9828–9832.

Moroldo M, Paillard S, Marconi R, Fabrice L, Canaguier A, Cruaud C, De Berardinis V, Guichard C, Brunaud V, Le Clainche I, Scalabrin S, Testolin R, Di Gaspero G, Morgante M, Adam-Blondon AF (2008) A physical map of the heterozygous grapevine 'Cabernet Sauvignon' allows mapping candidate genes for disease resistance. BMC Plant Biol 8: 66.

Mouille G, Witucka-Wall H, Bruyant MP, Loudet O, Pelletier S, Rihouey C, Lerouxel O, Lerouge P, Höfte H, Pauly M (2006) Quantitative trait loci analysis of primary cell wall composition in Arabidopsis. Plant Physiol 141: 1035–1044.

Mur LAJ, Kenton P, Lloyd AJ, Ougham H, Prats E (2008) The hypersensitive response; the centenary is upon us but how much do we know? J Exp Bot 59: 501–520.

Neu C, Stein N, Keller B (2002) Genetic mapping of the Lr20-Pm1 resistance locus reveals suppressed recombination on chromosome arm 7AL in hexaploid wheat. Genome 45: 737–744.

Olmo HP (1986) The potential role of (*Vinifera* X *Rotundifolia*) hybrids in grape variety improvement. Experientia 42: 921–926.

Osoegawa K, Vessere GM, Li Shu C, Hoskins RA, Abad JP, de Pablos B, Villasante A, de Jong PJ (2007) BAC clones generated from sheared DNA. Genomics 89: 291–299.

Pauquet J, Bouquet A, This P, Adam-Blondon A-F (2001) Establishment of a local map of AFLP markers around the powdery mildew resistance gene *Run1* in grapevine and assessment of their usefulness for marker assisted selection. Theor Appl Genet 103: 1201–1210.

Peterson DG, Tomkins JP, Frisch DA, Wing RA, Paterson AH (2000) Construction of plant bacterial artificial chromosome (BAC) libraries: An illustrated guide. J Agri Genom 5: Published with permission from CABI Publ. Full text available at *http: //www.cabi-publishing.org/JAG*

Phytowelt GmbH (2003) Study on the use of the varieties of interspecific vines: final report (Contract No. AGR 30881 of 30/12/2002). Phytowelt for the European Commission: *http: //ec.europa.eu/agriculture/markets/wine/studies/vine_en.pdf*

Planchon JE (1887) Monographie des *Ampelidae* vraies. In: *Monographia Phanerogamarum* A et C de Candolle 5: 305–364.

Riaz S, Dangl GS, Edwards KJ, Meredith CP (2004) A microsatellite marker based framework linkage map of *Vitis vinifera* L. Theor Appl Genet 108: 864–872.

Riaz S, Doligez A, Henri RJ, Walker MA (2007) Grape. In: C Kole (ed) Genome Mapping and Molecular Breeding in Plants, vol 4: Fruit and Nuts. Springer Berlin, Heidelberg, Germany, pp 63–101.

Riaz S, Tenscher AC, Rubin J, Graziani R, Pao SS, Walker MA (2008) Fine-scale genetic mapping of two Pierce's disease resistance loci and a major segregation distortion region on chromosome 14 of grape. Theor Appl Genet 117: 671–681.

Santos-Rosa M, Poutaraud A, Merdinoglu D, Mestre P (2008) Development of a transient expression system in grapevine via agro-infiltration. Plant Cell Rep 27: 1053–1063.

Savocchia S, Stummer BE, Wicks TJ, van Heeswijck R, Scott ES (2004) Reduced sensitivity of *Uncinula necator* to sterol demethylation inhibiting fungicides in southern Australian vineyards. Aust Plant Pathol 33: 465–473.

Shen KA, Meyers BC, Islam-Faridi MN, Chin DB, Stelly DM, Michelmore RW (1998) Resistance gene candidates identified by PCR with degenerate oligonucleotide primers map to clusters of resistance genes in lettuce. Mol Plant-Microbe Interact 11: 815–823.

Shizuya H, Birren B, Kim UJ, Mancino V, Slepak T, Tachiiri Y, Simon M (1992) Cloning and stable maintenance of 300-kilobase-pair fragments of human DNA in *Escherichia Coli* using an F-factor-based vector. Proc Natl Acad Sci USA 89: 8794–8797.

Song J, Dong F, Lilly JW, Stupar RM, Jiang J (2001) Instability of bacterial artificial chromosome (BAC) clones containing tandemly repeated DNA sequences. Genome 44: 463–469.

Stirling B, Newcombe G, Vrebalov J, Bosdet I, Bradshaw HD (2001) Suppressed recombination around the *MXC3* locus, a major gene for resistance to poplar leaf rust. Theor Appl Genet 103: 1129–1137.

Suetsugu Y, Minami H, Shimomura M, Sasanuma S, Narukawa J, Mita K, Yamamoto K (2007) End-sequencing and characterization of silkworm (*Bombyx mori*) bacterial artificial chromosome libraries. BMC Genom 8: 314.

Takken FLW, Albrecht M, Tameling WIL (2006) Resistance proteins: molecular switches of plant defence. Curr Opin Plant Biol 9: 383–390.

van Ooijen G, van den Burg HA, Cornelissen BJC, Takken FLW (2007) Structure and function of resistance proteins in Solanaceous plants. Annu Rev Phytopathol 45: 43–72.

Velasco R, Zharkikh A, Troggio M, Cartwright DA, Cestaro A, Pruss D, Pindo M, Fitzgerald LM, Vezzulli S, Reid J, Malacarne G, Iliev D, Coppola G, Wardell B, Micheletti D, Macalma T, Facci M, Mitchell JT, Perazzolli M, Eldredge G, Gatto P, Oyzerski R, Moretto M, Gutin N, Stefanini M, Chen Y, Segala C, Davenport C, Demattè L, Mraz A, Battilana J, Stormo K, Costa F, Tao Q, Si-Ammour A, Harkins T, Lackey A, Perbost C, Taillon B, Stella A, Solovyev V, Fawcett JA, Sterck L, Vandepoele K, Grando SM, Toppo S, Moser C, Lanchbury J, Bogden R, Skolnick M, Sgaramella V, Bhatnagar SK, Fontana P, Gutin A, Van de Peer Y, Salamini F, Viola R (2007) A high quality draft consensus sequence of the genome of a heterozygous grapevine variety. PLoS ONE 19: e1326.

Wei F, Gobel-Werner K, Morroll SM, Kurth J, Mao L, Wing RA, Leister D, Schulze-Lefert P, Wise RP (1999) The *Mla* (powdery mildew) resistance cluster is associated with three NBS-LRR gene families and suppressed recombination within a 240-kb DNA interval on chromosome 5S (1HS) of barley. Genetics 153: 1929–1948.

Welter LJ, Göktürk-Baydar N, Akkurt M, Maul E, Eibach R, Töpfer R, Zyprian EM (2007) Genetic mapping and localization of quantitative trait loci affecting fungal disease resistance and leaf morphology in grapevine (*Vitis vinifera* L). Mol Breed 20: 359–374.

Wiedemann-Merdinoglu S, Prado E, Coste P, Dumas V, Butterlin G, Bouquet A, Merdinoglu D (2006) Genetic analysis of resistance to downy mildew resistance from *Muscadinia rotundifolia* 9th Int Conf on Grape Genetics and Breeding. Udine, Italy, 2–6 July 2006.

Woo SS, Jiang JM, Gill BS, Paterson AH, Wing RA (1994) Construction and characterization of a bacterial artificial chromosome library of *Sorghum bicolor*. Nucl Acids Res 22: 4922–4931.

Wu CC, Nimmakayala P, Santos FA, Springman R, Scheuring C, Meksem K, Lightfoot DA, Zhang HB (2004) Construction and characterization of a soybean bacterial artificial chromosome library and use of multiple complementary libraries for genome physical mapping. *Theor Appl Genet* 109: 1041–50.

Zhou F, Kurth J, Wei F, Elliott C, Valè G, Yahiaoui N, Keller B, Somerville S, Wise R, Schulze-Lefert P (2001) Cell-autonomous expression of barley *Mla1* confers race-specific resistance to the powdery mildew fungus via a *Rar1*-independent signaling pathway. Plant Cell 13: 337–50.

Zipfel C, Kunze G, Chinchilla D, Caniard A, Jones JD, Boller T, Felix G (2006) Perception of the bacterial PAMP EF-Tu by the receptor EFR restricts *Agrobacterium*-mediated transformation. Cell 125: 749–760.

9

Genome Sequence Initiatives

Anne-Françoise Adam-Blondon,[1,*] *Olivier Jaillon,*[2] *Silvia Vezzulli,*[3] *Andrey Zharkikh,*[4] *Michela Troggio*[3] and *Riccardo Velasco*[3]

ABSTRACT

The possibility to access a complete genome sequence allows the development of new resources for high-throughput functional analysis without *a priori*, whole genome syntenic approaches for the detection of badly predicted genes or of conserved regulatory boxes. It also facilitates the identification of key genes by combined genetic and genomics approaches. Two parallel projects using different strategies for the sequencing of the grapevine genome were initiated. The first one aimed at the development of a reference genome sequence using a near homozygous genotype and the second one at the sequencing of a highly heterozygous high quality cultivar. Beside giving access to two annotated draft sequences of the genome of *Vits vinifera* L., the two projects developed genomic resources such as BAC libraries, BAC end sequences, sequenced full length cDNA libraries, physical and genetic maps. They offer very good material to assess the influence of heterozygosity on the development of physical maps or whole genome shotgun sequences and a first rough view on the polymorphism between the two haplotypes of a grapevine variety. Finally, they provided a first picture of the grapevine highly heterozygous genome, with a structure not much rearranged from the genome of its ancient hexaploid ancestor and the expansion of specific gene families.

Keywords: genome sequence, grapevine, *Vitis vinifera*, heterozygosity, annotation, hexaploid

[1]UMR INRA UEVE ERL CNRS Génomique Végétale, 2 rue Gaston Crémieux, BP 5708, 91 057 Evry cedex, France.
[2]CEA, DSV, Institut de Génomique, Genoscope, 2 rue Gaston Crémieux, CP5706, 91057 Evry, France.
[3]IASMA Research and Innovation Centre, Fondazione Edmund Mach, Via E. Mach 1, 38010, S. Michele all'Adige, Trento, Italy.
[4]Myriad Genetics Inc., 320 Wakara Way, Salt Lake City, 84108 UT, USA.
*Corresponding author: *adam@evry.inra.fr*

9.1 Introduction

The possibility to access a complete genome sequence allows the development of new resources for (i) high-throughput functional analysis without *a priori* (see for instance *http://www.catma.org/*), (ii) whole genome syntenic approaches for the detection of badly predicted genes or of conserved regulatory boxes (see for instance Roest Crollius et al. 2000). Finally, it greatly facilitates the identification of key genes by combined genetic and genomics approaches (Morgante and Salamini 2003).

Three main strategies can be developed for the sequencing of complex genomes: a clone-by-clone approach, a whole genome shotgun approach and a mix of both (for review, see Green 2001). The BAC by BAC approach consists of building an integrated genetic and physical map of the genome, then to select the minimum set of large insert clones that allows to cover the genome (minimum tiling path) and to obtain the finished sequence of these clones. The quality of such a genome sequence is the highest possible among these three approaches, and the number of remaining gaps in the sequence depends on the quality of the integrated genetic and physical map used. This usually necessitates many exchanges between the sequencing teams and the mapping teams during the sequencing process to improve the tiling path. In plants, the *Arabidopsis* genome sequence (Arabidopsis Genome Initiative 2000) and the rice genome sequence (International Rice Genome Sequencing Project, IRGSP 2005) were obtained using this approach. The second method, called Whole Genome Shotgun (WGS) is now the most widely used because it is less labor intensive, faster, and thus much cheaper than the BAC by BAC approach. In this case, different libraries are prepared with contrasting average insert size (plasmid, cosmid and BAC). The extremities of these clones are sequenced and the pairs of reads are assembled in sequence contigs and super-contigs. The super-contigs can then be aligned along an integrated genetic and physical map. The resulting genomic sequence contains sizeable gaps between the contigs of a super-contig, gaps of unknown size between the super-contigs and the total gap number is heavily dependent on the genome coverage. In Angiosperms, the poplar, rice and papaya genomes were sequenced using this approach (Goff et al. 2002; Yu et al. 2002; Tuskan et al. 2006; Ming et al. 2008). The recent development of ultra-high throughput new technologies for sequencing (Shendure et al. 2004; see also Chapter 10) has definitely lowered the price of WGS approaches and are now increasingly used for such projects (see for instance Huang et al. 2009). A comparison was made between the clone-by-clone rice sequence and a 6 × WGS sequence of the same subspecies (IRGSP 2005). The WGS draft sequence covered 78% of the rice genome and 75.3% of the full-length cDNA supported genes models, probably due to local bad quality of the WGS sequence or of the cDNA sequence, and also to contamination. Another issue is the level of heterozygosity of the genome sequenced. In WGS approaches, it may cause a bad assembly of the sequence, leading to artificial duplications of variable size (Vinson et al. 2005).

Two parallel projects using different strategies for the sequencing of the grapevine genome were initiated. The first one was established within the frame of the International Grapevine Genome Program (IGGP; *www.vitaceae.org*) and the second one resulted from collaboration between the Istituto Agrario di San Michele all'Adige and Myriad Genetics.

9.2 Strategies for Sequencing of the Grapevine Genome

9.2.1 Development of BAC Library Resources

A growing resource of grapevine BAC libraries is available (Table 9-1), mainly for *Vitis vinifera* L. well known cultivars but also for a *V. vinifera* genotype introgressed with disease resistance genes. The two *V. vinifera* cv. Cabernet Sauvignon BAC libraries were chosen as reference libraries by the IGGP, as the initial plan was to use Cabernet Sauvignon as the model genotype for grapevine, due to its wide distribution around the world. Now, the development of new BAC resources are more targeted to other species of the *Vitis* or *Muscadinia* genus, aiming at identifying efficient alleles for resistance to diseases or for the adaptation to abiotic stresses through comparative genomics approaches.

9.2.2 Development of Physical Maps of the Grapevine Genome

Physical maps consist of the assembly into contigs of overlapping large insert clones based on fingerprint similarities and on the presence of common markers. These assemblies can be produced using the software FPC (Nelson and Soderlund 2005) and the distance is in base pairs. Most of the physical maps developed in plants are now based on the assembly of BAC clones (see Meyers et al. 2004 for review).

Two physical maps of the *V. vinifera* genome have been recently constructed: one using a Pinot Noir ENTAV 115 BAC library (Scalabrin et al. 2010) and one using the Cabernet Sauvignon reference BAC library (Moroldo et al. 2008). The fingerprinting method that was used for the construction of both maps is based on the fluorescent labeling of four different restriction sites and was adapted by Moroldo (2008) from the original paper of Luo et al. (2003). 49,104 BAC clones from the Pinot Noir library representing 11 times the grape genome were fingerprinted (Scalabrin et al. 2010) whereas 30,828 clones, representing eight times the genome were processed from the Cabernet Sauvignon BAC library (Moroldo et al. 2008; Table 9-2).

Table 9-1 Grape BAC libraries published in grapevine.

Genotype	Species	Number of clones	Genome coverage	Publicly available BES	Reference	Distribution[$]
Syrah	*V. vinifera*	55296	16.5	No	Tomkins et al. 2001	CUGI
Syrah cl.ENTAV 173	*V. vinifera*	23040	7.7		Adam-Blondon et al. 2005	
Cabernet-Sauvignon cl.412	*V. vinifera*	44544	13.2	Yes	Adam-Blondon et al. 2005	CNRGV
Pinot Noir cl.777	*V. vinifera*	48384	14.8	No	Adam-Blondon et al. 2005	
Pinot Noir cl.115	*V. vinifera*	23040	4.6	Yes	Adam-Blondon et al. 2005	
Pinot Noir cl. 115	*V. vinifera*	26064	6.8	Yes	Velasco et al. 2007	
VHR3294-R23	*V. vinifera**	55295	9.4	No	Barker et al. 2005	
PN40024	*V. vinifera*	70656	~14	Yes	Jaillon et al. 2007	CNRGV

**V. vinifera* introgressed with a resistance gene carrying genomic region from *Muscadinia rotundifolia*
$CUGI: *www.genome.clemson.edu*; CNRGV: *http://cnrgv.toulouse.inra.fr/*

Table 9-2 Features of the Cabernet Sauvignon (Moroldo et al. 2008) and of the Pinot Noir (Scalabrin et al. 2010) physical maps. The average band size (1.1 kbp) was calculated by dividing the average insert size of the BAC clones by the average number of bands *per* clone. This value was used to estimate the physical length of the contigs. The physical map of the Cabernet Sauvignon and its links with the grapevine genetic map and the reference genome sequence can be queried at *http://urgi.versailles.fr/cmap*. The physical map of the Pinot Noir and its links with the Pinot noir genome assembly and the grapevine genetic maps developed at IASMA can be viewed at *http://genomics.research.iasma.it*.

Whole physical map	Cabernet S	Pinot Noir
Number of clones fingerprinted	44,544	44,544
Number of clones used for map assembly	30,828	38,983
Number of singletons	1,111	3,372
Number of contigs	1,770	1,804
Physical length of the contigs (kbp)	715,684	888,000
Contigs anchored on the genetic map:		
No. of contigs	395	436
Coverage (kbp)	255,476	341,621

Collections of BAC end sequences (BES) were also developed for both genotypes allowing the development of new genetic markers to improve the integration between the genetic and the physical maps (Lamoureux et al. 2006; Troggio et al. 2007): in particular, 77,231 Cabernet Sauvignon BES contained 3,601 SSR loci (V. Tharreau, unpublished results), which are very efficient molecular markers for mapping in grapevine.

A common feature of both maps is that the total length of the contigs is larger than the estimated size of the grape genome (1.5–1.6 fold) and that this expansion could mainly be attributed to the effects of heterozygosity (Moroldo et al. 2008; Scalabrin et al. 2010). Indeed, genotyping with simple sequence repeat (SSR) markers of a *V. vinifera* germplasm collection showed a high level of heterozygosity in *V. vinifera* (Aradhya et al. 2003). Moreover, it was later shown that an important part of the sequence variation is due to insertion/deletion events and that the frequency of Single Nucleotide Polymorphisms (SNPs) is uneven, by comparing the two haplotype sequences in "Cabernet Sauvignon" over two different genomic regions encompassing 182 kbp (F. Chopin et al. unpublished) or as a result of the WGS sequencing of the cultivar Pinot Noir (Velasco et al. 2007). These features affect the banding pattern of the fingerprints produced from two allelic regions and their assignment to overlapping regions by the FPC software. Even when the BAC clones corresponding to the two different haplotypes are assembled within the same contig, their local order has to be considered with caution as many false duplications are observed (Moroldo et al. 2008; Scalabrin et al. 2010).

Up to now, only three fingerprinting-based physical maps were assembled for heterozygous plants other than grape: *Prunus* (Jung et al. 2004), apple (Han et al. 2007) and poplar (Kelleher et al. 2007). The peach map was constructed using the same fingerprinting method as the one used for grapevine, but some other differences impair a homogeneous comparison of the two assemblies (Moroldo et al. 2008). The poplar and apple maps both showed an 1.2-fold genome size expansion (Han et al. 2007; Kelleher et al. 2007). This value is lower compared to the ones found in grapevine and it may be related to the use of agarose gels in fingerprinting (Nelson et al. 2007; discussed in Moroldo et al. 2008), as at least in poplar, the sequence variation was shown to have the same features as in grapevine (Kelleher et al. 2007). Another concern was an unusually high percentage of chimeric contigs that was found compared to physical maps of other species despite a stringent filtering of possible plate cross contaminations before the physical map assembly (Moroldo et al. 2008). Such chimerism may arise from partial sequence similarity within the BAC clones caused by shared fingerprint bands corresponding to repetitive sequences or large-scale duplications.

Despite these limitations, in absence of a whole genome sequence, such integrated maps are invaluable resources for the quick development of new markers in targeted regions using BAC-end sequences (see Barker et al. 2005; Castellarin et al. 2006; Troggio et al. 2007; Moroldo et al. 2008 for examples in grapevine), for candidate gene approaches by establishing links between genetic maps where quantitative trait loci (QTLs) for traits of interest have been located and gene-containing BACs and to prepare and accelerate map base cloning projects. For instance, it was proved to be very useful to understand the genome organization of a gene family like the flavonoid hydroxylase ones (F3′H and F3′5′H) at a fine scale. Three groups recently cloned and characterized these genes in grape, based on available expressed sequence tags (ESTs) and sequence homologies with known F3′H and F3′5′H genes in other plants and analyzed their expression in relation with the composition in particular classes of flavonoids like cyanidin and delphinidin-based anthocyanins (Jeong et al. 2005; Bogs et al. 2006; Castellarin et al. 2006). Two groups showed that both F3′H and F3′5′H belong to a multigene family, either by sequencing several fragments amplified with degenerate primers (Jeong et al. 2005), or by a physical mapping of sequences obtained with the same approach combined with a single-strand conformational polymorphism (SSCP) based evidence of the different members of the gene family (Castellarin et al. 2006). The second approach revealed bigger and tandemly duplicated gene families (Castellarin et al. 2006). The fact that different forms of the F3′5′H genes showed different patterns of transcription (Jeong et al. 2005; Castellarin et al. 2006), underlines the importance of obtaining the complete family of sequences, which is now easy to implement with the availability of the grapevine whole genome sequence. Finally, the establishment of finely integrated local genetic and physical maps will remain the key step of map-based cloning projects for genes of interest. The long generation time and the space needed for large progenies make such approaches much more difficult in perennial species than in annual species, even though it has already been proved possible (see for instance Patocchi et al. 1999 in apple tree and Claverie et al. 2004 in plum) and especially for target genes expressed in plantlets. The first local fine map in grapevine was published by Barker et al. (2005) for a region containing a major gene for resistance to powdery mildew (see Chapter 8).

9.2.3 Strategy for the Development of a Reference Genome Sequence for Grapevine by the French-Italian Public Consortium for the Grapevine Genome Characterization

In order to have a WGS sequence of the highest quality possible, the international consortium decided to sequence a nearly homozygous genotype, the PN40024 with a 12 × coverage. The near homozygous line was derived from Pinot Noir at the INRA station of Colmar (Bronner and Oliveira 1990) by nine successive selfing steps. Its analysis with SSR markers led Jaillon et al. (2007) to suspect that an outcross might have occurred in any of the three first generations of selfing. A rough paternity search showed that the most probable parent would be Helfensteiner (Jaillon et al. 2007). Even if the number of selfing generations the PN40024 was derived from may be lower than nine, its level of homozygosity was showed to be quite good: 7/102 tested SSR loci were heterozygous and the heterozygous genome sequence contigs did not represent more than 7% of the genome size (Jaillon et al. 2007).

In 2005, an agreement was signed between France and Italy to share the costs of this public initiative (*http://www.agriculture.gouv.fr/spip/IMG/pdf/cp_vigne170505-1.pdf*). The necessary funds have been provided in France by the French Ministry of Education and Research, the Genoscope (*www.genoscope.cns.fr*), INRA and the French National Agency for Research (*www.agence-nationale-recherche.fr*) and in Italy by the Italian Ministry for Agriculture and Forestry, the Friuli Venezia Giulia County and a consortium of private companies and banks. The sequence production was shared between Genoscope, IGA (Istituto di Genomica Applicata; *www.appliedgenomics.org*) and CRIBI (Centro Ricerca Interdipartimentale Biotechnologie Innovative; *www.cribi.unipd.it*) and the rest of the work was shared by other French (Institut National de Recherche Agronomique; INRA; *www.inra.fr*) and Italian laboratories (VIGNA consortium; *www.vitisgenome.it*).

9.2.4 Strategy for Sequencing a Highly Heterozygous Genome: The Pinot Noir Genome Project

The Grapevine Genome Initiative, collaboration between the Fondazione Edmund Mach-Istituto Agrario di San Michele all'Adige–(FEM-IASMA) and two private companies, Myriad Genetics, Inc. and the 454 Life Science, has focused on sequencing the elite cultivar Pinot Noir to provide insight into the structural nature of heterozygosity in an outcrossing species known to be highly heterozygous. The Sanger sequences (7x) and the 454 sequences (4.2x) were assembled by a Myriad proprietary software able to handle heterozygous sequences, revealing a large amount of polymorphism. Of special interest to biologists and breeders are polymorphisms in and around the coding regions representing a substantial resource for molecular breeding programs, as well as trait and quantitative trait loci (QTL) marker association. This project adopted a new approach for the sequencing and assembly of a large heterozygous eukaryotic genome, which consisted of a whole genome shotgun combining paired reads produced by Sanger sequencing (Sanger et al. 1977) and unpaired reads obtained through sequencing by synthesis (SBS, Margulies et al. 2005). The coverage of SBS reads aimed at identifying polymorphic sites and at closing most gaps between DNA contigs. A novel approach to genomic alignment was developed to generate a single consensus sequence from both chromosomes of Pinot Noir.

The project was funded entirely by the Province of Trento (Italy) while the sequence production was shared by Myriad Genetics, Inc., IASMA Research Centre and 454 Life Science.

Both genome sequences were recently published by Jaillon et al. (2007) for the 8 × version of the reference genome sequence and by Velasco et al. (2007) for the Pinot Noir genome sequence (10.7×), revealing interesting features of the grapevine genome sequence.

9.3 The Reference Grapevine Genome Sequence: The Grapevine Genome Descends from an Ancient Event of Polyploidization

All the results so far generated (maps, markers, shotgun reads, genome sequence and its annotations), once analyzed and published, were submitted to the relevant database at NCBI and can also be viewed in the two genome browsers of the project: *http://www.genoscope.cns.fr/externe/GenomeBrowser/Vitis/* and *http://www.appliedgenomics.org/*.

9.3.1 Assembling the Reads of the WGS

The first version of the reference sequence of the grapevine genome has been obtained from a set of sequences covering 8.4 times the genome size, which would *a priori* lead to a relatively high quality sequence compared with other similar projects (Table 9-3, Goff et al. 2002; Tuskan et al. 2006; Velasco et al. 2007; Ming et al. 2008). Several assemblies were produced along the project and the total size of the genome sequence (487 Mb) was obtained at a six-fold coverage. After this stage, the improvements that were observed concerned an increase of the length of the contigs and super-contigs and the diminution of their respective numbers (Table 9-3). The objective of a 12 × coverage of the grapevine genome is now reached, a new assembly has been produced with an improvement of the assembly method and is now available to the public with its annotation at NCBI.

The anchorage of the sequence scaffolds along the grapevine genome was realized in two steps (Jaillon et al. 2007): (i) some super-contigs were joined together into 45 ultra-contigs using paired BAC-end sequences (BES) from Cabernet-Sauvignon and BAC contigs of the same BACs from our Cabernet-Sauvignon physical map and (ii) the ultra-contigs and remaining scaffolds were then aligned along the *V. vinifera* genetic map published by Doligez et al. (2006). This resulted in 61% of the assembly being anchored and orientated and 69% anchored (Jaillon et al. 2007). This aspect is being improved in the 12 × version of the genome sequence by a combination of improvements of the genetic map and of the assembly: currently, around 85% of the assembly is anchored and orientated and 91% anchored (The French-Italian Public Consortium for the Grapevine Genome Characterization, unpublished results).

9.3.2 Annotation of the Grapevine Genome Sequence

Around 41% of the grapevine genome sequence was found to correspond to repetitive sequences, which are higher in proportion to in rice and *Arabidopsis*, but surprisingly lower than in papaya, which has also a genome smaller in size than grapevine (Ming et al. 2008). A striking feature in grapevine was that introns are rich in repeats, with 12.4% of intron sequence that contains TE, mainly LINEs (Jaillon et al. 2007). Unlike in Arabisopsis, but as in poplar, in grapevine the TE density is non-homogeneous along the sequence and is opposite to gene density (Arabidopsis Genome Initiative 2000; Tuskan et al. 2006; Jaillon et al. 2007). Recently, an in-depth re-annotation of the Class II elements was carried out based on the two published genome sequences and the authors reported 1,160 complete elements and 2,086 degenerated elements distributed in four families: *hAT*, *PIF*, *Mutator* and *CACTA* (Benjak et al. 2008). The authors also found transcription evidence of Class II elements in the EST databases and showed that they contribute to inter-varietal polymorphism (Benjak et al. 2008). They also suggested that the percentage of genome corresponding to TE by Jaillon et al. (2007) and Velasco et al. (2007) may be underestimated (Benjak et al. 2008).

Table 9-3 Comparison of WGS assemblies in different plant genome sequencing projects.

	Grapevine[1]	Grapevine[1]	Grapevine[1]	Grapevine[1]	Rice[2]	Poplar[3]	Papaya[4]
Number reads produced	-	-	-	8,200,000	5,500,000	7,600,000	2,800,000
Coverage	2.9	3.7	5	8.4	6	7.5	3
Number of contigs	170,000	150,000	125,000	19,557	42,109	45,970	47,483
N50 contigs[6]	2.9 kb	4.3 kb	7.7 kb	65.9 kb	20 kb	2-29-126 kb[5]	11 kb
Number of super-contigs	140,000	100,000	77,000	3,514	-	-	17,764
Final size	-	-	477 Mb	487 Mb	390 Mb	410 Mb	370 Mb

[1]Near inbred line; Jaillon et al. 2007.
[2]Inbred line; Goff et al. 2002.
[3]Heterozygous individual; Tuskan et al. 2006.
[4]Near inbred line; Ming et al. 2008.
[5]The 26,363 contigs of 1–20 kb have a N50 = 2 kb; the 8,245 contigs with a size > 20 kb by unanchored on the genetic map have a N50 = 29 kb and the 11,362 contigs anchored on the genetic map have a N50 = 126 kb (Tuskan et al. 2006).
[6]N50 contigs is the median of the contig size.

The non-coding RNAs were also analyzed, showing that the tRNA, snRNA and ribosomal RNA sequences have similar numbers and distributions as in *Arabidopsis*, rice and poplar (Jaillon et al. 2007). As for the miRNA, 22 families were detected by MicroHarvester (Dezulian et al. 2006), with the noticeable specific expansion of two of them in grapevine (miR845 and miR395) and the first record of the miR535 and miR1213 families in a dicotyledon. As the miR395 family is thought to be involved into sulfate metabolism, these observations may lead to interesting research in relation with agronomical traits. Similar findings were derived from the Pinot Noir genome project, yet with grapevine specific miRNA families and targets (Velasco et al. 2007).

Automatic genome structural and functional annotations rely on the integration of different sets of data: *de novo* gene prediction, homologies with ESTs, cDNA, proteins, or sequences conserved across evolution (see for instance Town 2006 for the Medicago genome project and Jaillon et al. 2007 for the grapevine genome project). With regard to the latter point, in animals, the comparison of the human and of *Tetraodon* genome sequences allowed to improve or define gene models (Roest-Crolius et al. 2000; Jaillon et al. 2004). But the key resources with regard to the quality of the output of an automatic annotation are EST and full length cDNA collections. Full length cDNA sequences allow globally improving the quality of the protein coding gene models proposed by the process of automatic annotation. For instance, in *Arabidopsis*, although the genome had been sequenced four years ago and several rounds of annotation had already been done, a collection of 31,558 flcDNAs allowed to discover 326 unpredicted genes and to correct the structure of 45% models with no previous full length cDNA support (Castelli et al. 2004). Finally, it is an important resource for functional genomics (see as an example *http://www.evry.inra.fr/public/projects/orfeome/orfeome.html*). Large sets of public *Vitis* ESTs were available to Jaillon et al. (2007; around 190,000) and the consortium produced in addition around 48,000 sequences from libraries enriched in full length cDNAs. These sequences, mapped on available versions of the genome assembly, also allowed to quickly develop a set of 650 genes for which both the coding and genomic sequences were fully known and manually inspected, necessary for the training of the software used for *de novo* gene prediction (S. Aubourg and C. Clepet unpublished results).

The current number of proposed protein coding gene models in grapevine is 30,434 (Jaillon et al. 2007). The next version of the annotation should be improved by several means: an increase of the genome coverage will reduce the number of gaps and errors that still can affect gene predictions (IRGSP 2005); over 100,000 full length cDNA sequences have now been produced and finally the deep resequencing of grapevine transcriptome has also been included in the process (Denoeud et al. 2008; The French-

Italian Public Consortium for the Grapevine Genome Characterization, unpublished results).

9.3.4 Whole Genome Comparison between the Grapevine Genome and Itself or Other Angiosperm Genomes

One of the striking features of the grapevine genome is that it was found to be an ancient hexaploid with each genomic region always aligning (at the protein level) with two others (Jaillon et al. 2007; Velasco et al. 2007; Fig. 9-1), which confirmed and improved previous conclusions derived from cytological observations of F_1 hybrids between *V. vinifera* ($2n = 38$) and *Muscadinia rotundifolia* ($2n = 40$) and suggesting a polyploid origin of the *Vitis* genome (Chapter 1 Section 5.4; Patel and Olmo 1955). Moreover, the blocks of synteny between the three ancestral genomes have undergone few rearrangements (Jaillon et al. 2007).

A comparative analysis of the grapevine genome with the four angiosperm genomes available to date, showed that the hexaploid ancestor of grapevine is also the ancestor of *Arabidopsis*, papaya and poplar, but would not be at the origin of the monocot lineage (Jaillon et al. 2007; Ming et al. 2008). This result was only based on the observation of blocks of synteny between species like those represented in Fig. 9-1. A different timing of the whole genome duplication or hybridization events in the dicotyledon lineage, based on studies of substitution rates between pairs of othologous genes was proposed by Velasco et al. (2007), stressing the difficulty to date gene divergence between species where substitutions rates may be different (e.g., comparing annual and perennial species). Altogether these results confirmed the importance of whole genome duplications/polyploidy in genome evolution (Ohno 1970; Jaillon et al. 2004; Aury et al. 2006) and especially in plants (Bowers et al. 2003, Arabidopsis Genome Initiative 2000; IRGSP 2005; Tuskan et al. 2006; Ming et al. 2008; Rensing et al. 2008).

9.4 The DNA of Pinot Noir Discloses Features of a Highly Heterozygous Genome

9.4.1 Sequencing and Assembling a Heterozygous Genome: Problems and Solutions

As already discussed, the assembly of a whole genome shotgun sequence is more problematic for a highly heterozygous genotype than for an inbred genotype. For the Pinot Noir project, two sequencing techniques were adopted. The Sanger method was used to generate 6.5 × coverage of the genome. This was integrated with a 4.2 × coverage of sequence reads generated by a scalable, highly parallel sequencing by synthesis (SBS) method.

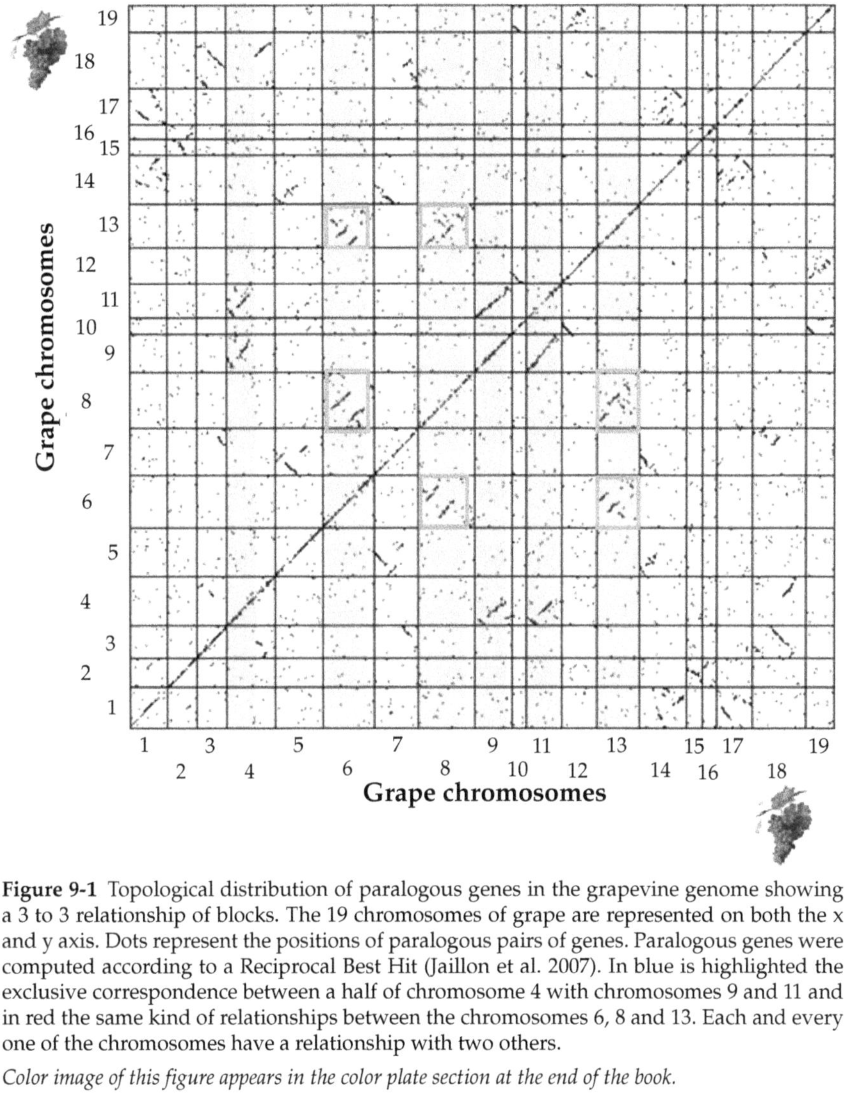

Figure 9-1 Topological distribution of paralogous genes in the grapevine genome showing a 3 to 3 relationship of blocks. The 19 chromosomes of grape are represented on both the x and y axis. Dots represent the positions of paralogous pairs of genes. Paralogous genes were computed according to a Reciprocal Best Hit (Jaillon et al. 2007). In blue is highlighted the exclusive correspondence between a half of chromosome 4 with chromosomes 9 and 11 and in red the same kind of relationships between the chromosomes 6, 8 and 13. Each and every one of the chromosomes have a relationship with two others.

Color image of this figure appears in the color plate section at the end of the book.

Existing software and strategies were not adequate for the assembly of this highly heterozygous genome. A modified version of the assembly pipeline developed for rice genome sequencing (Goff et al. 2002) was developed to handle the high level of heterozygosity in the Pinot Noir genome (Zharkikh et al. 2008). As a result, most of the assembled contigs were represented by a consensus sequence derived from the alignment of the two existing haplotypes. In addition to the regions in which such a merge was possible, many regions were found to be chromosome-specific, i.e., either segments with different DNA sequences flanked by orthologous regions of the two homologous chromosomes (Fig. 9-2a) or gaps corresponding to sequences missing in one chromosome but not in the other (Fig. 9-2b).

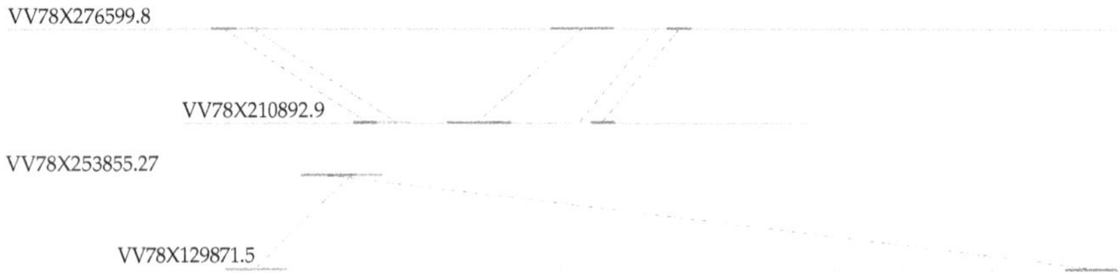

Figure 9-2 Two examples of chromosome-specific regions of the Pinot Noir genome. (a) segments with different DNA sequences (in green) flanked by orthologous regions of the two homologous chromosomes (same color for both chromosomes); (b) Gaps corresponding to a sequence missing in one chromosome but not in the other.

Color image of this figure appears in the color plate section at the end of the book.

The assembly started with aligning the unique sequences and then progressively included sequences with a higher degree of repetitiveness. Applying the procedure to about 6.6 M reads from Sanger sequencing, 90.6% of which represented paired clone ends, 211,374 initial seed contigs of unique sequences were generated. By using long clone links with non-repetitive clone ends, seed contigs were ordered into metacontigs (ordered assembly of contigs, referred to as supercontigs or scaffolds in other publications) and the overlapping members of the metacontigs were iteratively merged. After merging the sequences into 120,000 contigs, the data were combined with 4.2 genome-equivalents of SBS sequences. This helped identifying polymorphic sites in the assembled sequences and closing about 25% of the gaps between contigs. After removal of 10,847 contigs composed only of tandemly repeated sequences and disposal of 7,003 contigs shorter than 1,000 bp, the iterative assembly produced 58,611 contigs corresponding to 530.9 Mb of genomic DNA. 44,179 of the 58,611 contigs were assembled into 2,093 metacontigs covering 477.1 Mb of genomic DNA and the remaining 14,432 contigs were singletons.

9.4.2. Metaconting Alignment on a Dense Genetic Map

The next phase of the assembly involved positioning metacontigs along the chromosomes using a genetic map. A sample of validated single nucleotide polymorphisms (SNPs) in non-repetitive regions of the genome was used to develop 799 markers (Pindo et al. 2008; Vezzulli et al. 2008) for the genotyping of a Syrah x Pinot Noir full sib family (Troggio et al. 2007). This genetic map currently includes 1,767 molecular markers arranged in 19 linkage groups and covering 1,276 cM (Velasco et al. 2007).

By matching the sequences of 1,356 markers to the assembled contigs, the 397 largest metacontigs were anchored to the genetic map, thus assigning to the linkage groups about 435.1 Mb of sequence or more than 91% of the metacontig sequences. Over 81% of these sequences have been anchored by two or more markers and, therefore, are oriented. The largest metacontig, mapped to linkage group 18, was anchored with 21 markers and covered almost 32.3 cM of the map. Large metacontigs were mostly associated with non-centromeric regions of the chromosomes with a relatively high recombination rate: about 150–200 kb per cM. Pericentromeric regions, abundant in TEs and tandem repeats, were covered with multiple short metacontigs, most of which were poorly anchored to the genetic map (not oriented or ambiguous order). The recombination rate in these regions was proved to be low and the DNA content per cM to increase from 300 to 1,000 kb. Sequences which are still unmapped are mostly represented by short metacontigs and singleton contigs containing multiple tandem repetitive sequences (Zharkikh et al. 2008).

SNP-based markers were essential in improving the metacontig assembly. Many adjacent metacontigs that were not initially merged because of non-significant links between them were associated based on neighboring genetic markers and could therefore be successfully merged into a single larger metacontig. On the other hand, if a metacontig was associated with several markers from different linkage groups or with distant markers from the same linkage group, it was considered chimeric and was split into separate metacontigs by a semi-automated procedure (Velasco et al. 2007). The anchored sequence of the cultivated grapevine genome together with the large arsenal of SNP loci are available at the public databases and at *http://genomics.research.iasma.it* (Velasco et al. 2007).

9.4.3 The Pinot Noir Genome Structure

The set of Pinot Noir chromosome pairs included a considerable number of haplotype-specific gaps (sequences present in one haplotype but not in the other). The total length of the 1,042,174 identified gaps corresponded to 48.9 Mb (Velasco et al. 2007). In some chromosomal regions, the two alternative haplotypes were too different for the algorithm employed during assembly to combine them into a single contig and were considered by Velasco et al. (2007) as hemizygous DNA (22,061 contigs with a total length of 65.1 Mb). All these data allowed the authors to conclude that the homologous chromosomes of Pinot Noir differed on average by 11.2% of their DNA sequences and that the grapevine genome exists in a dynamic state, mediated at least in part by TE activity, as reported for other species (Morgante et al. 2005). Indeed, the large grapevine genomic gaps are frequently bordered by 5 bp direct repeats, reminiscent of a type of

DNA excision mediated by a precise process of transposition (Chandler and Mahillon 2002). A total of 2 million SNPs (1,751,176 of which are identified in the anchored contigs) were discovered along the grapevine genome together with more than a million of insertion/deletion events (Velasco et al. 2007) for an overall rate of 6.3 polymorphisms per kilobase. The level of heterozygosity observed in Pinot Noir was much higher compared to that reported for the heterogozygous Nisqually-1 genome in poplar (Tuskan et al. 2006) where 1,241,251 SNPs or indels were identified for an overall rate of 2.6 polymorphisms per kilobase. An average value of 4.0 SNP per kb was observed, with coding and non-coding regions demonstrating different degrees of polymorphism, 2.5 and 5.5 SNPs per kb respectively (Velasco et al. 2007), while in poplar 83% of the polymorphisms discovered occurred in non-coding portions of the genome. One or more SNPs were found in 86.7% of anchored genes and 71.4% of genes had more than four SNPs (Table 9-4). Those gene-based markers are valuable tools, as SNPs present in functional genes may cause natural phenotypic variation (Fridman et al. 2000; Thornsbery et al. 2001) and help in genetic diagnosis.

A reduction of SNP frequency in gene desert regions was observed as described for the dog genome (Lindblad-Toh et al. 2005). Independent of the gene density, SNP frequency was found not homogeneous along the grapevine genome, with regions where SNP frequency peaked between 5 and 7.5 per 1 kb, and others displaying dramatically reduced frequencies (Fig. 9-3). However, the sparseness of putative quasi-homozygous haplotypic blocks indicates that heterozygosity prevails (Velasco et al. 2007).

The existence of structural diversity between homologous chromosomes within plant species involving different contents in coding regions has been reported in maize (Brunner et al. 2005; Clark et al. 2007) as well as in allogamous plants (Rafalski 2002) such as grapevine (Velasco et al. 2007).

9.4.4 Comparison with the Nearly Homozygous Genotype PN40024 Assembly

The 58,611 contigs obtained from sequencing the heterozygous genome of grape (HTA = heterozygous assembly) were compared with the 57,634 contigs obtained from sequencing the near homozygous line of grapevine (PN40024) (HMA = homozygous assembly based on 8.4 × read coverage). About 57% of contigs from each set showed at least 1,000 bp non-repetitive sequence overlaps with contigs from the alternative set. More than 50% of the overlaps could be perfectly aligned with the number of sequence discrepancies corresponding to the estimated level of heterozygosity. The remaining overlaps contained multiple large heterozygous insertions and small (0.5–1.5 kb) sequence inversions which prevent reliable alignment of contigs. 15,851 HTA contigs and 14,148 HMA contigs were completely

Table 9-4 SNPs in exons and non-coding DNA (A) and percentage of anchored genes tagged with SNPs (B).

A.

Region	SNPs No.	SNPs No./kb
In non-coding	998,149	3.1
Intron	642,219	5.5
Exon	110,808	2.5

B.

SNPs/gene	No. anchored Genes	(%)
0	3,773	13.3
1-3	4,362	15.3
≥ 4	20,292	71.4
Total	28,427	100.0

overlapped by the corresponding alternative contigs, whereas 26,193 HTA contigs and 21,613 HMA contigs closed gaps in the alternative assembly. Adding all these numbers, the combined HTA+HMA assembly closed about 63.3% of HTA gaps and 70.0% of HMA gaps (Zharkikh et al. 2008).

9.5 Perspectives

The automatic annotation has then to be progressively refined by an expert-based manual annotation, which is made by a community of scientists and ideally should be centralized and organized (See as an example Aubourg et al. 2005; Menda et al. 2008). A first glance on the protein-coding sequences in grapevine showed an expansion of some gene families involved into the biosynthesis in terpen and flavonoid compounds, both important for the berry taste, before and after vinification. For instance, 89 genes and 27 pseudogenes were found in grapevine, arranged in clusters, sometimes of important size (33 genes in a 672 kb tandem array on chromosome 18; Jaillon et al. 2007) to be compared to around 40 genes in *Arabidopsis*, poplar and rice (Aubourg et al. 2002; Jaillon et al. 2007). Similarly, the stilbene synthase gene family is arranged in 9 clusters in grapevine, totaling 43 genes and 9 pseudogenes while in *Arabidopsis*, only 4 genes can be found and in rice and poplar, no more than 30 genes (Jaillon et al. 2007). It is now possible to set up a global strategy to decipher the individual biochemical function of each family member and its physiological role. In a more general point of view it is now possible to set up tools for the genome wide analysis of gene transcription and translation in grapevine.

Having a reference genome sequence allows the high throughput resequencing of varieties or wild species to extend the first insight on the molecular basis of polymorphism in grapevine given by the Pinot Noir genome sequence (Velasco et al. 2007). This was recently started by Myles et al. (2010) using a Solexa/Illumina platform and is necessary to fully

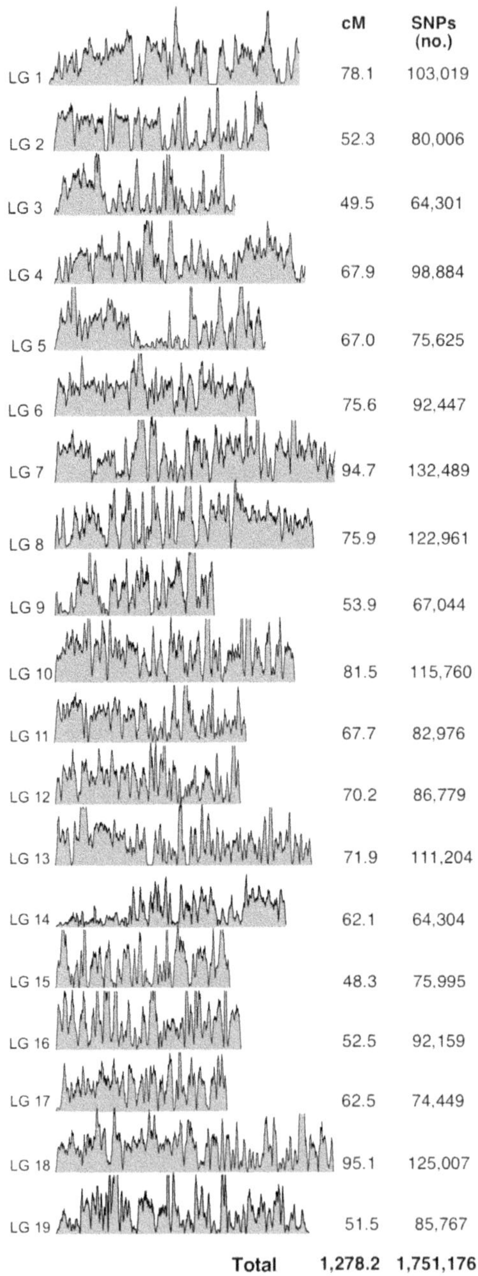

Figure 9-3 SNP density along the 19 chromosomes (LG) of *V. vinifera*. Left and right of the figure correspond respectively to top and bottom of chromosomes according to Troggio et al. (2007). The SNP values reported do not consider gaps in and among metacontigs.

understand the role of different structural variations (transposition, unequal recombination, mutations, etc.) to the phenotype variation (Hurles et al. 2008). It is also the ground for the development of efficient tools for the management and use of clonal variation (Benjak et al. 2008; Schellenbaum et al. 2008). The recent genome wide analysis of the class II TEs present in the grapevine genome showed a contribution of these elements to the high polymorphism observed among grapevine varieties (Benjak et al. 2008); these elements are therefore good candidates for scavenging the clonal variation within *V. vinifera* varieties. Finally, it allows studying the epigenetic regulation of grapevine at a genome scale and its interplay with the structural variations to shape the polymorphism in grapevine (Vaughn et al. 2007; Beck and Rakhyan 2008).

References

Adam-Blondon A-F, Bernole A, Faes G, Lamoureux D, Pateyron S, Grando MS, Caboche M, Velasco R, Chalhoub B (2005) Construction and characterization of BAC libraries from major grapevine cultivars. Theor Appl Genet 110: 1363–1371.

Arabidopsis Genome Initiative (2000) Analysis of the genome sequence of the flowering plant Arabidopsis thaliana. Nature 408: 796–815.

Aradhya MK, Dangl GS, Prins BH, Boursiquot J-M, Walker MA, Meredith CP, Simon CJ (2003) Genetic structure and differentiation in cultivated grape, *Vitis vinifera* L. Genet Res Camb 81: 179–192.

Aubourg S, Lecharny A, Bohlman J (2002) Genomic analysis of the terpenoid synthase (AtTPS) gene family of Arabidopsis thaliana. Mol Genet Genom 267: 730–747.

Aubourg S, Brunaud V, Bruyère C, Cock M, Cooke R, Cottet A, Couloux A, Déhais P, Deléage G, Drouard L, Duclert A, Echeverria M, Eschbach A, Falconet D, Filippi G, Gaspin C, Geourjon C, Grienenberger J-M, Guermann B, Houlné G, Jamet E, Lechauve F, Leleu O, Leroy P, Mache R, Meyer C, Nedjari H, Negrutiu I, Orsini V, Peyretaillade E, Pommier C, Raes J, Risler J-L. Rivière S, Rombauts S, Rouzé P, Schneider M, Schwob P, Small I, Soumayet-Kampetenga G, Stankovski D, Toffano C, Tognolli M, Caboche M, Lecharny A (2005) The GENEFARM project: structural and functional annotation of Arabidopsis gene and protein families by a network of experts. Nucl Acids Res 33: D641–D646.

Aury JM, Jaillon O, Duret L, Noel B, Jubin C, Porcel BM, Ségurens B, Daubin V, Anthouard V, Aiach N, Arnaiz O, Billaut A, Beisson J, Blanc I, Bouhouche K, Câmara F, Duharcourt S, Guigo R, Gogendeau D, Katinka M, Keller AM, Kissmehl R, Klotz C, Koll F, Le Mouël A, Lepère G, Malinsky S, Nowacki M, Nowak JK, Plattner H, Poulain J, Ruiz F, Serrano V, Zagulski M, Dessen P, Bétermier M, Weissenbach J, Scarpelli C, Schächter V, Sperling L, Meyer E, Cohen J, Wincker P (2006) Global trends of whole-genome duplications revealed by the ciliate Paramecium tetraurelia. Nature 444: 171–178.

Barker CL, Donald T, Pauquet J, Ratnaparkhe A, Bouquet A, Adam-Blondon A-F, Thomas MR, Dry I (2005) Genetic and physical mapping of the grapevine powdery mildew resistance gene, *Run1*, using a bacterial artificial chromosome library. Theor Appl Genet 111: 370–377.

Beck S, Rakyan VK (2008) The Methylome: approaches for global DNA methylation profiling. Trends Genet 24: 231–237.

Benjak A, Forneck A, Casacuberta JM (2008) Genome-wide analysis of the 'Cut-and-Paste" transposons of grapevine. PLOSone 3: e3107.

Bogs J, Ebadi A, McDavid D, Robinson SP (2006) Identification of the Flavonoid Hydroxylases from grapevine and their regulation during fruit development. Plant Physiol 140: 279–291.

Bowers JE, Chapman BA, Rong J, Paterson AH (2003) Unravelling Angiosperm genome evolution by phylogenetic analysis of chromosomal duplication events. Nature 422: 433–438.

Bronner A, Oliveira J (1990) Creation and study of the Pinot Noir variety lineage. In: Proc 5th Int Symp Grape Breeding, St Martin/Pflaz, Germany, Sept 1989, Vitis, spl issue, pp 69–80.

Brunner S, Fengler K, Morgante M, Tingey S, Rafalski A (2005) Evolution of DNA sequence non homologies among maize inbreds. Plant Cell 17: 343–360.

Castellarin SD, Di Gaspero G, Marconi R, Nonis A, Peterlunger E, Paillard S, Adam-Blondon A-F, Testolin R (2006) Colour variation in red grapevines (*Vitis vinifera* L.): genomic organisation, expression of flavonoid 3′-hydroxylase, flavonoid 3′, 5′-hydroxylase genes and related metabolite profiling of red cyanidin-/blue delphinidin-based anthocyanins in berry skin. BMC Genom 7: 12.

Castelli V, Aury JM, Jaillon O, Wincker P, Clepet C, Menard M, Cruaud C, Quetier F, Scarpelli C, Schachter V, Temple G, Caboche M, Weissenbach J, Salanoubat M (2004) Whole genome sequence comparisons and "full-length" cDNA sequences: a combined approach to evaluate and improve Arabidopsis genome annotation. Genome Res 14: 406–413.

Chandler M, Mahillon J (2002) Mobile DNA II. In: NL Craig, R Craigie, M Gellert, AM Lambowitz (eds) Insertion Sequences revisited. Am Soc Microbiol. Washington DC, USA, pp 305–366.

Clark RM, Schweikert G, Toomajian C, Ossowski S, Zeller G, Shinn P, Warthmann N, Hu TT, Fu G, Hinds DA, Chen H, Frazer KA, Huson DH, Schölkopf B, Nordborg M, Rätsch G, Ecker JR, Weigel D (2007) Common sequence polymorphisms shaping genetic diversity in *Arabidopsis thaliana*. Science 317: 338–342.

Claverie M, Dirlewanger E, Cosson P, Bosselut N, Lecouls AC, Voisin R, Kleinhentz M, Lafargue B, Caboche M, Chalhoub B, Esmenjaud D (2004) High-resolution mapping and chromosome landing at the root-knot nematode resistance locus Ma from Myrobolan plum using a large-insert BAC DNA library. Theor Appl Genet 109: 1318–1327.

Denoeud F, Aury J-M, Da Silva C, Noel B, Rogier O, Delledonne M, Morgante M, Valle G, Wincker P, Scarpelli C, Jaillon O, Artiguenave F (2008) Annotating genomes with massive-scale RNA-sequencing. Genome Biol 9: R175.

Dezulian T, Remmert M, Palatnik JF, Weigel D, Huson DH (2006) Identification of plant microRNA homologs. Bioinformatics 22: 359–360.

Doligez A, Adam-Blondon A-F, Cipriani G, Di Gaspero G, Laucou V, Merdinoglu D, Meredith CP, Riaz S, Roux C, This P (2006) An integrated SSR map of grapevine based on five mapping populations. Theor Appl Genet 113: 369–382.

Fridman E, Pleban T, Zamir D (2000) A recombination hotspot delimits a wild-species quantitative trait locus for tomato sugar content to 484 bp within an invertase gene. Proc Natl Acad Sci USA 97: 4718–4723.

Goff SA, Ricke D, Lan TH, Presting G, Wang R, Dunn M, Glazebrook J, Sessions A, Oeller P, Varma H, Hadley D, Hutchison D, Martin C, Katagiri F, Lange B.M, Moughamer T, Xia Y, Budworth P, Zhong J, Miguel T, Paszkowski U, Zhang S, Colberrt M, Sun WL, Chen L, Cooper B, Park S, Wood TC, Mao L, Quail P, Wing R, Dean R, Yu Y, Zharkikh A, Shen R, Sahasrabudhe S, Thomas A, Cannings R, Gutin A, Pruss D, Reid J, Tavtigian S, Mitchell J, Eldredge G, Scholl T, Miller RM, Batnagar S, Adey N, Rubano T, Tusneem N, Robinson R, Feldhaus J, Macalma T, Oliphant A, Briggs S (2002) A draft sequence of the rice genome (*Oryza sativa* L. ssp. *japonica*). Science 296: 92–100.

Green ED (2001) Strategies for the systematic sequencing of complex genomes. Nat Rev 2: 573–583.

Han Y, Gasic K, Marron B, Beever JE, Korban SS (2007) A BAC-based physical map of the apple genome. Genomics 89: 630–637.

Huang S, Li R, Zhang Z, Li L, Gu X, Fan W, Lucas WJ, Wang X, Xie B, Ni P, Ren Y, Zhu H, Li J, Lin K, Jin W, Fei Z, Li G, Staub J, Kilian A, van der Vossen EAG, Wu Y, Guo J, He J, Jia Z, Ren Y, Tian G, Lu Y, Ruan J, Qian W, Wang M, Huang Q, Li B, Xuan Z, Cao J, Asan, Wu Z, Zhang J, Cai Q, Bai Y, Zhao B, Han Y, Li Y, Li X, Wang S, Shi Q, Liu S, Cho WK, Kim J-Y, Xu Y, Heller-Uszynska K, Miao H, Cheng Z, Zhang S, Wu J, Yang Y, Kang H, Man Li, Liang H, Ren X, Shi Z, Wen M, Jian M, Yang H, Zhang G, Yang Z, Chen R, Liu S, Li J, Ma L, Liu H, Zhou Y, Zhao J, Fang X, Li G, Fang L, Li Y, Liu D, Zheng H, Zhang Y, Qin N, Li Z, Yang G, Yang S, Bolund L, Kristiansen K, Zheng H, Li S, Zhang X, Yang H, Wang J, Sun R, Zhang B, Jiang S, Wang J, Du Y, Li S (2009) The genome of the cucumber, *Cucumis sativus* L. Nat Genet 41: 1275–1282.

Hurles ME, Dermitzakis ET, Tyler-Smith C (2008) The functional impact of structural variation in humans. Trends Genet 24: 238–245.

International Rice Genome Sequencing Project -IRGSP-(2005) The map-based sequence of the Rice genome. Nature 436: 793–800.

Jaillon O, Aury JM, Brunet F, Petit JL, Stange-Thomann N, Mauceli E, Bouneau L, Fischer C, Ozouf-Costaz C, Bernot A, Nicaud S, Jaffe D, Fisher S, Lutfalla G, Dossat C, Segurens B, Dasilva C, Salanoubat M, Levy M, Boudet N, Castellano S, Anthouard V, Jubin C, Castelli V, Katinka M, Vacherie B, Biemont C, Skalli Z, Cattolico L, Poulain J, De Berardinis V, Cruaud C, Duprat S, Brottier P, Coutanceau JP, Gouzy J, Parra G, Lardier G, Chapple C, McKernan KJ, McEwan P, Bosak S, Kellis M, Volff JN, Guigo R, Zody MC, Mesirov J, Lindblad-Toh K, Birren B, Nusbaum C, Kahn D, Robinson-Rechavi M, Laudet V, Schachter V, Quetier F, Saurin W, Scarpelli C, Wincker P, Lander ES, Weissenbach J, Roest Crollius H (2004) Genome duplication in the teleost fish Tetraodon nigroviridis reveals the early vertebrate proto-karyotype. Nature 431: 946–957.

Jaillon O, Aury J-M, Noel B, Policriti A, Clepet C, Casagrande A, Choisne N, Aubourg S, Vitulo N, Jubin C, Vezzi A, Legeai F, Hugueney P, Dasilva C, Horner D, Mica E, Jublot D, Poulain J, Bruyere C, Billault A, Segurens B, Gouyvenoux M, Ugarte E, Cattonaro F, Anthouard V, Vico V, Del Fabbro C, Alaux M, Di Gaspero G, Dumas V, Felice N, Paillard S, Juman I, Moroldo M, Scalabrin S, Canaguier A, Le Clainche I, Malacrida G, Durand E, Pesole G, Laucou V, Chatelet P, Merdinoglu D, Delledonne M, Pezzotti M, Lecharny A, Scarpelli C, Artiguenave F, Pé E, Valle G, Morgante M, Caboche M, Adam-Blondon A-F, Weissenbach J, Quétier F, Wincker P (2007) The grapevine genome sequence suggests ancestral hexaploidization in major angiosperm phyla. Nature 449: 463–468.

Jeong ST, Goto-Yamamoto N, Hashizume K, Esaka M (2005) Expression of the flavonoid 3′-hydroxylase and flavonoid composition in grape (*Vitis vinifera*). Plant Sci 170: 61–69.

Jung S, Jesudurai C, Staton M, Du Z, Ficklin S, Cho I, Abbott A, Tomkins J, Main D (2004) GDR (Genome Database for *Rosaceae*): integrated web resources for *Rosaceae* genomics and genetics research. BMC Bioinformat 5: 130.

Kelleher CT, Chiu R, Shin H, Bosdet IE, Krzywinski MI, Fjell CD, Wilkin J, Yin T, DiFazio SP, Ali J, Asano JK, Chan S, Cloutier A, Girn N, Leach S, Lee D, Mathewson CA, Olson T, O'connor K, Prabhu AL, Smailus DE, Stott JM, Tsai M, Wye NH, Yang GS, Zhuang J, Holt RA, Putnam NH, Vrebalov J, Giovannoni JJ, Grimwood J, Schmutz J, Rokhsar D, Jones SJ, Marra MA, Tuskan GA, Bohlmann J, Ellis BE, Ritland K, Douglas CJ, Schein JE (2007) A physical map of the highly heterozygous *Populus* genome: integration with the genome sequence and genetic map and analysis of haplotype variation. Plant J 50: 1063–1078.

Lamoureux D, Bernole A, Le Clainche I, Tual, S, Thareau V, Paillard S, Legeai F, Dossat C, Wincker P, Oswald M, Merdinoglu D, Vignault C, Delrot S, Caboche M, Chalhoub B, Adam-Blondon A-F (2006) Anchoring a large set of markers onto a BAC library for the development of a draft physical map of the grapevine genome. Theor Appl Genet 113: 344–356.

Lindblad-Toh K, Wade CM, Mikkelsen TS, Karlsson EK, Jaffe DB, Kamal M, Clamp M, Chang JL, Kulbokas III EJ, Zody MC, Mauceli E, Xie X, Breen M, Wayne RK, Ostrander EA, Ponting CP, Galiber F, Smith DR, deJong PJ, Kirkness E, Alvarez P, Biagi T, Brockman W, Butler J, Chin CW, Cook A, Cuff J, Daly MJ, DeCaprio D, Gnerre S, Grabherr M, Kellis

M, Kleber M, Bardeleben C, Goodstadt L, Heger A, Hitte C, Kim L, Koepfli KP, Parker HG, Pollinger JP, Searle SMJ, Sutter NB, Thomas R, Weber C, Broad Sequencing Platform members, Lander ES. (2005) Genome sequence, comparative analysis and haplotype structure of the domestic dog. Nature 438: 803–819.

Luo MC, Thomas C, You FM, Hsiao J, Ouyang S, Buell CR, Malandro M, McGuire PE, Anderson OD, Dvorak J (2003) High-throughput fingerprinting of bacterial artificial chromosomes using the snapshot labeling kit and sizing of restriction fragments by capillary electrophoresis. Genomics 82: 378–389.

Margulies M, Egholm M, Altman WE, Attiya S, Bader JS, Bemben LA, Berka J, Braverman MS, Chen YJ, Chen Z, Dewell SB, Du L, Fierro JM, Gomes XV, Godwin BC, He W, Helgesen S, Ho CH, Irzyk GP, Jando SC, Alenquer ML, Jarvie TP, Jirage KB, Kim JB, Knight JR, Lanza JR, Leamon JH, Lefkowitz SM, Lei M, Li J, Lohman KL, Lu H, Makhijani VB, McDade KE, McKenna MP, Myers EW, Nickerson E, Nobile JR, Plant R, Puc BP, Ronan MT, Roth GT, Sarkis GJ, Simons JF, Simpson JW, Srinivasan M, Tartaro KR, Tomasz A, Vogt KA, Volkmer GA, Wang SH, Wang Y, Weiner MP, Yu P, Begley RF, Rothberg JM (2005) Genome sequencing in microfabricated high-density picolitre reactors. Nature 437: 376–380.

Menda N, Buels RM, Tecle I, Mueller LA (2008) A community-based annotation framework for linking Solanaceae genomes with phenomes. Plant Physiol 147: 1788–1799.

Meyers BC, Scalabrin S, Morgante M (2004) Mapping and sequencing complex genomes: let's get physical! Nat Rev Genet 5: 578–588.

Ming R, Hou S, Feng Y, Yu Q, Dionne-Laporte A, Saw JH, Senin P, Wang W, Ly BV, Lewis KL, Salzberg SL, Feng L, Jones MR, Skelton RL, Murray JE, Chen C, Qian W, Shen J, Du P, Eustice M, Tong E, Tang H, Lyons E, Paull RE, Michael TP, Wall K, Rice DW, Albert H, Wang ML, Zhu YJ, Schatz M, Nagarajan N, Acob RA, Guan P, Blas A, Wai CM, Ackerman CM, Ren Y, Liu C, Wang J, Wang J, Na JK, Shakirov EV, Haas B, Thimmapuram J, Nelson D, Wang X, Bowers JE, Gschwend AR, Delcher AL, Singh R, Suzuki JY, Tripathi S, Neupane K, Wei H, Irikura B, Paidi M, Jiang N, Zhang W, Presting G, Windsor A, Navajas-Pérez R, Torres MJ, Feltus FA, Porter B, Li Y, Burroughs AM, Luo MC, Liu L, Christopher DA, Mount SM, Moore PH, Sugimura T, Jiang J, Schuler MA, Friedman V, Mitchell-Olds T, Shippen DE, dePamphilis CW, Palmer JD, Freeling M, Paterson AH, Gonsalves D, Wang L, Alam M (2008) The draft genome of the transgenic tropical fruit tree papaya (*Carica papaya* Linnaeus). Nature 452: 991–996.

Morgante M, Salamini F (2003) From plant genomics to breeding practice. Curr Opin Biotechnol 14: 214–219.

Morgante M, Brunner S, Pea G, Fengler K, Zuccolo A, Rafalski A (2005) Gene duplication and exon shuffling by helitron-like transposons generate intraspecies diversity in maize. Nat Genet 37: 997–1002.

Moroldo M, Paillard S, Marconi R, Legeai F, Canaguier A, Cruaud C, De Berardinis V, Guichard C, Brunaud V, Le Clainche I, Scalabrin S, Testolin R, Di Gaspero G, Morgante M, Adam-Blondon AF (2008) A physical map of the heterozygous grapevine 'Cabernet Sauvignon' allows mapping candidate genes for disease resistance. BMC Plant Biol 8: 66.

Myles S, Chia J-M, Hurwitz B, Simon C, Zhong GY, Buckler E, Ware D (2010) Rapid Genomic Characterization of the Genus Vitis. PLOS ONE 5: e8219.

Nelson W, Soderlund C (2005) Software for restriction fragment physical maps. In: K Meksem, G Kahl (eds) The Handbook of Genome Mapping: Genetic and Physical Mapping. (eds K. Meksem and G. Kahl), Wiley-VCH Verlag GmbH & Co. KGaA, Weinheim, FRG pp 285–306 .

Nelson WM, Dvorak J, Luo MC, Messing J, Wing RA, Soderlund C (2007) Efficacy of clone fingerprinting methodologies. Genomics 89: 160–166.

Ohno S (1970) Evolution by Gene Duplication. Springer-Verlag, New York, USA.

Patel GI, Olmo HP (1955) Cytogenetics of Vitis 1 The hybrid *V. vinifera* x *V. rotundifolia*. Am J Bot 42: 141–159.

Patocchi A, Vinatzer BA, Gianfranceschi L, Tartarini S, Zhang HB, Sansavini S, Gessler C (1999) Construction of a 550 kb BAC contig spanning the genomic region containing the apple scab resistance gene Vf. Mol Gen Genet 4–5: 884–891.

Pindo M, Vezzulli S, Coppola G, Cartwright D, Zharkikh A, Velasco R, Troggio M (2008) SNP High throughput screening in grapevine using the SNPlexTM genotyping system. BMC Plant Biol 8: 12.

Rafalski A (2002) Applications of single nucleotide polymorphisms in crop genetics. Curr Opin Plant Biol 5: 94–100.

Rensing RA, Lang DD, Zimmer AD, Terry A, Salamov A, Shapiro H, Nishiyama T, Perroud P-F, Lindquist EA, Kamisugi Y, Tanahashi T, Sakakibara K, Fujita T, Oishi K, Shin-IT, Kuroki Y, Toyoda A, Suzuki Y, Hashimoto S-I, Yamaguchi K, Sugano S, Kohara Y, Fujiyama A, Anterola A, Aoki S, Ashton N, Barbazuk WB, Barker E, Bennetzen JL, Blankenship R, Cho SH, Dutcher SK, Estelle M, Fawcett JA, Gundlach H, Hanada K, Heyl A, Hicks KA, Hughes J, Lohr M, Mayer K, Melkozernov A, Murata T, Nelson DR, Pils B, Prigge M, Reiss B, Renner T, Rombauts S, Rushton PJ, Sanderfoot A, Schween G, Shiu S-H, Stueber K, Theodoulou FL, Tu H, Van de Peer Y, Verrier PJ, Waters E, Wood A, Yang L, Cove D, Cuming AC, Hasebe M, Lucas S, Mishler BD, Reski R, Grigoriev IV, Quatrano RS, Boore JL (2008) The *Physcomitrella* genome reveals evolutionary insights into the conquest of the land by plants. Science 319: 64–69.

Roest Crollius H, Jaillon O, Bernot A, Dasilva C, Bouneau L, Fischer C, Fizames C, Wincker P, Brottier P, Quétier F, Saurin W, Weissenbach J (2000) Human gene number estimate provided by genome wide analysis using *Tetradon nigroviridis* genomic DNA. Nature Genet 25: 235–238.

Sanger F, Nicklen S, Coulson AR (1977) DNA sequencing with chain-terminating inhibitors. Proc Natl Acad Sci USA 74(12): 5463–5467.

Scalabrin S, Troggio M, Moroldo M, Pindo M, Felice N, Coppola G, Prete G, Malacarne G, Marconi R, Faes G, Jurman I, Grando S, Jesse T, Segala C, Valle G, Policriti A, Fontana P, Morgante M, Velasco R (2010) Physical mapping in highly heterozygous genomes: a physical contig map of the Pinot Noir grapevine cultivar BMC Genom 11: 204.

Schellenbaum P, Mohler V, Wenzel G, Walter B (2008) Variation in DNA methylation patterns of grapevine somaclones (*Vitis vinifera* L.). BMC Plant Biol 8: 78.

Shendure J, Mitra RD, Varma C, Church GM (2004) Advanced sequencing technologies: methods and goals. Nat Rev Genet 5: 335–344.

Tomkins JP, Peterson DG, Yang TJ, Ablett ER, Henry RJ, Lee LS, Holton TA, Waters D, Wing RA (2001) Grape (*Vitis vinifera* L.) BAC library construction, preliminary STC analysis, and identification of clones associated with flavonoid and stilbene biosynthesis. Am J Enol Vitic 52: 287–291.

Town CD (2006) Annotating the genome of *Medicago truncatula*. Curr Opin Plant Biol 9: 122–127.

Thornsberry JM, Goodman MM, Doebley J, Kresovich S, Nielsen D, Buckler IV ES (2001) *Dwarf8* polymorphisms associate with variation in flowering time. Nat Genet 28: 286–289.

Troggio M, Malacarne G, Coppola G, Segala C, Cartwright DA, Pindo M, Stefanini M, Mank R, Moroldo M, Morgante M, Grando MS, Velasco R (2007) A dense single-nucleotide polymorphism-based genetic linkage map of grapevine (*Vitis vinifera* L.) anchoring Pinot Noir bacterial artificial chromosome contigs. Genetics 176: 2637–2650.

Tuskan GA, Difazio S, Jansson S, Bohlmann J, Grigoriev I, Hellsten U, Putnam N, Ralph S, Rombauts S, Salamov A, Schein J, Sterck L, Aerts A, Bhalerao RR, Bhalerao RP, Blaudez D, Boerjan W, Brun A, Brunner A, Busov V, Campbell M, Carlson J, Chalot M, Chapman J, Chen GL, Cooper D, Coutinho PM, Couturier J, Covert S, Cronk Q, Cunningham R, Davis J, Degroeve S, Déjardin A, Depamphilis C, Detter J, Dirks B, Dubchak I, Duplessis S, Ehlting J, Ellis B, Gendler K, Goodstein D, Gribskov M, Grimwood J, Groover A, Gunter L, Hamberger B, Heinze B, Helariutta Y, Henrissat B, Holligan D, Holt R, Huang W, Islam-Faridi N, Jones S, Jones-Rhoades M, Jorgensen R, Joshi C, Kangasjärvi J, Karlsson J, Kelleher C, Kirkpatrick R, Kirst M, Kohler A, Kalluri U, Larimer F, Leebens-Mack

J, Leplé JC, Locascio P, Lou Y, Lucas S, Martin F, Montanini B, Napoli C, Nelson DR, Nelson C, Nieminen K, Nilsson O, Pereda V, Peter G, Philippe R, Pilate G, Poliakov A, Razumovskaya J, Richardson P, Rinaldi C, Ritland K, Rouzé P, Ryaboy D, Schmutz J, Schrader J, Segerman B, Shin H, Siddiqui A, Sterky F, Terry A, Tsai CJ, Uberbacher E, Unneberg P, Vahala J, Wall K, Wessler S, Yang G, Yin T, Douglas C, Marra M, Sandberg G, Van de Peer Y, Rokhsar D (2006) The genome of black cottonwood, *Populus trichocarpa* (Torr.& Gray). Science 313: 1596–1604.

Vaughn MW, Tanurd IcM, Lippman Z, Jiang H, Carrasquillo R, Rabinowicz PD, Dedhia N, McCombie WR, Agier N, BulskiA, Colot V, Doerge RW, Martienssen RA (2007) Epigeneic natural variation in Arabidopsis thaliana. PLOS Biol 5: e174.

Velasco R, Zharkikh A, Troggio M, Cartwright DA, Cestaro A, Pruss D, Pindo M, Fitzgerald LM, Vezzulli S, Reid J, Malacarne G, Iliev D, Coppola G, Wardell B, Micheletti D, Macalma T, Facci M, Mitchell JT, Perazzolli M, Eldredge G, Gatto P, Oyzerski R, Moretto M, Gutin N, Stefanini M, Chen Y, Segala C, Davenport C, Demattè L, Mraz A, Battilana J, Stormo K, Costa F, Tao Q, Si-Ammour A, Harkins T, Lackey A, Perbost C, Taillon B, Stella A, Solovyev V, Fawcett JA, Sterck L, Vandepoele K, Grando SM, Toppo S, Moser C, Lanchbury J, Bogden R, Skolnick M, Sgaramella V, Bhatnagar SK, Fontana P, Gutin A, Van de Peer Y, Salamini F, Viola R (2007) A high quality draft consensus sequence of the genome of a heterozygous grapevine variety. PLoS ONE 2: e1326.

Vezzulli S, Micheletti D, Riaz R, Pindo M, Viola R, This P, Walker MA, Troggio M, Velasco R (2008) A SNP transferability survey within the genus *Vitis*. BMC Plant Biol 8: 128.

Vinson JP, Jaffe DB, O'Neill K, Karlsson EK, Stange-Thomann N, Anderson S, Mesirov JP, Satoh N, Satou Y, Nusbaum C, Birren B, Galagan JE, Lander ES (2005) Assembly of polymorphic genomes: algorithms and application to Ciona savignyi. Genome Res 15: 1127–1135.

Yu J, Hu S, Wang J, Wong GK-S, Li S, Liu B, Deng Y, Dai L, Zhou Y, Zhang X, Cao M, Liu J, Sun J, Tang J, Chen Y, Huang X, Lin W, Ye C, Wei Tong W, Cong L, Geng G, Han Y, Li L, Li W, Hu G, Huang X, Li W, Li J, Liu Z, Li L, Liu J, Qi Q, Liu J, Li L, Li T, Wang X, Lu H, Wu T, Zhu M, Ni P, Han H, Dong W, Ren X, Feng X, Cui P, Li X, Wang H, Xu X, Zhai W, Xu Z, Zhang J, He S, Zhang J, Xu J, Zhang K, Zheng X, Dong J, Zeng W, Tao L, Ye J, Tan J, Ren X, Chen X, He J, Liu D, Tian W, Tian C, Xia H, Bao Q, Li G, Gao H, Cao T, Wang J, Zhao W, Li P, Chen W, Wang X, Zhang Y, Hu J, Wang J,Liu S, Yang J, Zhang G, Xiong Y, Li Z, Mao L, Zhou C, Zhu Z, Chen R, Hao B, Zheng W, Chen S, Guo W, Li G, Liu S, Tao M, Wang J, Zhu L, Yuan L, Yang H (2002) A draft sequence of the rice genome (*Oryza sativa* L. ssp. *indica*). *Science* 296: 79–92.

Zharkikh A, Troggio M, Pruss D, Cestaio A, Eldrdge G, Pindo M, Mitchell JT, Vezzulli S, Bhatnagar S, Fontana P, Viola R, Gutin A, Salamini F, Skolnick M, Velasco R (2008) Sequencing and assembly of highly heterozygous genome of *Vitis vinifera*: problems and solutions. J Biotechnol 36: 38–43.

10

Vitis Functional Genomics: Open Systems for Transcriptome Analysis

Richard L. Tillett and *John C. Cushman**

ABSTRACT

Technological advances in the last decade have resulted in a dramatic increase in the volume of genomic and transcriptomic sequence data available for *Vitis vinifera* L. and related *Vitis* species. Until recently, the vast majority of transcript data has been obtained using traditional Sanger sequencing with over 350,000 expressed sequence tags (ESTs) being generated from a limited number of varieties and species. Such EST data has been instrumental in the fabrication of "closed" microarray platforms for assessing mRNA abundance patterns in a wide variety of grapevine tissues and developmental states. Traditional Sanger EST sequencing and massively parallel signature sequencing (MPSS) have now been largely displaced by the emergence of high-throughput "next generation" or "second generation" sequencing technologies that provide "open" deep-sequencing platforms for the characterization of transcriptomes also known as "RNA-Seq". In this chapter, we review research progress obtained using traditional EST and MPSS as well as summarize more recent RNA-Seq studies that are revolutionizing our understanding of the *Vitis* transcriptome. We also summarize so-called "third generation" sequencing technologies that promise to improve the efficiency and lower the cost of genome and transcriptome sequencing in the near future.

Keywords: *Vitis vinifera*, expressed sequence tag (EST), open platform, next generation sequencing, transcriptome, RNA-Seq.

Department of Biochemistry and Molecular Biology, MS 330, University of Nevada, Reno, Nevada 89557-0330, USA.

*Corresponding author: *jcushman@unr.edu*

10.1 Introduction

The berries of grapevine (*Vitis vinifera* L.) and related species are among the most widely grown and economically most import fruit crops in the world. Since the initial domestication of the grapevine more than 7000 years ago (Arroyo-Garcia et al. 2006; This et al. 2006), its berries have been used for wine production, as well as grape juice, table grapes, raisins, and more recently for leaf, seed, and skin extracts by the nutraceutical and cosmetic industries (Iriti and Faoro 2006; Monagas et al. 2006). Given its international importance as a fruit crop, intensive functional genomics investigations of grapevine have focused mainly on understanding the complex transcriptional regulatory hierarchy controlling developmental and tissue-specific gene expression patterns at nearly all stages of berry development, and on biotic and abiotic stress-specific responses in leaf, root, and berry tissues.

Recent developments in DNA sequencing technology are rapidly changing the ability of researchers to perform in-depth gene expression profiling using so-called "open" transcriptome technologies. The term "open" refers to the direct characterization, in most cases, by sequencing of a small to medium size region or "tag" of a messenger RNA (mRNA). Until the last few years, only two technological approaches dominated "open transcriptome" analysis in grapevine: 1) large-scale expressed sequence tag (EST) sampling and 2) massively parallel signature sequencing (MPSS). These now largely obsolete technologies resulted in the rapid proliferation of functional genomic resources, mainly in the form of large, publicly available EST or MPSS databases, for many varieties of *V. vinifera* and related species over the last decade. The available information has permitted large-scale mRNA expression profiling studies using cDNA or oligonucleotide microarrays or so called "closed" transcriptome technologies (see Chapter 11). However, the recent development of an assortment of next-generation short-read sequencing technologies will not only vastly accelerate tag-based studies of gene expression in grapevine, but also greatly facilitate *de novo* transcript discovery that has displaced traditional EST sequencing. Here we review the major studies that have been conducted in grapevine to date using EST and MPSS approaches. We then survey the emerging next-generation sequencing (NGS) technologies that are rapidly revolutionizing our understanding of the *Vitis* transcriptome.

10.2 Traditional Expressed Sequence Tag (EST) Sequencing

The most basic approach towards transcriptome analysis has been to perform traditional Sanger sequencing to collect expressed sequence tags (ESTs). ESTs are typically automatically curated, single-read sequences of cDNA (Rudd 2003). Libraries of cDNA can be constructed from a single tissue or a specific developmental stage and then sampled by sequencing of thousands of randomly selected clones in order to create a snapshot of the gene expression pattern within that sample. The relatively inexpensive nature of EST sequencing has made this approach a very attractive and popular means to sample the transcriptome (Rudd 2003) and provide the raw material for fabrication of cDNA microarrays (Alba et al. 2004). EST projects typically target either the 5′ end of directionally cloned cDNA, which this favors obtaining sequence from the protein-coding region, or the 3′ end, which improves the prospect of obtaining sequence from the unique untranslated region (UTR) of a transcript. Alternatively, both 5′ and 3′ sequences are collected in a so-called "paired end" approach in order to better define the ends of cDNAs, which are important for genome annotation. In the absence of complete genome sequence data, the availability of EST data provides a low-cost resource for gene discovery and accessible means to sample the actively transcribed portions of a genome (Rudd 2003). Comparison of available EST resources between diverse taxa (e.g., tomato and grape) can also provide novel insights into molecular processes, such as fruit ripening, and can result in the identification of conserved transcription factors not previously associated with ripening (Fei et al. 2004). Furthermore, now that complete genome sequences have become available for *Vitis vinifera* (Jaillon et al. 2007; Velasco et al. 2007; see also Chapter 9), EST data provide an invaluable resource for genome annotation as well as bioinformatic databases and visualization tools (see Chapter 13).

Despite its utility, EST sequencing has a number of weaknesses. First, the sampling depth of most traditional Sanger EST projects is often limited to a few hundred or thousand sequenced cDNAs. Although cDNA libraries should accurately reflect the relative abundance of a particular transcript, because they are typically made from a single organ or tissue or a mixture of developmental stages of a single organ or tissue, some low abundance transcripts will be poorly represented, and genes not expressed in that particular organ or tissue will be absent (Rudd 2003). Comprehensive sampling of all tissues at all developmental stages and under exposure to all possible biotic agents and environmental stress conditions would be needed to develop a comprehensive gene expression encyclopedia or atlas. However, such limited sampling of mRNA populations is inadequate for accurate and statistically significant determination of the relative transcript abundance of weakly to moderately expressed mRNAs. Inadequate sampling can be partially overcome by enrichment strategies, such as normalization and subtraction (Bonaldo et al. 1996), which can dramatically improve the sequence diversity within a particular cDNA library by equalizing the relative occurrence of abundant versus rare transcripts. However, such strategies may not overcome inadequate sampling. Fortunately, new high-throughput sampling technologies have been developed that allow for more cost-effective, in-depth quantitative mRNA expression profiling.

A second weakness is that the quality of cDNA-derived ESTs can be poor and sequences may not accurately reflect the genomic sequences from which they were reverse-transcribed and cloned (Rudd 2003). Traditional EST data derived from Sanger sequencing are typically about 97% accurate when compared to cognate genomic reference sequences, considering all types of errors including base-calling inaccuracies (Li et al. 2004; Prosdocimi et al. 2007) or insertions, deletions, and substitutions (Hillier et al. 1996) arising from the fidelity of the reverse transcriptase and sequencing polymerase (Arezi and Hogrefe 2007). Furthermore, EST sequences, if not properly cleansed, can contain xenocontaminants (e.g., vector, polylinker and primer-adaptor sequences), contaminant sequences derived not from the target organism, but from foreign organisms (e.g., *E. coli*, fungi), and abundant structural or regulatory RNAs (e.g., rRNAs, organellar transcripts). Typically, such contaminants are present in 1–3% of all ESTs (Lee and Shin 2009).

As a third limitation, ESTs do not necessarily provide complete representations of gene models (or full-length cDNAs) from which they were derived. To overcome this limitation, full-length cDNA collections have been developed for many important model plant species including *Arabidopsis* (Seki et al. 2002, 2004), soybean (Umezawa et al. 2008), and the model halophyte, *Thellungiella halophila* (Taji et al. 2008). Such full-length cDNA collections serve as a useful resource for gene discovery and a means to provide precise annotation of sequenced genomes. As such, full-length cDNA collections should also be a key component of the *Vitis vinifera* genome annotation effort and are being pursued in some laboratories. In addition, such full-length cDNA collections provide a useful tool for the exploration of plant gene function in heterologous hosts and for plant improvement using heterologous genetic resources (Ichikawa et al. 2006; Kondou et al. 2008).

10.2.1 Vitis EST Sequencing

One of the earliest reports of cloning and sequencing large numbers of cDNA clones from wine grape (*Vitis vinifera* L. cv. Shiraz) used a differential screening approach to isolate and characterize ESTs from a ripening grape berry cDNA library (Davies and Robinson 2000). Differentially expressed cDNAs were selected on the basis of their hybridization to nylon membranes using cDNA isolated either two weeks before or two weeks after véraison (6 or 10 weeks after flowering), respectively. In this study, a total of 17 unique grape ripening-induced protein (GRIP) cDNA clones were sequenced that encoded a diverse group of putative cell wall proteins and stress response proteins (Davies and Robinson 2000).

One of the first reports of ESTs from *V. vinifera* L. (cv. Chardonnay) generated a total of 2479 ESTs from fully expanded leaves and 2438 ESTs from pre-véraison berries; however, these data were not released to the public (Ablett et al. 2000). The first publicly deposited set of 275 ESTs representing just 215 unique transcripts was reported from cDNA libraries generated from green, softening and ripening *Vitis vinifera* L. cv. Shiraz berries (Terrier et al. 2001). A much larger dataset of 8647 ESTs representing 4203 unique transcripts from field grown vines including young root, bud, juvenile leaf, and berry of *V. vinifera* L. (cv. Pinot Noir) and developing inflorescence (pre-flowering) and growing shoot tips of *V. vinifera* L. (cv. Regent) was reported (Moser et al. 2005). The juvenile leaf cDNA library was normalized, whereas the other cDNA libraries described were not.

One of the major motivations for the generation of EST collections has been to provide physical probe collections for cDNA microarrays. One of the earliest EST libraries was generated with RNA isolated from grape buds (*V. vinifera* L. cv. Purple Cornichon) harvested at weekly intervals over a 10-week period from nine weeks before bud burst to just after shoot emergence (Pacey-Miller et al. 2003). A total of 4270 randomly selected cDNA clones were sequenced. The EST clone inserts were then used to construct a printed cDNA microarray that was used to survey mRNA expression patterns over the same 10-week time course (Pacey-Miller et al. 2003).

Using a single mixed cDNA library prepared from *V. vinifera* L. (cv. Shiraz) whole berries collected before, during and after véraison, a total of 4608 randomly selected cDNA clones were used to fabricate a printed cDNA insert microarray (Waters et al. 2005). Of these cDNAs, 3206 were sequenced in order to annotate the microarray. The microarray was used to survey mRNA abundance changes in flowers, whole berries at two, five, eight, and 10 weeks post flowering, and in pigmented and unpigmented berry skin at véraison and two and three weeks after véraison in *V. vinifera* L. (cv. Shiraz) (Waters et al. 2005). This same cDNA library was used as a clone source to fabricate a second, larger, printed cDNA insert microarray containing 9200 features that was used to survey mRNA abundance changes in post-véraison berry skins (13 weeks after flowering) from a collection of red and white wine varieties including cv. Shiraz (at weeks 12 and 13), Cabernet Sauvignon, Pinot Noir, Shiraz, Chardonnay, Riesling, Sauvignon Blanc, and Semillon (Waters et al. 2006).

More recently, cDNA collections have been used to develop printed cDNA insert microarrays containing 2,712 cDNAs that were then used to investigate differential mRNA expression between resistant (cv. Regent) and susceptible (cv. Trincadeira) grapevine cultivars in response to fungal pathogen infection under field conditions (Figueiredo et al. 2008). A total of 121 differentially expressed transcripts representing 29 unique genes were identified, some of which were predicted to play roles in defense against fungal pathogen infection.

A much larger custom, printed cDNA microarray containing 9061 probes was used to survey gene expression changes associated with the chilling requirement associated for bud break in endodormant grapevines (Mathiason et al. 2009). Probe cDNA content was derived from six different non-normalized bud cDNA libraries. Four libraries were constructed from either control or corresponding hydrogen-cyanamide-induced dormancy release treated vines, from buds sampled at several time points during the natural dormancy cycle, or from young paradormant buds of *Vitis vinifera* L. cv. Perlette. Two other cDNA libraries were constructed from paradormant and endodormant buds of *Vitis riparia* L. Michx. Microarray analysis identified 1,469 significantly differentially expressed genes following 500–1500 hours of chilling, with the number of such genes increasing as chilling progressed. Pearson correlation hierarchical clustering revealed eight major clusters of genes that were either repressed (e.g., genes involved in carbohydrate and energy metabolism) or activated by chilling (e.g., signaling and cell growth) (Mathiason et al. 2009).

The sequence data derived from EST collections has also been used to develop oligonucleotide microarrays. The first oligonucleotide array designed for a *Vitis* species contained 50-mer oligonucleotide probes designed preferentially from 3′ non-coding regions of 3175 Unigenes based on ESTs from cDNA libraries generated from green, softening, or ripening grape berries *Vitis vinifera* L. cv. Shiraz berries (Terrier et al. 2001), and on ESTs in the public National Center for Biotechnology Information (NCBI, Bethesda, MD, USA) repository at the time (Terrier et al. 2005). Patterns of mRNA expression patterns were compared among berry pericarp tissues at pre-véraison, véraison, and post-véraison from various *V. vinifera* L. cultivars including Cabernet Sauvignon, Chardonnay and from nine stages of berry development for cv. Shiraz (Terrier et al. 2005). K-means clustering was used to identify 15 distinct mRNA expression profiles for 2593 genes during berry development. The oligonucleotide microarray developed by Terrier et al. (2005) was also used to identify a set of nine genes with functions in flavonoid (anthocyanin) metabolic pathways associated specifically with red berry tissues (Ageorges et al. 2006) and to investigate differentially expressed genes associated with early berry development in the *fleshless berry* (*flb*) mutant (*V. vinifera* L. cv. Ugni Blanc) (Fernandez et al. 2007).

By the end of 2003, EST resources for *Vitis vinifera* and related species began to proliferate rapidly. Research groups at the University of Nevada, Reno and the University of California, Davis generated more than 80%

of the 146,075 ESTs available at that time (Da Silva et al. 2005). More than 51,000 ESTs were generated from leaf, root, and berry tissue from *V. vinifera* L. (cv. Chardonnay) vines subjected to mixed abiotic stresses including salinity, cold, and water deficit stresses at the University of Nevada, Reno. More than 61,000 ESTs were generated from leaf, stem, bud, flower, and berry tissue from healthy *V. vinifera* L. (cv. Cabernet Sauvignon) vines or vines infected with the bacterial pathogen *Xylella fastidiosa* at the University of California, Davis. These EST data were used for gene discovery and to create *in silico* digital expression profiles of 665 differentially expressed transcripts across 29 different cDNA libraries to representing a range of developmental expression profiles. These data also provided the data set for the first-generation, commercially available Affymetrix GeneChip® *Vitis vinifera* (Grape) Genome Array containing approximately 14,000 distinct probe sets together with approximately 1700 probe sets from other *Vitis* species and hybrids, capable of surveying an estimated 40% of the *Vitis* genome (*http://www.affymetrix.com/products_services/arrays/specific/vitis.affx#1_1*) (Da Silva et al. 2005). These data were also the data resource for the synthesis of a commercially available Array-Ready Oligo Set for Grape (*V. vinifera*) Genome version 1.0 containing 14,562 70-mer probes representing a similar number of transcripts from The Institute for Genomics Research (TIGR) Grape Gene Index (VvGI) release 3.

These high-density oligonucleotide microarray designs have been used for a number of "closed" transcriptome studies (see Chapter 11 for more details) of various aspects of wine grape biology including 1) the effects of water-deficit and isoosmotic salinity stress on *V. vinifera* L. (cv. Cabernet Sauvignon) shoot tissues (Cramer et al. 2007), 2) tissue-specific expression patterns within mature *V. vinifera* L. (cv. Cabernet Sauvignon) berry tissues under normal and water deficit stress conditions (Grimplet et al. 2007), 3) short-term responses to chilling, polyethylene glycol, and salinity stress treatment in wine *V. vinifera* L. (cv. Cabernet Sauvignon) leaves (Tattersall et al. 2007), 4) transcriptional changes over seven stages of berry development in *V. vinifera* L. (cv. Cabernet Sauvignon) (Deluc et al. 2007), or only three stages of berry development in *V. vinifera* L. (cv. Pinot Noir) (Pilati et al. 2007), or around véraison in two greenhouse-grown grape clusters of *V. vinifera* cv. Cabernet Sauvignon in which approximately 50% of the berries had begun to accumulate anthocyanins at 4 stages (e.g., green hard, green soft, pink soft, and red soft) closely proximate to the initiation of ripening (Lund et al. 2008) and 5) induced or repressed steady-state transcript abundance changes associated with natural viral infection in leaves of red wine varieties (cv. Carménère and Cabernet-Sauvignon (Espinoza et al. 2007). The Array-Ready Oligo Set for Grape (*V. vinifera*) Genome version 1.0 was used to fabricate microarrays slides and then employed to identify

73 out of 14,562 genes whose expression was significantly modulated in *V. vinifera* L. (cv. Cabernet Sauvignon) berry clusters at véraison (eight weeks after flowering) by 24 hour exposure to ethylene (Chervin et al. 2008).

Since the landmark 2005 EST study (Da Silva et al. 2005), several additional large-scale EST collections have been generated. Because *Vitis* bud tissues were not yet well-represented in the EST databases available at the time, four different cDNA libraries prepared from *V. vinifera* L. (cv. Superior) control buds, or buds treated with dormancy releasing compounds (e.g., hydrogen cyanamide, HCL) sampled at two week intervals during dormancy and dormancy release to develop a collection of 9065 unigenes derived from sequencing both 5′ and 3′ ends of 14,001 ESTs (Keilin et al. 2007). Comparison of EST frequencies between control and dormancy release-treated vines revealed altered mRNA abundance for a wide variety of genes whose functions might be associated with grape bud dormancy release (Keilin et al. 2007).

To gain insight into the intricacy of the transcriptome with a focus on genes expressed in pericarp tissues, 77,583 EST (3′ only) were generated from two non-normalized and nine normalized cDNA libraries prepared from wine grape cultivar Cabernet Sauvignon, and from normalized cDNA libraries from the table grape variety Muscat Hamburg (Peng et al. 2007). Field-grown Cabernet Sauvignon vines were used to collect leaf, flower and root tissues, and pericarp and seed tissues from eight different stages throughout berry development. Field-grown vines of Muscat Hamburg were used to collect post-anthesis flowers, berries from fruit set through véraison (with seeds), and berries from véraison to full maturity (without seeds). ESTs with functions associated with metabolite movement and cell wall structural changes were more highly represented in pericarp tissues than in other tissues studied (Peng et al. 2007). Using public EST data and these normalized cv. Cabernet Sauvignon cDNA libraries, a second-generation, proprietary Affymetrix GeneChip® *Vitis vinifera* (GrapeGen) Genome Array, representing approximately 23,000 *Vitis vinifera* unigene probe sets was prepared. This GrapeGen GeneChip® has been used to survey transcriptional changes over nine stages of berry development and ripening in *V. vinifera* L. (cv. Muscat Hamburg) (Lijavetzky et al. 2008).

Currently, more than 353,000 ESTs have been made publicly available in EST databases (Table 10-1). Approximately 50% of all ESTs have been sequenced from the *V. vinifera* cv. Cabernet Sauvignon and 16.2% from cv. Chardonnay, in balance with their relative importance for the production of red and white wine, respectively. When evaluated according to tissue, the most popular source of ESTs is berries (with or without seeds), from which approximately 42% of all *Vitis* ESTs have been derived. Within cv. Cabernet Sauvignon, both EST and MPSS data have been generated, with approximately 185,000 ESTs and 30,000 MPSS tags, respectively (Table

Table 10-1 Distribution of *Vitis vinifera* ESTs.

Cultivar	ESTs (%)	Organ	ESTs (%)	Development	ESTs (%)	Treatment	ESTs (%)
Cabernet Sauvignon	184,885 (52.2)	Berry	81,738 (23.1)	Ripening—Pre-Véraison	57,133 (16.1)	Abiotic stress	81,483 (23)
Chardonnay	57,497 (16.2)	Berry (seeds removed)	64,870 (18.3)	Ripening—Post-Véraison	34,677 (9.8)	GA3 treatment	13,496 (3.8)
Thompson-seedless	34,093 (9.6)	Leaf	43,323 (12.2)	Pre-bloom	34,494 (9.7)	Biotic stress	9,767 (2.8)
Muscat Hamburg	26,750 (7.6)	Flower	32,970 (9.3)	Ripening—Véraison	23,089 (6.5)	n.d.[a]	243,810 (70.4)
Perlette	15,796 (4.5)	mixed	30,561 (8.6)	Flowering	18,157 (5.1)		
Carmenere	8,857 (2.5)	Bud	26,000 (7.4)	Seedling	16,452 (4.7)		
Shiraz	8,431 (2.4)	Root	23,857 (6.7)	Ripening—other[b]	16,440 (4.6)		
Pinot Noir	5,681 (1.6)	Seed	12,366 (3.5)	Mature	11,957 (3.4)		
Ugni blanc	2,406 (0.7)	Inflorescence	11,753 (3.3)	Senescent	6,859 (1.9)		
Regent	1,755 (0.5)	Cell culture	6,296 (1.8)	Vegetative	5,623 (1.6)		
Lacryma christii	1,106 (0.3)	Stem	4,731 (1.3)	Sapling	5,394 (1.5)		
Gamay NJB	876 (0.2)	Petiole	4,491 (1.3)	Ripening—Ripe[c]	4,363 (1.2)		
Mixture of four cultivars[d]	704 (0.2)	Pedicels	4,011 (1.1)	Bud break	1,750 (0.5)		
Chasselas	454 (<0.1)	Skins	1,399 (0.4)	Mixed Stages	10,5991 (29.9)		
Regent & Trincadeira	64 (<0.1)	Shoot Tip	857 (0.2)	Unknown Stage	6,821 (1.9)		
Razegui	60 (<0.1)	Mixed	170 (<0.1)				
Zinfandel	31 (<0.1)						
Merlot	7 (<0.1)						
Total	353,936 (100)						

Footnotes: [a]n.d. : not described. [b]Ripening—other category includes libraries described as "Ripening" in dbEST, where it is unknown whether pre-, post-, at véraison, or ripe. [c]Ripening –ripe category includes tissue described as "ripe", "at harvest" or stage 38 of the modified E-L system (Coombe 1995), [d]Mixture is Pinot Noir, Moscato Bianco, Teroldego and Freiburg 933/60 rootstock.

10-2). Most ESTs have been generated from whole berry or pericarp tissues collected at various stages throughout berry development. However, extensive collections of ESTs have also been generated from floral tissues, vegetative tissues (e.g., leaf, petiole, stem, and bud), and roots (Table 10-2).

10.2.2 ESTs for Molecular Marker Development

Another major motivation for the development of EST resources for *Vitis vinifera* and related species has been the development of molecular markers. A cDNA library from *V. vinifera* L. cv. Cabernet Sauvignon roots was enriched for microsatellites by hybridization to CA- and GA-bound nylon membranes, and randomly picked inserts were sequenced. A total of eight EST simple sequence repeat (SSR) loci with various di- and trinucleotide repeats (e.g., CA, GT, GA, GGA, etc.) were characterized (Decroocq et al. 2003). Primer sets were developed to sequence across these SSRs and all were found to be highly polymorphic within the family Vitaceae and were more highly conserved across the Vitaceae compared with similar microsatellite markers tested within the Rosaceae (Decroocq et al. 2003). A total of 405 SSR markers were derived from a total of 8647 ESTs produced from various tissues of *V. vinifera* L. (cv. Pinot Noir and Regent) resulting in 179 dimeric, 208 trimeric, and 16 repeats of tetramers or greater microsatellites (Moser et al. 2005). An integrated SSR map for grapevine based on five different mapping populations was constructed from 502 SSR markers (mainly $(AC)_n$ and $(AG)_n$ microsatellites) (Doligez et al. 2006).

To identify single nucleotide polymorphisms (SNPs) across the grapevine genomes, researchers are using a SNP discovery approach that involves resequencing EST-derived sequence tags in a set of selected grapevine genotypes. This approach is the most direct approach to identifying SNPs and has been widely used for many other plant species. PCR primers were designed for 451 randomly selected ESTs from public EST data. Of these, 267 primer pairs were found to generate PCR fragments for a set of 11 genotypes, which, following sequencing resulted in 230 DNA fragments (> 86%) with an average amplicons size of 437 bp, that were able to identify 1,573 SNPs at an average frequency of 1 per 64 bp (Lijavetzky et al. 2007). Library-derived (Moser et al. 2005) and publicly deposited EST data were used to generate and map 151 EST/BAC-end derived marker loci (SSR or SNP), furthering progress towards a transcript map for the *Vitis* genome (Salmaso et al. 2008). A similar approach has added 444 SNP-based markers from 332 BAC-end sequences, and 112 EST-derived marker loci in three different mapping populations (Vezzulli et al. 2008). These SNP-based markers were also validated across five different *Vitis* genotypes providing a foundation for the analysis of genome organization and evolution in

Table 10-2 ESTs and MPSS signature tags generated from *V. vinifera* cv. Cabernet Sauvignon.

Source	Stresses or Pathogens	#Libraries	Sequences	Reference
EST Libraries		60	184,885	
Berry tissue			43,721	
Whole Berries			19,748	
Ripening—pre-véraison		4	11,761	da Silva et al. 2005; Iocca et al. 2003 (unpublished)
Ripening—véraison		1	4,429	da Silva et al. 2005
Post-véraison		1	3,558	da Silva et al. 2005
Ripening—other		1	35	Hughes et al. 2004 (unpublished)
Fruit (seeds removed)			34,178	
Ripening—pre-véraison		2	1,996	Terrier et al. 2005; Iocca et al. 2003 (unpublished)
Ripening—véraison		3	1,737	Terrier et al. 2005; Iocca et al. 2003 (unpublished)
Post-véraison	Water deficit	1	13,439	Iocca et al. 2003 (unpublished)
Post-véraison		3	3,557	Terrier et al. 2005; Iocca et al. 2003 (unpublished)
Ripening—Ripe		2	1,399	Terrier et al. 2005; Iocca et al. 2003 (unpublished)
Mixed Dev		4	12,050	Peng et al. 2007
Seeds			12,366	
Ripening—véraison		1	478	Terrier et al. 2005
Mixed Dev		3	11,888	Peng et al. 2007
Skins			1,399	
Ripening—pre-véraison		2	584	Terrier et al. 2005
Ripening—Ripe		2	815	Terrier et al. 2005
Floral tissues			35,944	
Inflorescence				
Pre-bloom		3	2,174	Iocca et al. 2003 (unpublished)

Table 10-2 contd....

Table 10-2 contd....

Source	Stresses or Pathogens	#Libraries	Sequences	Reference
Ripening—other		1	967	Iocca et al. 2003 (unpublished)
Flowers				
Pre-bloom		2	22,360	da Silva et al. 2005; Iocca et al. 2003 (unpublished)
Flowering		2	10,443	da Silva et al. 2005; Iocca et al. 2003 (unpublished)
Vegetative tissues			31,430	
Leaf			17,877	
Vegetative development	Powdery mildew	3	146	Kovacs 2006 (unpublished)
Ripening/mid-season	*Xylella fastidiosa*	2	6,125	da Silva et al. 2005
Ripening/mid-season		2	4,720	da Silva et al. 2005
Senescent	*Xylella fastidiosa*	2	3,465	da Silva et al. 2005
Senescent		2	3,394	da Silva et al. 2005
Unknown Development		1	27	Hughes et al. 2004 (unpublished)
Petiole				
Ripening—véraison		1	4,491	da Silva et al. 2005
Stem				
Pre-bloom		1	4,700	da Silva et al. 2005
Bud				
Pre-bloom		1	4,362	da Silva et al. 2005
Root			16,594	
Seedling	Mixed abiotic stresses[a]	1	16,452	Ergul et al. 2007 (unpublished)

Unknown Development		1	102	Hughes et al. 2004 (unpublished)
Unknown Dev	Iron and bicarbonate	1	40	Gogorcena et al. 2006 (unpublished)
Other tissue				
Cell culture		2	6,296	Iocca et al. 2003 (unpublished); Decroocq et al. 2003
Mixed organs[b]		2	26,895	Peng et al. 2007
MPSS Libraries				
Whole Berries				
Pre-véraison		2	30,737[c]	Iandolino et al. 2008

[a]Mixture of unstressed, cold-, water deficit and 150 mM NaCl-stressed leaves.
[b]Flowers, leaves and roots.
[c]Confirmed unique sequences derived from 2,635,293 17-mer and 2,259,286 20-mer tags.

grapevine (Vezzulli et al. 2008). Finally, a total of 35 consensus EST-derived markers were generated from selected genes to genotype segregating progeny from a cross between two table grape varieties (*V. vinifera* L. cv. Italian x Big Perlon) as part of a larger study to map quantitative trait loci (QTLs) for berry size and seed content (Costantini et al. 2008).

10.2.3 Subtractive Suppression Hybridization (SSH)

Most EST sequencing is conducted on randomly selected clones from cDNA libraries made from a specific tissue or developmental stage. As such, this random sampling approach lacks the ability to efficiently and selectively target a subpopulation of transcripts within a particular cDNA library. In contrast, the SSH approach allows cDNA libraries to be constructed that are specifically enriched for a subpopulation of transcripts using a process that combines normalization and subtraction into a single process (Diatchenko et al. 1996). SSH involves the use of two hybridizations, which normalize then enrich, respectively, for differentially expressed target genes in "tester" versus "driver" cDNA populations. The differentially expressed target cDNAs are then preferentially amplified, whereas the amplification of nontarget cDNA amplification is suppressed during PCR. The SSH procedure can achieve a greater than 1000-fold enrichment in differentially expressed cDNAs (Diatchenko et al. 1996). This enrichment reduces the cloning of abundantly expressed genes common to both the control (driver) and treatment (tester) plants, and thereby increases the chance of cloning differentially expressed genes of interest. Many genes of interest are expressed at relatively low abundance or in specific cell or tissue types making selective enrichment by SSH an important and useful strategy.

The SSH approach was first used in wine grape to isolate genes differentially expressed between two red-skinned cultivars with either white flesh (*V. vinifera* L. cv. Gamay) or red flesh (cv. Lacryma) (Ageorges et al. 2006). Reciprocal SSH cDNA libraries were prepared and a total of 1406 ESTs were generated representing 96 unigenes, 60 of which were derived from the Gamay library and 36 from the Lacryma library. The cDNA library from the red-fleshed variety Lacryma was found to be significantly enriched for sequences with functions related to secondary metabolism. Subsequent qRT-PCR and oligonucleotide microarray analysis (Terrier et al. 2005) revealed that red berry tissue was associated specifically with a set of nine genes with functions in flavonoid (anthocyanin) metabolic pathways (Ageorges et al. 2006).

The SSH approach was also used to enrich for genes involved in early berry development in the *fleshless berry* (*flb*) mutant (*V. vinifera* L. cv. Ugni Blanc) (Fernandez et al. 2007). The *flb* mutant is unable to undergo division and cell fate-specification in the cell layers dedicated to cell enlargement in the ovary immediately following fertilization resulting in a 20-fold reduction in the inner mesocarp or fleshy part of the berry (Fernandez et al. 2006). Reciprocal SSH cDNA libraries were constructed using RNA isolated from young berries (3, 6, and 15 days after anthesis) from both wild type and the *flb* mutant, and 812 and 918 ESTs were generated, respectively (Fernandez et al. 2007). Of these, 24 and 42 ESTs were expressed more abundantly (> 2-fold) in wild type and *flb* mutant, respectively. The *flb* mutation was found to promote the expression of several genes associated with ripening or stress responses and to impair the expression of at least five genes associated with the regulation of growth processes that occur during early stages of fruit development (Fernandez et al. 2007).

A much larger study compared the response of resistant and susceptible sibling genotypes (*V. rupestris* x *V. arizonica*, respectively) to *Xylella fastidiosa* (Xf), a xylem-limited, Gram-negative bacterium and the causal agent of Pierce's disease (PD) in grape. SSH was employed to identify differences in gene expression associated with host-pathogen interactions (Lin et al. 2007). PD has been a major threat to the California wine industry since the 1880s and strategies for the selective enrichment of genes associated with PD resistance were lacking. Therefore, SSH was used to compare resistant and susceptible responses to Xf infection in a mapping population derived from a *V. rupestris* x *V. arizonica* hybrid (Krivanek et al. 2005; Riaz et al. 2006). A total of 5794 ESTs were sequenced from 12 different SSH cDNA libraries (average of 482 ESTs per library) resulting in 1942 unique sequences (993 contigs and 949 singlets) (Lin et al. 2007). Four states were compared, with each state comprising three SSH cDNA libraries from leaf, stem, and shoot tissue: resistant infected (RI), resistant noninfected or control (RC), susceptible infected (SI), and susceptible non-infected or control (SC). Comparison of the resistant and susceptible stem cDNA libraries showed that the RI stem library contained ESTs encoding known pathogenesis-related (PR) proteins, such as β-1-3 glucanase, chitinase, proline-rich proteins, and primary cell wall-modifying enzymes, such as xyloglucan endotransglycosylase, which modifies xyloglucan by cleavage and rejoining of the β-(1-4)-xyloglucan backbone. These patterns of expression indicated that the RI lines might limit pathogen invasion via the production of antimicrobial proteins or alterations in cell wall structure to limit further pathogen invasion (Lin et al. 2007).

10.2.4 *cDNA-Amplified Fragment Length Polymorphism*

Quantitative cDNA-amplified fragment length polymorphism (cDNA-AFLP) is a powerful technique that has been used for gene discovery and large-scale discovery of genes with differential expression patterns (Bachem et al. 1996; Breyne et al. 2003). The basic approach involves digesting cDNA with a combination of restriction enzymes to generate a fragment or tag. Then, the resulting fragments are subjected to a series of selective amplifications to produce less complex subsets of transcript tags that are then separated by electrophoresis on high-resolution polyacrylamide gels and visualized by autoradiography or an automated LI-COR system (Vuylsteke et al. 2007). The original cDNA-AFLP method has undergone several modifications that allow it to provide for genome-wide, quantitative screening of differentially expressed genes. These improvements include 1) reducing the number of tags per transcript to one by careful selection of 3′ end restriction fragments, which significantly reduces the total number of AFLP tags that must be screened, 2) selecting the most appropriate combination of restriction enzymes (e.g., *Bst*YI/MseI or *Apo*I/*Mse*I) based on *in silico* analysis of full-length cDNAs from a model or target organism, and 3) determining and using the optimal number of selective nucleotides and primer combinations to screen and visualize a majority of both abundantly and weakly expressed genes (Breyne et al. 2003).

To date, cDNA-AFLP analysis has been applied to the study of grapevine biology in only two major studies. The first large-scale investigation into the molecular basis of compatibility between wine grape and the strictly biotrophic pathogen *Plasmopara viticola* (Oomycetes), the causative agent of downy mildew, was conducted using artificially infected leaves of susceptible *V. vinifera* L. (cv. Riesling) to identify grapevine and *P. viticola* genes associated with the infection process (Polesani et al. 2008). RNA samples isolated from infected leaves at the oil spot stage, from healthy control leaves, and from pure sporangia were used for cDNA-AFLP analysis. For each of 128 primer combinations, 55–75 transcript-derived fragments (TDFs) were visualized (over a 25–760 bp size range) representing about 7000 transcripts. From this data set, a total of 1197 differentially expressed TDFs corresponding to about 17% of all visualized transcripts were excised, eluted, re-amplified, purified, and subjected to direct sequencing, resulting in 982 cDNA fragments with useful sequence data. Sixty-one percent of the TDFs matched known tentative consensus or ESTs. Surprisingly, about 20% of the TDFs were found to be homologous to *Vitis* genomic DNA sequences (Jaillon et al. 2007), but were not represented in any EST database. A total of 96 (9.7%) TDFs were homologous to known *Phytophthora* spp. sequences derived from the *Phytophthora* genome sequencing projects (Tyler et al. 2006). This cDNA-AFLP analysis showed that *P. viticola* infection resulted in the overall down-regulation of grapevine transcripts, particularly genes encoding photosynthesis-related functions, but an increase in transcript abundance for selected genes associated with downy mildew infection at the oil spot stage (Polesani et al. 2008).

The second major cDNA-AFLP analysis in wine grape was performed to investigate gene expression profiles in the post-harvest phases of fruit ripening know as "withering". Post-harvest withering of berries is used in the production of dessert and fortified wines to alter must characteristics, namely to increase the concentration of simple sugars and pigments (Zamboni et al. 2008). RNA samples for cDNA-AFLP analysis was prepared from berry skin and flesh harvested at eight different time points ranging from early fruit set, pre-véraison, véraison, full ripeness and 22, 41, 74 or 99 days after ripening with the first three post-ripening samples being collected from clusters maintained either on or off the vines. Using 128 primer combinations, an average of 75 transcript-derived fragments (TDFs) were visualized representing about 9600 transcripts. From this data set, a total of 2093 TDFs were found to be differentially expressed during berry development or withering, corresponding to about 21.8% of all visualized transcripts. Of these, 1829 TDFs were successfully excised, eluted, re-amplified, purified, and sequenced resulting in 1267 cDNA fragments with good-quality sequence data (Zamboni et al. 2008). Twelve hierarchical mRNA expression pattern clusters were identified from among the 2093 TDFs using a Pearson correlation. Sequence information combined with expression patterns of the TDFs indicated that berry withering is characterized by increased expression of transcripts associated with flavonoid and terpenoid biosynthetic pathways, ethylene biosynthesis, and hexose catabolism via the pyruvate pathway. Comparison of the on- and off-vine withering treatments revealed that only 252 TDFs were differentially expressed (167 TDFs for off-vine; 86 TDFs for on-vine) between the two treatments. However, off-vine withering induced a greater dehydration stress and greater induction of transcripts encoding stress response genes, such as dehydrins and osmolyte biosynthetic enzymes (Zamboni et al. 2008).

10.2.5 *Vitis EST Databases*

The rapid accumulation of EST data in recent years has resulted in the creation of several different databases that feature *Vitis* EST data. Historically, the most detailed functional annotation for *Vitis* EST data has been provided by the Dana Farber Cancer Institute (DFCI) Grape Gene Index (*http://compbio.dfci.harvard.edu/tgi/cgi-bin/tgi/gimain.pl?gudb=grape*), which is part of the Gene Index Project (Quackenbush et al. 2001). The current release (Release 7.0; April 17, 2010) of the DFCI Grape Gene Index contains 352,730 ESTs and 25,497 ETs (curated, non-redundant, mature transcripts) organized into 34,154 tentative consensus sequences (TCs; virtual transcripts derived from EST and ET assemblies), 31,813 singleton ESTs or ESTs not contained in an assembly, and 14,811 singleton ETs. The DFCI Grape Gene

Index, like all gene indices, can be searched by gene identifiers (e.g., ET numbers, GenBank Accession numbers, EST identifiers and clone names), by keyword(s) in annotation, or with a query sequence using BLAST (Basic local alignment search tool) (Altschul et al. 1990). Functional annotation features include details on predicted alternative splice variants (5306 splice variant clusters composed at least one TC with at least one additional singleton or TC contained in Release 7.0), comparison of EST digital expression patterns between different cDNA libraries or tissues, classifications of TCs by Gene Ontology (GO) vocabularies, associations of TCs with metabolic and signaling pathways, and predictions of unique 70-mer predictions for the design of non-redundant oligonucleotide microarrays.

Another useful resource is the *Vitis* UniGene database (*http://www.ncbi.nlm.nih.gov/UniGene/UGOrg.cgi?TAXID=29760*) curated at the NCBI. The current UniGene database (build #10) contains 344,597 total sequences including ESTs (329,788), mRNAs (513) and supplemented with predicted transcript models (14,296) from which 22,083 gene clusters have been generated, each containing one or more EST or mRNA (Pontius et al. 2003).

The Vitis Gene Expression Database (VitisExpDB) is an online MySQL-PHP-driven relational database that houses annotated EST and gene expression data (*http://cropdisease.ars.usda.gov/vitis_at/main-page.htm*) (Doddapaneni et al. 2008). Currently, the database stores ~ 320,000 EST sequences, from different *Vitis* species and cultivars, including *V. vinifera* L. and their BLAST (top hit) annotation details and GO-based structured vocabulary. Furthermore, putative homologs in other species or varieties for each EST are designated along with information on their nucleotide identity percentages and, phylogenetic relationships, and there is the ability to design common primers to amplify and retrieve these putative homologous sequences. The same site also hosts microarray data obtained from a 60-mer oligonucleotide microarray containing 191,450 probes (nine probes per EST spotted in duplicate) from a non-redundant set of 20,020 ESTs derived from various sources (Doddapaneni et al. 2008). Microarray expression profiles surveyed leaf, stem and shoot tissues of highly resistant and the highly susceptible sibling genotypes selected from a *V. rupestris* x *V. arizonica* population (Lin et al. 2007) at 11 time points following infection with *Xylella fastidiosa* (Xf) or control.

To facilitate the identification of proteins from berry pericarp and mesocarp tissues isobarically tagged for relative and absolute quantitation by iTRAQ MS/MS, a predicted peptide database was created from EST data using clustering and trimming approaches (Lücker et al. 2009). An automated EST data processing system was developed and used to process both in house-generated EST datasets and publicly available EST data from the NCBI. The assembled datasets were then annotated and translated according to the top hit from BLASTX searches. Then, the top-scoring protein sequence identified in the BLASTX search was subjected to BLASTP analysis using the UniProt Knowledgebase (UniProtKB; *http://www.uniprot.org/*) in order to determine its sequence identity. The number of high confidence peptides identified with this custom *Vitis* peptide database was 1.8- and 1.9-fold higher for exocarp and mesocarp tissues, respectively, than a standard MS database from the Hammersmith Campus of the Imperial College, London (Lücker et al. 2009). Other useful databases and links to *Vitis* genomic resources can be found at the International Grape Genome Program website (*http://www.vitaceae.org*).

10.3 High Throughput Sequencing Approaches

Although EST sequencing and its variations described above have been commonly used for gene discovery, the depth of coverage is typically limited for digital measurements of gene expression in which the relative abundance of each sequence tag is used to infer the relative abundance of its corresponding transcript within a particular tissue or condition (Audic and Claverie 1997). Therefore, a number of methods have been developed to provide greater depth of sequencing of short tags, and thereby provide a more accurate, quantitative measure of the relative abundance of a transcript. High-throughput, sequence-based methods such as serial analysis of gene expression (SAGE) or massively parallel signature sequencing (MPSS) can be regarded as complementary to hybridization-based or "closed" transcriptome approaches provided that a sufficient number of biological replicates are performed (Liu et al. 2007; Vega-Sanchez et al. 2007) and generally provide similar results at high transcript abundance ranges (Nygaard et al. 2008). However, high-throughput sequencing-based methods exhibit increased sensitivity to the detection of low abundance transcripts and because they are "open" platforms, they have the added advantage of measuring new transcripts (Liu et al. 2007). In a related study comparing two different oligonucleotide microarray platforms with MPSS using identical mRNA samples from *Arabidopsis*, the two microarray platforms showed more consistent results with each other than with MPSS. This study also showed that measurements of relative transcript abundance, particularly small differences in expression ratios, can be masked by the platform used rather than by biological differences between samples and tissues (Chen et al. 2007).

10.3.1 Serial Analysis of Gene Expression

The serial analysis of gene expression (SAGE) method relies on Sanger sequencing of short (10–14 bp), concatenated cDNA tags derived from a cDNA region immediately 3′ to the 3′-most restriction site present in a double-stranded cDNA synthesized on magnetic beads (Velculescu et al. 1995). The original method generated 10–14 bp tags and represented an up to 30-fold improvement in tag coverage for the same number of Sanger sequencing reads. However, such short tags did not always permit the unambiguous assignment of tags to corresponding transcripts. Subsequently, improved versions of the original method called LongSAGE, which generates 17-19-mer tags (Saha et al. 2002), and SuperSAGE, which generates 26-mer tags (Matsumura et al. 2003; Matsumura et al. 2008), were developed. Despite the improved throughput and reduced cost compared to EST sequencing and its widespread use in diverse species (Anisimov 2008), no SAGE-related profiling studies using *V. vinifera* or related species have been reported to date.

10.3.2 Massively Parallel Signature Sequencing

Massively parallel signature sequencing (MPSS) involves cloning a cDNA library on a collection of 5 μm diameter microbeads such that one transcript from the original RNA sample is represented on a single bead (Brenner et al. 2000a). Each bead is then subjected to successive rounds of sequencing using digestion with a four base pair-recognizing restriction endonuclease (e.g., *Bbv*I) to expose four nucleotides followed by a complex series of adapter ligation and decoding reactions that generate 17 to 20-base signature tags (Brenner et al. 2000b; Reinartz et al. 2002). A typical MPSS experiment yields 250,000–400,000 high quality signature sequences, which can be used for tag-based measurements of gene expression as with SAGE. However, a direct comparison between LongSAGE and MPSS platforms revealed that MPSS libraries were significantly less complex than smaller LongSAGE libraries (Hene et al. 2007). The complexity of MPSS limited its use to an on-site service performed by Lynx Therapeutics and when Solexa, Inc. (now Illumina, Inc.) developed sequencing-by-synthesis (SBS) technology, which is simpler and provides much longer sequence read lengths, MPSS became obsolete (Simon et al. 2009).

MPSS was used to generate two signature tag libraries from green, hard stage II berries sampled from field-grown *Vitis vinifera* L. (cv. Cabernet Sauvignon) (Iandolino et al. 2008). A total of 2,635,293 17-base and 2,259,286 20-base signatures were produced that corresponded to 30,737 and 26,878 distinct sequences, respectively (Table 10-2). At the time these data were generated, the *Vitis* genome had not yet been completed, so the tags were related to available grape EST collections and compiled into a Grape MPSS Database (*http://mpss.udel.edu/grape*) (Nakano et al. 2006; Iandolino et al. 2008). This comparison showed that 6430 distinct EST contigs and 2190 EST singletons had perfect matches to at least one MPSS signature (Iandolino et al. 2008). Thus, the database contains a substantial number of highly abundant MPSS signatures that lack any match to known ESTs. This data will be increasingly useful for the annotation of previously unidentified gene models within the grape genome (Nakano et al. 2006). Interestingly, the MPSS dataset also revealed the presence of antisense transcription not reported previously in grape, but comparable with that characterized in many other plant species (Iandolino et al. 2008).

10.4 Next-Generation Sequencing Technologies

Recent advances in high-throughput NGS sequencing, made possible by the commercial introduction of "second generation" sequencing instrumentation, capable of producing millions of DNA sequence reads in a single run, are rapidly transforming the pace and scope of functional genomic analyses (Mardis 2008a; Simon et al. 2009). Newer, so-called "third generation", single-molecular DNA sequencing technologies are also emerging, but have not yet been fully commercialized (Gupta 2008). Such NGS systems not only drive down the cost of gene discovery by transcriptome or genome sequencing, but also provide genome-wide sequence readouts as endpoints for a wide variety of applications including mutation or polymorphism discovery, comparative genomics, analysis of chromatin structural and epigenetic regulation, and discovery of noncoding RNAs (Mardis 2008b). When used for transcriptome profiling, RNA-Seq provides a far more precise measurement of the relative abundance of transcript and their isoforms than traditional EST or MPSS sequencing methods (Wang et al. 2009; Wall et al. 2009).

10.4.1 Pyrosequencing

The first "next generation" Roche (454) Genome Sequencer (GS) FLX sequencer, commercialized in 2004 by 454 Life Sciences/Roche Applied Science (*http://www.454.com*), is based on pyrosequencing (Ronaghi et al. 1998). This approach mixes the library of DNA or cDNA fragments to be sequenced with a population of agarose beads whose surfaces carry oligonucleotides complementary to the 454-specific adapter sequence on the fragment library, so that each bead is associated with a single fragment. Emulsion PCR (Dressman et al. 2003) is then used to produce approximately 1 million copies of each DNA fragment on the surface of each bead. The beads are arrayed in a picotiter plate (PTP), which holds each bead in a fixed location and acts as a flow cell as pyrosequencing reagents are introduced in a step-wise fashion. The amount of light released by the firefly enzyme luciferase at the location of each bead as the amplified single molecules are sequenced *en masse* is proportional to the number of nucleotides incorporated. Light emitted after each nucleotide incorporation step is recorded by a CCD camera opposed to the PTP (Margulies et al. 2005; Droege and Hill 2008). The system is capable of producing 100–200 Mb during a typical run with an average read length of up to 500 bp making this platform best–suited for *de novo* sequencing (Table 10-3).

10.4.2 Sequencing-by-synthesis

Illumina's SBS Genome Analyzer II system introduced in 2006 uses an approach similar to pyrosequencing, called sequencing-by-synthesis (SBS; *http://www.illumina.com*). In this sequencing platform, cDNA or size-fractionated DNA is ligated to an Illumina-specific adapter library on both ends. The DNA fragments are then loaded into an eight-channeled flow cell coated with covalently linked oligonucleotides on its surface. Single-molecule amplification takes place on this surface through the formation of double-stranded bridges between the adapter-ligated DNA fragment and the surface oligos, which build up by repeated cycles of solid-phase amplification resulting in the production of dense clusters of template DNA. Following this cluster generation, DNA polymerase and all four nucleotides are added simultaneously to the flow cell channels for incorporation into the oligo-primed cluster fragments. Each nucleotide carries a base-unique fluorescent label, and as it is incorporated, fluorescence imaging detects each base incorporation step. The blocked 3′ terminus and the fluorophore are chemically removed and the process is repeated until 25–35 bp are read for each cluster, however, longer read lengths will be possible in the near future (Mardis 2008a). A typical run will yield 30–50 million sequence tags. The major applications for this platform include transcriptome profiling in organisms with fully sequenced genomes, SNP detection (genome resequencing), and genome-wide detection of protein-DNA interactions using ChIP-sequencing (Bentley 2006; Smith et al. 2008; Simon et al. 2009).

10.4.3 Sequencing by Oligo Ligation and Detection

The Applied Biosystems, Inc. sequencing by oligo ligation and detection (SOLiD) System (*http://www.appliedbiosystems.com*) generates an adapter-ligated fragment library using emulsion PCR to amplify the fragments on the surface of small magnetic beads similar to the Roche 454 system. However, after amplification, the beads are covalently attached to the surface of a specially treated glass slide, which is placed inside a fluidics cassette inside the sequencer. Two slides are prepared and processed in two separate flow cells so that one slide can receive sequencing reactants as the second slide is being imaged in order to double the speed and throughput of the system

Table 10-3 Comparison of current and next-generation sequencing (NGS) high-throughput sequencing technologies.

Platform	Date	Sequencing Chemistry	Read length (nt)	Time/run	Throughput (Mb/Run)	Comments
EST	1990	Sanger	650–800	5–10 days	Low (1.5 Mb)[a]	Labor intensive; 97% accuracy
SAGE	1995	Sanger	10–23	5–10 days	Low (1.5 Mb)[a]	Labor intensive; 97% accuracy
MPSS	2000	Ligation-based sequencing using encoder/decoder adapters	16–20	2 days	Medium (2.5–3.0 Mb)	Site-specific biases; 97% accuracy. obsolete
Roche 454 GS FLX (TITANIUM upgrade)	2004 (2008)	Pyrosequencing	250 (500)	8 hours (16 hours)	High (100–200 Mb)	*De novo* and resequencing; 99.5% accuracy
Illumina Genome Analyzer	2006	Sequencing-by-synthesis	32–40	4 days	Very High (1300 Mb)	Resequencing; 99.5% accuracy
Applied Biosystems SOLiD	2007	Ligation-based sequencing	25–35	5 days	Extremely High (3000 Mb)	Resequencing; 99.94% accuracy
Helicos Biosciences tSMS	2008	Single molecule sequencing-by-synthesis	~ 30–55	14 days	Extremely High (7500 Mb)	Resequencing; 99.4% accuracy
Pacific Biosciences SMRT	2010	Single molecule sequencing-by-synthesis	~ 100,000?	Real-time	Extremely High	*De novo* and resequencing; 99.3% accuracy

[a]Throughput of an ABI 3730 Automated DNA Sequencing System programmed to sequence 24 x 96-well plates per day with an average read length of 650 bp per sample.

(Mardis 2008a). The SOLiD system uses a unique DNA sequence process catalyzed by DNA ligase involving a complex series of reiterative ligation cycles in which a limited set of semi-degenerate 8-mer oligonucleotides are hybridized to the DNA fragment sequence adjacent to a universal primer 3′ end. When a matching 8-mer hybridizes to the DNA fragment, DNA ligase seals the phosphate backbone and a fluorescent readout identifies the fixed base of the 8-mer, which corresponds to either the fifth or the second position depending on the cycle number. Part of the 8-mer and its fluorescent tag are then removed and the process is repeated to identify the sequence at each fragment at five nucleotide intervals. The synthesized fragments that end at base 25 or 35 are removed by denaturation and are washed away. Then, four subsequent rounds of sequencing are initiated at n-1 to n-4-positioned universal primers. A unique attribute of the SOLiD system is that it contains an inherent quality check feature, called "2 base encoding", to distinguish miscalled bases from correct base calls during the data analysis step (Mardis 2008a; Simon et al. 2009). Each SOLiD run takes five days, but produces 2–3 Gb of sequence data (Table 10-3). The major applications for SOLiD include very deep transcriptome profiling, SNP characterization, dissection of chromatin architecture (nucleosome positioning) with ChIP sequencing, and resequencing of bacterial genomes (Simon et al. 2009).

10.4.4 Next-Generation Transcriptome Studies in Wine Grape

Application of new "open platform" transcriptome sequencing technologies to *Vitis vinifera* are having a profound influence on our understanding of grapevine gene expression dynamics. In the short time period since NGS technologies have been introduced, they have been used to generate and analyze vast amounts of transcript sequence data from *Vitis vinifera* (Table 10-4). In one of the first studies published in which NGS was applied to wine grape transcriptomics, the power and utility of these emerging platforms was demonstrated (Bellin et al. 2009). In that study, 454 sequencing was performed on two similar 3′-targeted berry cDNA libraries, one normalized (N) and one non-normalized (NN). Each library was allotted one half run of 454 GS FLX sequencing, generating 556,742 high quality reads (~ 290k from NN; ~ 266k from N). Reads from N and NN were assembled independently, identifying 1,468 (NN) and 3709 (N) putative new genes that mapped to unannotated regions of the genome. From the two assembled libraries, custom CombiMatrix microarray probesets were designed, hybridized to berry RNA, and used to confirm the expression of 1251 (NN) and 3098 (N) novel genes that had not been identified by any prior expression data.

Grapevine RNA-Seq has also been used for SNP discovery, for identification and quantititation of differential gene expression during berry development, and for the identification of alternative splicing events,

Table 10-4 Next-generation studies in wine grape.

NGS Platform	Sequences (millions)	Topic	Select Outcomes	Cultivar(s)	Reference
Illumina	175	*de novo* gene modeling	Prediction of 28,399 CDS, 12,341 loci	Pinot Noir PN40024	Denoeud et al. 2008
454	0.56	Gene discovery and annotation	1,251 and 3,098 novel loci confirmed 19K CombiMatrix berry microarray	Corvina	Bellin et al. 2009
Illumina	24 + 7.8	miRNAs + targets	21 new miRNAs 44 new targets	Pinot Noir ENTAV115	Pantaleo et al. 2010
llumina	13 + 135	miRNAs + targets	36 miRNAs from 65 loci	Pinot Noir PN40024	Mica et al. 2010
454	0.56		12K CombiMatrix miRNA microarray used to detect 56 mature miRNAs	Corvina	
SOLiD Illumina	328 205	Mitochondrial RNA editing	445 C-to-U edits found	Pinot Noir PN40024	Picardi et al. 2010
Illumina	59	Alternative splicing	385 alternative splicing events detected	Corvina	Zenoni et al. 2010
		SNP detection	85,870 SNPs		
		Gene expression in berry development	6,695 differentially expressed transcripts from 5, 10, 15 weeks after flowering		

all in a single set of experiments (Zenoni et al. 2010). Using the Illumina platform, Zenoni et al. generated ~59 million reads from *V. vinifera* cv. Corvina berries at fruit set, from berries at véraison, and from berries that had undergone further ripening (> 17 million reads from each stage). Reads were then annotated against the Pinot Noir 40024 genome (Jaillon et al. 2007) and compared to a database of theoretical splice junctions derived from prior transcript (EST) annotation of gene loci. RNA-Seq evidence of exon-exon junctions was mapped conservatively, with 5 independent tags required for consideration; nonetheless, 385 alternative splicing events were found. Next, the Corvina Illumina sequence reads were compared with the reference 42004 genome for SNP discovery. From ~ 50 million successfully aligned reads, 85,870 polymorphisms were identified with 9.1% located in coding regions. Throughout the course of berry development, transcripts of 17,324 genes were identified in one or more of the sampled developmental stages. Lastly, by the reads per exon kilobase per million (RPKM) density metric, and Fisher's exact testing, 6695 genes were found to be expressed in a stage-specific manner.

RNA-Seq has also been employed to generate genomic annotations of sites of transcription entirely *de novo*, that is, without reference to prior EST- or model-based gene predictions (Denoeud et al. 2008). In this study, 173 million Illumina reads from leaf, root, stem and callus were assembled into genomic coverage contigs ("covtigs") in a procedure designated G-Mo.R-Se (for Gene Modeling using RNA-Seq). Using G-Mo.R-Se, 28,399 transcript models with plausible coding sequences were clustered into 12,341 loci.

The depth of sequencing afforded by RNA-Seq has also been used to facilitate the detection of mitochondrial RNA editing in grapevine (Picardi et al. 2010). In the organelles of higher plants, RNA editing results almost exclusively in C-to-U substitutions, and in coding regions is further restricted to the 1st and 2nd positions of a codon, generating non-synonymous mutations, and occasionally new start and stop codons (Takenaka et al. 2008). In non-coding RNAs, similar C-to-U substitutions are thought to promote the stability of tRNA secondary structure and group II intron self-splicing (Gray 2003). Of 328 million SOLiD and 205 million Illumina transcript reads, Picardi et al. mapped over 6 million uniquely to mtDNA. With two NGS sources, platform-specific errors were minimized, and 401 C-to-U conversions in mitochondrial coding sequences, 38 in tRNAs, and six in group II introns were unambiguously identified. Further analysis identified tissue specificity in some editing sites.

To date, two studies utilizing NGS have been performed to identify microRNAs (miRNAs), small (19–24 nt) non-coding RNAs that post-transcriptionally inhibit the expression of their target genes via complementary sequences, selecting them for degradation or translational arrest (Jones-Rhoades et al. 2006; Brodersen et al. 2008). In the first study, by comparing the 8.4X *V. vinifera* genome (Jaillon et al. 2007) against the known genomic sequences of miRNAs in other dicotyledenous plants, 140 loci were predicted to encode pre-miRNAs that would mature into 79 unique miRNA sequences, which could be grouped into 28 conserved miRNA families (Mica et al. 2010). To test these predictions, Mica et al. generated a 13M read Pinot Noir 40024 short RNA Illumina run and detected 36 of the 79 predicted mature miRNA tags for one or more unique miRNA in 23 of the 28 predicted families. A 12 K CombiMatrix custom array was designed and tested against four small RNA tissue isolates, again detecting miRNAs from 23 predicted families, with up to 56 of the 79 unique sequences identified along with strong tissue-specific expression of miRNAs apparent.

A second study combining NGS sequencing of 24 million short RNAs with NGS degradome sequencing of the 5′ ends of uncapped poly-adenylated RNAs, supported and extended the findings of Mica et al. by identifying short NGS RNA reads corresponding to 24 of the 28 miRNA families, in addition to another 26 non-conserved miRNAs (i.e., shared among some, but not all, plants studied) and an additional set of putative grapevine-specific miRNAs (Pantaleo et al. 2010). From sequencing 7.7 million uncapped poly-adenylated RNAs and screening for complementarity to the 10th nucleotide of an miRNA, the authors were able to identify 112 targets of miRNA degradation by known miRNAs and 44 additional targets to provide supporting evidence for the existence of 21 of their hypothesized 42 grapevine-specific miRNAs.

10.4.5 Single Molecule Sequencing

Two emerging, although not yet commercialized, "third generation" sequencing approaches involve the direct sequencing of single DNA molecules (Gupta 2008; Simon et al. 2009). However, neither of these technologies have been applied to wine grape to date. One proprietary method called True single-molecule sequencing (tSMS™) developed by Helicos Biosciences (*http://www.helicosbio.com*) uses a form of sequencing-by-synthesis in which a DNA sample to be sequenced is fragmented and denatured to produce individual single-stranded templates, which are then sequenced in a flow cell using DNA polymerase and fluorescent nucleotides (Table 10-3). The templates are anchored to the flow cell by polyTs and their site of attachment is recorded by a CCD camera connected to a microscope that repeatedly images the growing DNA chain as each of four different nucleotides are added sequentially. This approach was used to resequence the bacteriophage M13 genome (Harris et al. 2008). Helicos claims that 30-55 bp reads are possible and up to 7500 Mb of sequence can be collected in 14 days with 99.4% accuracy (Simon et al. 2009).

A second proprietary approach called single-molecule, real-time (SMRT) sequencing developed by Pacific Biosciences (PacBio) (*http://www.pacificbiosciences.com*) resembles the Helicos system in that it employs real-time observation of a single molecule of DNA polymerase as it synthesizes DNA (Table 10-3). However, PacBio's approach involves the use of so-called SMRT chips, each made of 100-nm thick metal film containing thousands of "zero-mode-waveguide" (ZMW) nanostructures or cavities 10-50 nm in diameter (Korlach et al. 2008). A DNA polymerase molecule is attached to the bottom surface of the chamber, into which differentially fluorophore-labeled phospho-linked nucleotides are introduced in solution. Upon addition of a complementary nucleotide, the fluorophore of the incorporated nucleotide emits a light pulse in a wavelength corresponding to one of the four bases allowing identification of the base. Upon incorporation, the polymerase cleaves the fluorophore bond, allowing the dye of the fluorophore to diffuse away, returning the signal to baseline levels. The process is repeated through the length of the DNA fragment in real-time (Eid et al. 2009). If scaled up so that 14,000 ZMW nanostructures were in use simultaneously, the technology could potentially sequence the equivalent of a human genome (3 Gb) in a single day.

10.5 Conclusion

Technological advances in the last decade have resulted in a dramatic increase in the volume of transcriptomic and genomic sequence for *Vitis vinifera* L. and related *Vitis* species. To date, the vast majority of transcript data has been obtained using traditional Sanger sequencing of ESTs from only a very limited number of varieties and species. Such EST data has been instrumental in the fabrication of "closed" microarray platforms for assessing mRNA abundance patterns in a wide variety of grapevine tissues and developmental states. Although only limited MPSS sequencing was conducted, this technology is now obsolete. However, transcriptome analysis in grapevine is entering a new phase of discovery that has begun to embrace the use of a variety of NGS platforms described here. Large amounts of transcriptome data are now routinely being obtained using NGS platforms to yield a much more detailed and comprehensive picture of the complex RNA and miRNA expression patterns that control grapevine development and its interaction with an ever-changing environment being brought about by global climate change.

Acknowledgements

This work was support by grants from the National Science Foundation Plant Genome Program (DBI-0217653) to JCC and the University of Nevada

Agricultural Experiment Station. The authors wish to thank Mary Ann Cushman for critical reading of the manuscript. This publication was also made possible by NIH Grant Number P20 RR-016464 from the INBRE Program of the National Center for Research Resources through its support of the Nevada Genomics, Proteomics and Bioinformatics Centers.

References

Ablett E, Seaton G, Scott K, Shelton D, Graham M, Baverstock P, Lee L, Henry R (2000) Analysis of grape ESTs: global gene expression patterns in leaf and berry. Plant Sci 159: 87–95.

Ageorges A, Fernandez L, Vialet S, Merdinoglu D, Terrier N, Romieu C (2006) Four specific isogenes of the anthocyanin metabolic pathway are systematically co-expressed with the red colour of grape berries. Plant Sci 170: 372–383.

Alba R, Fei Z, Payton P, Liu Y, Moore S, Debbie P, Cohn J, D'Ascenzo M, Gordon J, Rose J, Martin G, Tanksley S, Bouzayen M, Jahn M, Giovannoni J (2004) ESTs, cDNA microarrays, and gene expression profiling: tools for dissecting plant physiology and development. Plant J 39: 697–714.

Altschul S, Gish W, Miller W, Myers E, Lipman D (1990) Basic local alignment search tool. J Mol Biol 403–410: 403–410.

Anisimov S (2008) Serial analysis of gene expression (SAGE): 13 years of application in research. Curr Pharm Biotechnol 9: 338–350.

Arezi B, Hogrefe H (2007) *Escherichia coli* DNA polymerase III epsilon subunit increases Moloney murine leukemia virus reverse transcriptase fidelity and accuracy of RT-PCR procedures. Anal Biochem 360: 84–91.

Arroyo-Garcia R, Ruiz-Garcia L, Bolling L, Ocete R, Lopez M, Arnold C, Ergul A, Soylemezoglu G, Uzun H, Cabello F, Ibanez J, Aradhya M, Atanassov A, Atanassov I, Balint S, Cenis J, Costantini L, Goris-Lavets S, Grando M, Klein B, McGovern P, Merdinoglu D, Pejic I, Pelsy F, Primikirios N, Risovannaya V, Roubelakis-Angelakis K, Snoussi H, Sotiri P, Tamhankar S, This P, Troshin L, Malpica J, Lefort F, Martinez-Zapater J (2006) Multiple origins of cultivated grapevine (*Vitis vinifera* L. ssp. sativa) based on chloroplast DNA polymorphisms. Mol Ecol 15: 3707–3714.

Audic S, Claverie J (1997) The significance of digital gene expression profiles. Genome Res 7: 986–995.

Bachem C, van der Hoeven R, de Bruijn S, Vreugdenhil D, Zabeau M, Visser R (1996) Visualization of differential gene expression using a novel method of RNA fingerprinting based on AFLP: analysis of gene expression during potato tuber development. Plant J 9: 745–753.

Bellin D, Ferrarini A, Chimento A, Kaiser O, Levenkova N, Bouffard P, Delledonne M (2009) Combining next-generation pyrosequencing with microarray for large scale expression analysis in non-model species. BMC Genom 10: 555.

Bentley D (2006) Whole-genome re-sequencing. Curr Opin Genet Dev 16: 545–552.

Bonaldo M, Lennon G, Soares M (1996) Normalization and subtraction: two approaches to facilitate gene discovery. Genome Res 6: 791–806.

Brenner S, Williams S, Vermaas E, Storck T, Moon K, McCollum C, Mao J, Luo S, Kirchner J, Eletr S, DuBridge R, Burcham T, Albrecht G (2000a) *In vitro* cloning of complex mixtures of DNA on microbeads: physical separation of differentially expressed cDNAs. Proc Natl Acad Sci USA 97: 1665–1670.

Brenner S, Johnson M, Bridgham J, Golda G, Lloyd D, Johnson D, Luo S, McCurdy S, Foy M, Ewan M, Roth R, George D, Eletr S, Albrecht G, Vermaas E, Williams S, Moon K, Burcham T, Pallas M, DuBridge R, Kirchner J, Fearon K, Mao J, Corcoran K (2000b) Gene expression analysis by massively parallel signature sequencing (MPSS) on microbead arrays. Nat Biotechnol 18: 630–634.

Breyne P, Dreesen R, Cannoot B, Rombaut D, Vandepoele K, Rombauts S, Vanderhaeghen R, Inzé D, Zabeau M (2003) Quantitative cDNA-AFLP analysis for genome-wide expression studies. Mol Genet Genom 269: 173–179.

Brodersen P, Sakvarelidze-Achard L, Bruun-Rasmussen M, Dunoyer P, Yamamoto YY, Sieburth L, Voinnet O (2008) Widespread translational inhibition by plant miRNAs and siRNAs. Science 320: 1185–1190.

Chen J, Agrawal V, Rattray M, West M, St Clair D, Michelmore R, Coughlan S, Meyers B (2007) A comparison of microarray and MPSS technology platforms for expression analysis of Arabidopsis. BMC Genom 8: 414.

Chervin C, Tira-Umphon A, Terrier N, Zouine M, Severac D, Roustan J (2008) Stimulation of the grape berry expansion by ethylene and effects on related gene transcripts over the ripening phase. Physiol Plant 134: 534–546.

Coombe B (1995) Adoption of a system for identifying grapevine growth stages. Aust J Grape Wine Res 1: 104–110.

Costantini L, Battilana J, Lamaj F, Fanizza G, Grando M (2008) Berry and phenology-related traits in grapevine (*Vitis vinifera* L.): from quantitative trait loci to underlying genes. BMC Plant Biol 8: 38.

Cramer G, Ergül A, Grimplet J, Tillett R, Tattersall E, Bohlman M, Vincent D, Sonderegger J, Evans J, Osborne C, Quilici D, Schlauch K, Schooley D, Cushman J (2007) Water and salinity stress in grapevines: early and late changes in transcript and metabolite profiles. Funct Integr Genom 7: 111–134.

da Silva F, Iandolino A, Al-Kayal F, Bohlmann M, Cushman M, Lim H, Ergul A, Figueroa R, Kabuloglu E, Osborne C, Rowe J, Tattersall E, Leslie A, Xu J, Baek J, Cramer G, Cushman J, Cook D (2005) Characterizing the grape transcriptome. analysis of expressed sequence tags from multiple *Vitis* species and development of a compendium of gene expression during berry development. Plant Physiol 139: 574–597.

Davies C, Robinson S (2000) Differential screening indicates a dramatic change in mRNA profiles during grape berry ripening. Cloning and characterization of cDNAs encoding putative cell wall and stress response proteins. Plant Physiol 122: 803–812.

Decroocq V, Favé M, Hagen L, Bordenave L, Decroocq S (2003) Development and transferability of apricot and grape EST microsatellite markers across taxa. Theor Appl Genet 106: 912–922.

Deluc L, Grimplet J, Wheatley M, Tillett R, Quilici D, Osborne C, Schooley D, Schlauch K, Cushman J, Cramer G (2007) Transcriptomic and metabolite analyses of Cabernet Sauvignon grape berry development. BMC Genom 8: 429.

Denoeud F, Aury JM, Da Silva C, Noel B, Rogier O, Delledonne M, Morgante M, Valle G, Wincker P, Scarpelli C, Jaillon O, Artiguenave F (2008) Annotating genomes with massive-scale RNA sequencing. Genome Biol 9: R175.

Diatchenko L, Lau Y, Campbell A, Chenchik A, Moqadam F, Huang B, Lukyanov S, Lukyanov K, Gurskaya N, Sverdlov E, Siebert P (1996) Suppression subtractive hybridization: a method for generating differentially regulated or tissue-specific cDNA probes and libraries. Proc Natl Acad Sci USA 93: 6025–6030.

Doddapaneni H, Lin H, Walker M, Yao J, Civerolo E (2008) VitisExpDB: a database resource for grape functional genomics. BMC Plant Biol 8: 23.

Doligez A, Adam-Blondon A, Cipriani G, Di Gaspero G, Laucou V, Merdinoglu D, Meredith C, Riaz S, Roux C, This P (2006) An integrated SSR map of grapevine based on five mapping populations. Theor Appl Genet 113: 369–382.

Dressman D, Yan H, Traverso G, Kinzler K, Vogelstein B (2003) Transforming single DNA molecules into fluorescent magnetic particles for detection and enumeration of genetic variations. Proc Natl Acad Sci USA 100: 8817–8822.

Droege M, Hill B (2008) The Genome Sequencer FLX System—longer reads, more applications, straight forward bioinformatics and more complete data sets. J Biotechnol 136: 3–10.

Eid J, Fehr A, Gray J, Luong K, Lyle J, Otto G, Peluso P, Rank D, Baybayan P, Bettman B, Bibillo A, Bjornson K, Chaudhuri B, Christians F, Cicero R, Clark S, Dalal R, Dewinter

A, Dixon J, Foquet M, Gaertner A, Hardenbol P, Heiner C, Hester K, Holden D, Kearns G, Kong X, Kuse R, Lacroix Y, Lin S, Lundquist P, Ma C, Marks P, Maxham M, Murphy D, Park I, Pham T, Phillips M, Roy J, Sebra R, Shen G, Sorenson J, Tomaney A, Travers K, Trulson M, Vieceli J, Wegener J, Wu D, Yang A, Zaccarin D, Zhao P, Zhong F, Korlach J, Turner S (2009) Real-time DNA sequencing from single polymerase molecules. Science 323: 133–138.

Espinoza C, Vega A, Medina C, Schlauch K, Cramer G, Arce-Johnson P (2007) Gene expression associated with compatible viral diseases in grapevine cultivars. Funct Integr Genom 7: 95–110.

Fei Z, Tang X, Alba R, White J, Ronning C, Martin G, Tanksley S, Giovannoni J (2004) Comprehensive EST analysis of tomato and comparative genomics of fruit ripening. Plant J 40: 47–59.

Fernandez L, Romieu C, Moing A, Bouquet A, Maucourt M, Thomas M, Torregrosa L (2006) The grapevine fleshless berry mutation. A unique genotype to investigate differences between fleshy and nonfleshy fruit. Plant Physiol 140: 537–547.

Fernandez L, Torregrosa L, Terrier N, Sreekantan L, Grimplet J, Davies C, Thomas M, Romieu C, Ageorges A (2007) Identification of genes associated with flesh morphogenesis during grapevine fruit development. Plant Mol Biol 63: 307–323.

Figueiredo A, Fortes A, Ferreira S, Sebastiana M, Choi Y, Sousa L, Acioli-Santos B, Pessoa F, Verpoorte R, Pais M (2008) Transcriptional and metabolic profiling of grape (*Vitis vinifera* L.) leaves unravel possible innate resistance against pathogenic fungi. J Exp Bot 59: 3371–3381.

Gray MW (2003) Diversity and evolution of mitochondrial RNA editing systems. IUBMB Life 55: 227–233.

Grimplet J, Deluc L, Tillett R, Wheatley M, Schlauch K, Cramer G, Cushman J (2007) Tissue-specific mRNA expression profiling in grape berry tissues. BMC Genom 8: 187.

Gupta P (2008) Single-molecule DNA sequencing technologies for future genomics research. Trends Biotechnol 26: 602–611.

Harris T, Buzby P, Babcock H, Beer E, Bowers J, Braslavsky I, Causey M, Colonell J, Dimeo J, Efcavitch J, Giladi E, Gill J, Healy J, Jarosz M, Lapen D, Moulton K, Quake S, Steinmann K, Thayer E, Tyurina A, Ward R, Weiss H, Xie Z (2008) Single-molecule DNA sequencing of a viral genome. Science 320: 106–109.

Hene L, Sreenu V, Vuong M, Abidi S, Sutton J, Rowland-Jones S, Davis S, Evans E (2007) Deep analysis of cellular transcriptomes—LongSAGE versus classic MPSS. BMC Genom 8: 333.

Hillier L, Lennon G, Becker M, Bonaldo M, Chiapelli B, Chissoe S, Dietrich N, DuBuque T, Favello A, Gish W, Hawkins M, Hultman M, Kucaba T, Lacy M, Le M, Le N, Mardis E, Moore B, Morris M, Parsons J, Prange C, Rifkin L, Rohlfing T, Schellenberg K, Bento Soares M, Tan F, Thierry-Meg J, Trevaskis E, Underwood K, Wohldman P, Waterston R, Wilson R, Marra M (1996) Generation and analysis of 280,000 human expressed sequence tags. Genome Res 6: 807–828.

Iandolino A, Nobuta K, da Silva F, Cook D, Meyers B (2008) Comparative expression profiling in grape (*Vitis vinifera*) berries derived from frequency analysis of ESTs and MPSS signatures. BMC Plant Biol 8: 53.

Ichikawa T, Nakazawa M, Kawashima M, Iizumi H, Kuroda H, Kondou Y, Tsuhara Y, Suzuki K, Ishikawa A, Seki M, Fujita M, Motohashi R, Nagata N, Takagi T, Shinozaki K, Matsui M (2006) The FOX hunting system: an alternative gain-of-function gene hunting technique. Plant J 48: 974–985.

Iriti M, Faoro F (2006) Grape phytochemicals: a bouquet of old and new nutraceuticals for human health. Med Hypoth 67: 833–838.

Jaillon O, Aury J, Noel B, Policriti A, Clepet C, Casagrande A, Choisne N, Aubourg S, Vitulo N, Jubin C, Vezzi A, Legeai F, Hugueney P, Dasilva C, Horner D, Mica E, Jublot D, Poulain J, Bruyère C, Billault A, Segurens B, Gouyvenoux M, Ugarte E, Cattonaro F, Anthouard V, Vico V, Del Fabbro C, Alaux M, Di Gaspero G, Dumas V, Felice N, Paillard S, Juman I,

Moroldo M, Scalabrin S, Canaguier A, Le Clainche I, Malacrida G, Durand E, Pesole G, Laucou V, Chatelet P, Merdinoglu D, Delledonne M, Pezzotti M, Lecharny A, Scarpelli C, Artiguenave F, Pè M, Valle G, Morgante M, Caboche M, Adam-Blondon A, Weissenbach J, Quétier F, Wincker P (2007) The grapevine genome sequence suggests ancestral hexaploidization in major angiosperm phyla. Nature 449: 463–467.

Jones-Rhoades MW, Bartel DP, Bartel B (2006) MicroRNAS and their regulatory roles in plants. Annu Rev Plant Biol 57: 19–53.

Keilin T, Pang X, Venkateswari J, Halaly T, Crane O, Keren A, Ogrodovitch A, Ophir R, Volpin H, Galbraith D, Or E (2007) Digital expression profiling of a grape-bud EST collection leads to new insight into molecular events during grape-bud dormancy release. Plant Sci 173: 446–457.

Kondou Y, Higuchi M, Takahashi S, Sakurai T, Ichikawa T, Kuroda H, Yoshizumi T, Tsumoto Y, Horii Y, Kawashima M, Hasegawa Y, Kuriyama T, Matsui K, Kusano M, Albinsky D, Takahashi H, Nakamura Y, Suzuki M, Sakakibara H, Kojima M, Akiyama K, Kurotani A, Seki M, Fujita M, Enju A, Yokotani N, Saitou T, Ashidate K, Fujimoto N, Ishikawa Y, Mori Y, Nanba R, Takata K, Uno K, Sugano S, Natsuki J, Dubouzet J, Maeda S, Ohtake M, Mori M, Oda K, Takatsuji H, Hirochika H, Matsui M (2008) Systematic approaches to using the FOX hunting system to identify useful rice genes. Plant J 57: 883–894.

Korlach J, Marks P, Cicero R, Gray J, Murphy D, Roitman D, Pham T, Otto G, Foquet M, Turner S (2008) Selective aluminum passivation for targeted immobilization of single DNA polymerase molecules in zero-mode waveguide nanostructures. Proc Natl Acad Sci USA 105: 1176–1181.

Krivanek A, Famula T, Tenscher A, Walker M (2005) Inheritance of resistance to *Xylella fastidiosa* within a *Vitis rupestris* x *Vitis arizonica* hybrid population. Theor Appl Genet 111: 110–119.

Lee B, Shin G (2009) CleanEST: a database of cleansed EST libraries. Nucl Acids Res 37: D686–689.

Li M, Nordborg M, Li L (2004) Adjust quality scores from alignment and improve sequencing accuracy. Nucl Acids Res 32: 5183–5191.

Lijavetzky D, Cabezas J, Ibáñez A, Rodríguez V, Martínez-Zapater J (2007) High throughput SNP discovery and genotyping in grapevine (*Vitis vinifera* L.) by combining a re-sequencing approach and SNPlex technology. BMC Genom 8: 424.

Lijavetzky D, Francisco R, Peng F, Bravo G, Ibanez A, Oliveros-Collazos JC, Lund ST, Martinez Zapater JM (2008) A new GeneChip for grapevine transcriptomic analyses 8th Int Symp on Grapevine Physiology and Biotechnology. Adelaide, Australia, 23–28 Nov 2008.

Lin H, Doddapaneni H, Takahashi Y, Walker M (2007) Comparative analysis of ESTs involved in grape responses to *Xylella fastidiosa* infection. BMC Plant Biol 7: 8.

Liu F, Jenssen T, Trimarchi J, Punzo C, Cepko C, Ohno-Machado L, Hovig E, Kuo W (2007) Comparison of hybridization-based and sequencing-based gene expression technologies on biological replicates. BMC Genom 8: 153.

Lücker J, Laszczak M, Smith D, Lund S (2009) Generation of a predicted protein database from EST data and application to iTRAQ analyses in grape (*Vitis vinifera* cv. Cabernet Sauvignon) berries at ripening initiation. BMC Genom 10: 50.

Lund S, Peng F, Nayar T, Reid K, Schlosser J (2008) Gene expression analyses in individual grape (*Vitis vinifera* L.) berries during ripening initiation reveal that pigmentation intensity is a valid indicator of developmental staging within the cluster. Plant Mol Biol 68: 301–315.

Mardis E (2008a) Next-generation DNA sequencing methods. Annu Rev Genom Hum Genet 9: 387–402.

Mardis E (2008b) The impact of next-generation sequencing technology on genetics. Trends Genet 24: 133–141.

Margulies M, Egholm M, Altman W, Attiya S, Bader J, Bemben L, Berka J, Braverman M, Chen Y, Chen Z, Dewell S, Du L, Fierro J, Gomes X, Godwin B, He W, Helgesen S, Ho C, Irzyk G, Jando S, Alenquer M, Jarvie T, Jirage K, Kim J, Knight J, Lanza J, Leamon J,

Lefkowitz S, Lei M, Li J, Lohman K, Lu H, Makhijani V, McDade K, McKenna M, Myers E, Nickerson E, Nobile J, Plant R, Puc B, Ronan M, Roth G, Sarkis G, Simons J, Simpson J, Srinivasan M, Tartaro K, Tomasz A, Vogt K, Volkmer G, Wang S, Wang Y, Weiner M, Yu P, Begley R, Rothberg J (2005) Genome sequencing in microfabricated high-density picolitre reactors. Nature 437: 376–380.

Mathiason K, He D, Grimplet J, Venkateswari J, Galbraith D, Or E, Fennell A (2009) Transcript profiling in *Vitis riparia* during chilling requirement fulfillment reveals coordination of gene expression patterns with optimized bud break. Funct Integr Genom 9: 81–96.

Matsumura H, Reich S, Ito A, Saitoh H, Kamoun S, Winter P, Kahl G, Reuter M, Kruger D, Terauchi R (2003) Gene expression analysis of plant host-pathogen interactions by SuperSAGE. Proc Natl Acad Sci USA 100: 15718–15723.

Matsumura H, Krüger D, Kahl G, Terauchi R (2008) SuperSAGE: a modern platform for genome-wide quantitative transcript profiling. Curr Pharm Biotechnol 9: 368–374.

Mica E, Piccolo V, Delledonne M, Ferrarini A, Pezzotti M, Casati C, Del Fabbro C, Valle G, Policriti A, Morgante M, Pesole G, Pè ME, Horner DS (2009) High throughput approaches reveal splicing of primary microRNA transcripts and tissue specific expression of mature microRNAs in Vitis vinifera. BMC Genom 10: 558.

Mica E, Piccolo V, Delledonne M, Ferrarini A, Pezzotti M, Casati C, Del Fabbro C, Valle G, Policriti A, Morgante M, Pesole G, Pè ME, Horner DS (2010) Correction: High throughput approaches reveal splicing of primary microRNA transcripts and tissue specific expression of mature microRNAs in Vitis vinifera. BMC Genomics 11: 109.

Monagas M, Hernandez-Ledesma B, Gomez-Cordoves C, Bartolome B (2006) Commercial dietary ingredients from *Vitis vinifera* L. leaves and grape skins: antioxidant and chemical characterization J Agri Food Chem 54: 319–327.

Moser C, Segala C, Fontana P, Salakhudtinov I, Gatto P, Pindo M, Zyprian E, Toepfer R, Grando M, Velasco R (2005) Comparative analysis of expressed sequence tags from different organs of *Vitis vinifera* L. Funct Integ. Genom 5: 208–217.

Nakano M, Nobuta K, Vemaraju K, Tej S, Skogen J, Meyers B (2006) Plant MPSS databases: signature-based transcriptional resources for analyses of mRNA and small RNA. Nucl Acids Res 34: D731–735.

Nygaard V, Liu F, Holden M, Kuo W, Trimarchi J, Ohno-Machado L, CL C, Frigessi A, Glad I, Wiel M, Hovig E, Lyng H (2008) Validation of oligoarrays for quantitative exploration of the transcriptome. BMC Genom 9: 258.

Pacey-Miller T, Scott K, Ablett E, Tingey S, Ching A, Henry R (2003) Genes associated with the end of dormancy in grapes. Funct Integr Genom 3: 144–152.

Pantaleo V, Szittya G, Moxon S, Miozzi L, Moulton V, Dalmay T, Burgyan J (2010) Identification of grapevine microRNAs and their targets using high throughput sequencing and degradome analysis. Plant J 62: 960–976.

Peng F, Reid K, Liao N, Schlosser J, Lijavetzky D, Holt R, Martínez Zapater J, Jones S, Marra M, Bohlmann J, Lund S (2007) Generation of ESTs in *Vitis vinifera* wine grape (Cabernet Sauvignon) and table grape (Muscat Hamburg) and discovery of new candidate genes with potential roles in berry development. Gene 402: 40–50.

Picardi E, Horner DS, Chiara M, Schiavon R, Valle G, Pesole G (2010) Large-scale detection and analysis of RNA editing in grape mtDNA by RNA deep-sequencing. Nucl Acids Res.

Pilati S, Perazzolli M, Malossini A, Cestaro A, Demattè L, Fontana P, Dal Ri A, Viola R, Velasco R, Moser C (2007) Genome-wide transcriptional analysis of grapevine berry ripening reveals a set of genes similarly modulated during three seasons and the occurrence of an oxidative burst at vèraison. BMC Genom 8: 428.

Polesani M, Desario F, Ferrarini A, Zamboni A, Pezzotti M, Kortekamp A, Polverari A (2008) cDNA-AFLP analysis of plant and pathogen genes expressed in grapevine infected with *Plasmopara viticola*. BMC Genom 9: 142.

Pontius JU, Wagner L, Schuler GD (2003) UniGene: a unified view of the transcriptome. In: The NCBI Handbook. National Center for Biotechnology Information, Bethesda (MD) USA.

Prosdocimi F, Lopes D, Peixoto F, Mourão M, Pacífico L, Ribeiro R, Ortega J (2007) Effects of sample re-sequencing and trimming on the quality and size of assembled consensus sequences. Genet Mol Res 6: 756–765.

Quackenbush J, Cho J, Lee D, Liang F, Holt I, Karamycheva S, Parvizi B, Pertea G, Sultana R, White J (2001) The TIGR Gene Indices: analysis of gene transcript sequences in highly sampled eukaryotic species. Nucl Acids Res 29: 159–164.

Reinartz J, Bruyns E, Lin J, Burcham T, Brenner S, Bowen B, Kramer M, Woychik R (2002) Massively parallel signature sequencing (MPSS) as a tool for in-depth quantitative gene expression profiling in all organisms. Brief Funct Genom Proteom 1: 95–104.

Riaz S, Krivanek A, Xu K, Walker M (2006) Refined mapping of the Pierce's disease resistance locus, PdR1, and Sex on an extended genetic map of *Vitis rupestris* x *V. arizonica*. Theor Appl Genet 113: 1317–1329.

Ronaghi M, Uhlén M, Nyrén P (1998) A sequencing method based on real-time pyrophosphate. Science 281: 363–365.

Rudd S (2003) Expressed sequence tags: alternative or complement to whole genome sequences? Trends Plant Sci 8: 321–329.

Saha S, Sparks A, Rago C, Akmaev V, Wang C, Vogelstein B, Kinzler K, Velculescu V (2002) Using the transcriptome to annotate the genome. Nat Biotechnol 20: 508–512.

Salmaso M, Malacarne G, Troggio M, Faes G, Stefanini M, Grando M, Velasco R (2008) A grapevine (*Vitis vinifera* L.) genetic map integrating the position of 139 expressed genes. Theor Appl Genet 116: 1129–1143.

Seki M, Narusaka M, Kamiya A, Ishida J, Satou M, Sakurai T, Nakajima M, Enju A, Akiyama K, Oono Y, Muramatsu M, Hayashizaki Y, Kawai J, Carninci P, Itoh M, Ishii Y, Arakawa T, Shibata K, Shinagawa A, Shinozaki K (2002) Functional annotation of a full-length Arabidopsis cDNA collection. Science 296: 141–145.

Seki M, Satou M, Sakurai T, Akiyama K, Iida K, Ishida J, Nakajima M, Enju A, Narusaka M, Fujita M, Oono Y, Kamei A, Yamaguchi-Shinozaki K, Shinozaki K (2004) RIKEN Arabidopsis full-length (RAFL) cDNA and its applications for expression profiling under abiotic stress conditions. J Exp Bot 55: 213–223.

Simon S, Zhai J, Nandety R, McCormick K, Zeng J, Mejia D, Meyers B (2009) Short-read sequencing technologies for transcriptional analyses. Annu Rev Plant Biol 60: 305–333.

Smith A, Xuan Z, Zhang M (2008) Using quality scores and longer reads improves accuracy of Solexa read mapping. BMC Bioinformat 9: 128.

Taji T, Sakurai T, Mochida K, Ishiwata A, Kurotani A, Totoki Y, Toyoda A, Sakaki Y, Seki M, Ono H, Sakata Y, Tanaka S, Shinozaki K (2008) Large-scale collection and annotation of full-length enriched cDNAs from a model halophyte, *Thellungiella halophila*. BMC Plant Biol 8: 115–.

Takenaka M, Verbitskiy D, van der Merwe JA, Zehrmann A, Brennicke A (2008) The process of RNA editing in plant mitochondria. Mitochondrion 8: 35–46.

Tattersall E, Grimplet J, DeLuc L, Wheatley M, Vincent D, Osborne C, Ergül A, Lomen E, Blank R, Schlauch K, Cushman J, Cramer G (2007) Transcript abundance profiles reveal larger and more complex responses of grapevine to chilling compared to osmotic and salinity stress. Funct Integr Genom 7: 317–333.

Terrier N, Ageorges A, Abbal P, Romieu C (2001) Generation of EST from grape berry at various developmental stages. J Plant Physiol. 158: 1575–1583.

Terrier N, Glissant D, Grimplet J, Barrieu F, Abbal P, Couture C, Ageorges A, Atanassova R, Leon C, Renaudin J, Dedaldechamp F, Romieu C, Delrot S, Hamdi S (2005) Isogene specific oligo arrays reveal multifaceted changes in gene expression during grape berry (*Vitis vinifera* L.) development. Planta 222: 832–847.

This P, Lacombe T, Thomas M (2006) Historical origins and genetic diversity of wine grapes. Trends Genet. 22: 511–519.

Tyler B, Tripathy S, Zhang X, Dehal P, Jiang R, Aerts A, Arredondo F, Baxter L, Bensasson D, Beynon J, Chapman J, Damasceno C, Dorrance A, Dou D, Dickerman A, Dubchak I, Garbelotto M, Gijzen M, Gordon S, Govers F, Grunwald N, Huang W, Ivors K, Jones R,

Kamoun S, Krampis K, Lamour K, Lee M, McDonald W, Medina M, Meijer H, Nordberg E, Maclean D, Ospina-Giraldo M, Morris P, Phuntumart V, Putnam N, Rash S, Rose J, Sakihama Y, Salamov A, Savidor A, Scheuring C, Smith B, Sobral B, Terry A, Torto-Alalibo T, Win J, Xu Z, Zhang H, Grigoriev I, Rokhsar D, Boore J (2006) Phytophthora genome sequences uncover evolutionary origins and mechanisms of pathogenesis. Science 313: 1261–1266.

Umezawa T, Sakurai T, Totoki Y, Toyoda A, Seki M, Ishiwata A, Akiyama K, Kurotani A, Yoshida T, Mochida K, Kasuga M, Todaka D, Maruyama K, Nakashima K, Enju A, Mizukado S, Ahmed S, Yoshiwara K, Harada K, Tsubokura Y, Hayashi M, Sato S, Ana T, Ishimoto M, Funatsuki H, Teraishi M, Osaki M, Shinano T, Akashi R, Sakaki Y, Yamaguch-Shinozaki K, Shinozaki K (2008) Sequencing and analysis of approximately 40,000 soybean cDNA clones from a full-length-enriched cDNA library. DNA Res 15: 333–346.

Vega-Sanchez M, Gowda M, Wang G (2007) Tag-based approaches for deep transcriptome analysis in plants. Plant Sci 173: 371–380.

Velasco R, Zharkikh A, Troggio M, Cartwright D, Cestaro A, Pruss D, Pindo M, Fitzgerald L, Vezzulli S, Reid J, Malacarne G, Iliev D, Coppola G, Wardell B, Micheletti D, Macalma T, Facci M, Mitchell J, Perazzolli M, Eldredge G, Gatto P, Oyzerski R, Moretto M, Gutin N, Stefanini M, Chen Y, Segala C, Davenport C, Demattè L, Mraz A, Battilana J, Stormo K, Costa F, Tao Q, Si-Ammour A, Harkins T, Lackey A, Perbost C, Taillon B, Stella A, Solovyev V, Fawcett J, Sterck L, Vandepoele K, Grando S, Toppo S, Moser C, Lanchbury J, Bogden R, Skolnick M, Sgaramella V, Bhatnagar S, Fontana P, Gutin A, Van de Peer Y, Salamini F, Viola R (2007) A high quality draft consensus sequence of the genome of a heterozygous grapevine variety. PLoS ONE 2: e1326.

Velculescu V, Zhang L, Vogelstein B, Kinzler K (1995) Serial analysis of gene expression. Science 270: 484–487.

Vezzulli S, Troggio M, Coppola G, Jermakow A, Cartwright D, Zharkikh A, Stefanini M, Grando M, Viola R, Adam-Blondon A, Thomas M, This P, Velasco R (2008) A reference integrated map for cultivated grapevine (*Vitis vinifera* L.) from three crosses, based on 283 SSR and 501 SNP-based markers. Theor Appl Genet 117: 499–511.

Vuylsteke M, Peleman J, van Eijk M (2007) AFLP-based transcript profiling (cDNA-AFLP) for genome-wide expression analysis. Nat Protoc 2: 1399–1413.

Wall PK, Leebens-Mack J, Chanderbali AS, Barakat A, Wolcott E, Liang H, Lendherr L, Tomsho LP, Hu Y, Carlson JE, Ma H, Schuster SC, Soltis DE, Soltis PS, Altman N, dePamphilis CW (2009) Comparison of next generation sequencing technologies for transcriptome characterization. BMC Genom 10: 347.

Wang Z, Gerstein M, Snyder M (2009) RNA-Seq: a revolutionary tool for transcriptomics. Nat Rev Genet 10: 57–63.

Waters D, Holton T, Ablett E, Lee L, Henry R (2005) cDNA microarrays analysis of developing grape (Vitis vinifera cv. Shiraz) berry skin Funct Integ Genom 5: 40–58.

Waters D, Holton T, Ablett E, Slade L, Henry R (2006) The ripening wine grape berry skin transcriptome. Plant Sci 171: 132–138.

Zamboni A, Minoia L, Ferrarini A, Tornielli G, Zago E, Delledonne M, Pezzotti M (2008) Molecular analysis of post-harvest withering in grape by AFLP transcriptional profiling. J Exp Bot 59: 4145–4159.

Zenoni S, Ferrarini A, Giacomelli E, Xumerle L, Fasoli M, Malerba G, Bellin D, Pezzotti M, Delledonne M (2010) Characterization of transcriptional complexity during berry development in *Vitis vinifera* using RNA-Seq. Plant Physiol 152: 1787–1795.

11

Functional Genomics: Closed System Approaches for Transcriptome Analyses

Charles Romieu,[1] *Christian Kappel*[2] and *Serge Delrot*[2,*]

ABSTRACT

This chapter describes the evolution of high throughput transcriptomic tools available for the study of gene expression during normal development and stress responses of grapevine. Differential screening of cDNA clones prepared from berries and blotted on nylon membranes provided the first attempt of "high" throughput transcriptomics. This was quickly followed by sequencing of expressed sequence tags (EST), and by the design of several generations of microarrays. The sequencing of the grape genome in 2007 finally resulted in arrays that now cover the whole genome. The use of microarrays allowed identifying gene families which were differentially expressed during grape berry ripening (transcription factors, secondary metabolism, cell wall metabolism) and networks of genes that are co-regulated. The approach has also been extended to various compartments of the berries (skin, flesh, seed) and to gene expression changes induced by water stress and berry microenvironment. Microarrays were also used to identify target genes of transcription factors that were overexpressed into grapevine hairy roots and plants. More recently, high throughput transcriptomics was used to obtain an overall view of changes in gene expression that were induced by fungal diseases. The current trends consist in the use of graphic tools that facilitate the interpretation of microarray data, and in the use of high throughput sequencing technologies in place of microarray hybridization.

Keywords: microarrays, gene expression, berry development, berry ripening, water stress, fungal diseases

[1]UMR INRA 1334 CIRAD Montpellier Supagro AGAP, 2, place P. Viala, 34060 Montpellier Cedex, France.
[2]Lab Ecophysiology and Grape Functional Genomics, Institute of Vine and Wine Sciences, University of Bordeaux, INRA, 210 Chemin de Leysotte, CS 2008, 33882 Villenave d'Ornon, France.
*Corresponding author: *serge.delrot@bordeaux.inra.fr*

11.1 Introduction

There have been impressive advances in grape genomics in the past few years, and a substantial number of genomic tools have been developed so far, or are under development. These include mapping populations, genetic and physical maps (Adam-Blondon et al. 2005; Marguerit et al. 2009), and an extensive expressed sequence tags (EST) collection (more than 353,000 sequences deposited at the (National Center for Bioinformatics) NCBI EST database as on February 8, 2009). Analysis of ESTs (Iolandino et al. 2008), cDNA-amplified fragment length polymorphism (AFLP) analysis (Polesani et al. 2008) and microarrays have been used for high-throughput studies of gene expression and to provide functional data for many genes simultaneously.

The advantages and limits of microarray analysis for plant genomics have been recently summarized (Rensink and Buell 2005). Structural annotation of genes relies on abundant experimental evidence such as ESTs and full-length complementaryDNA (cDNA) sequences. By contrast, functional annotation relies largely on the use of bioinformatics because experimental evidence that can be used to assign gene function is available for only a limited number of genes within a genome. On a whole genome level, microarray technology enables to monitor the expression level of thousands of genes simultaneously in one single experiment (Meyers et al. 2004). Microarrays only measure expression levels for those genes for which a probe, either a clone or sequence, is available and, therefore, should be considered a targeted approach to expression profiling.

The primary method used to determine gene function from expression data is correlative, using the temporal and spatial expression patterns that are associated with a specific phenotype or response. Access to a large collection of reference expression data from mutants, tissues or treatments can provide a tool for identifying the function of unknown genes (Hughes et al. 2000).

A second method involves association of unknown genes with "known" genes based on co-expression and co-regulation. An analysis of the co-expression relationships among a large quantity of expression data from humans, flies, worms and yeast allowed to identify relationships that were conserved across evolution (Stuart et al. 2003). With the growing availability of microarray expression data, similar approaches can also be used to analyze plants. For example, publicly available *Arabidopsis* arrays were used to identify genes required for cellulose synthesis through the analysis of co-regulation with cellulose synthase genes (Persson et al. 2005). This illustrates the potential of mining many expression profiles for functional discovery via co-regulation. A co-response database (CSB.DB) based on publicly available expression data for *Arabidopsis* (Steinhauser et al. 2004) was used for the identification of brassinosteroid-related genes by co-regulation analysis of known signaling components (Lisso et al. 2005).

EST and genome sequencing projects have fueled the progressive development of high throughput methods for measurement of transcript abundance in Vitaceae (see also Chapter 10). There are several microarrays in use for scientific research in the grapevine community. Most of them are designed for *Vitis vinifera* but microarrays for other organisms as *Arabidopsis thaliana* are also used (Espinoza et al. 2007a). *Vitis* microarrays can be divided in two categories, cDNA microarrays with deposited DNA probes and microarrays with *in situ* synthesis of the oligonucleotide probes. One or more labeled samples can be hybridized against them to measure absolute or relative mRNA expressions levels. It has been used to study the grapevine transcriptome related to berry development (Terrier et al. 2005; Waters et al. 2005; Ageorges et al. 2006), modulation of endogenous grapevine gene expression (Franks et al. 2006), water and salinity stress (Cramer et al. 2007) and responses to virus infection (Espinoza et al. 2007a,b).

This review focuses on recent advances in the application of microarrays in grape genomic research. After a brief description of the grapevine microarrays available, we will illustrate how they have been used to study major physiological processes such as berry development and ripening, and tolerance to abiotic and biotic stress.

11.2 Currently Available Microarrays

11.2.1 cDNA Microarrays

Differential screening of cDNA clones blotted on nylon membranes can be considered as the ancestor of the microarray technology. This initial "non hypothesis driven" screening method was reported nine years ago in grapevine (Davies and Robinson 2000). Before this date, most if not all previous work in the molecular biology of grapevine had been confined to the level of individual or small groups of candidate genes, and focused at structural genes of known functions, such as enzymes or transporters involved in primary metabolism, berry softening or anthocyanin synthesis, in addition to few regulatory genes. Differential screening of nylon membrane with ^{32}P labeled cDNAs obtained from, green and ripening stage berries helped to discover 16 genes among those that were the most strongly up-regulated at the onset of berry ripening, regardless of their functions (Davies and Robinson 2000). The number of clones that were spotted on the filter is not known, but transcript abundance of Grip3 and Grip4 was so high that a preliminary cleaning of the ripening stage cDNA sample (10 weeks post flowering) became mandatory in order to detect additional genes differentially hybridized on the membrane. In this respect, one should notice that, owing to their prevalence, the same mRNAs could have been reasonably flagged as "ripening related" simply upon sequencing two 96 well plates, one from the green stage cDNA library and the other for ripe stage (Terrier et al. 2001). 146,075 *Vitis* sequences have been deposited into GenBank (NCBI) before September 30, 2003. Clustering into contigs with The Institute for Genomic Research (TIGR) Gene Indices Clustering Tools (*http://www.tigr.org/tdb/tgi/software/*) yielded an unigene set of 25,746 unique contig and singleton sequences (da Silva et al. 2005). *In silico* analysis and EST counting across 29 cDNA libraries, representing a range of developmental and stress conditions helped to identify 665 differentially expressed transcripts. A more focused analysis of flower and berry development identified 87 differentially expressed transcripts and provided the basis for a compendium that relates gene expression and annotation to previously characterized aspects of berry development and physiology.

Preliminary sequence information is not an absolute prerequisite for the preparation of cDNA microarrays that are often considered as open systems allowing the identification of novel transcribed sequences. However, five years elapsed between this pioneering work (Davies and Robinson 2000) and the first report on the use cDNA microarray on grapevine (Waters et al. 2005). Microarray slides including 4,608 PCR-amplified cDNA clones were derived from a ripening grape berry cDNA library. Singlepass DNA sequences could be obtained from the 5′ends of 3,206 of these clones. The cDNA inserts within the full set of 4,608 plasmids were amplified by PCR using T3 and T7 primers, before being spotted on microarrays. Dual labeling with cy3 and cy5 allowed to screen 341 mRNAs displaying significant (three-fold) changes during berry skin development. Since the number of PCR products in the array was probably lower than the number of colonies plated by Davies and Robinson (2000), it may be reasonably assumed that the precise evaluation of the expression level with fluorescent dyes was determinant for such a marked improvement in the number of identified genes, when compared with visual inspection of autoradiography slides. The best matches in protein database and the putative functions of these genes were published, but unfortunately, the sequences of these clones were not released in public *Vitis* database, so that the most relevant information still remains inaccessible to the scientific community.

Dual label cDNAs microarrays including duplicate spots of 1,225 sequences were then designed in order to screen changes in gene expression triggered by the ectopic expression of different transgene in *Vitis* hairy roots, but the experiment failed in detecting differentially expressed cDNAs because the number of cDNAs spotted on the array was probably too low (Franks et al. 2006). The preparation of custom cDNA microarrays did not cease when 14K oligoarrays representing about 50% of the *vitis* transcriptome became commercially available in 2003–2004. Indeed, cDNA libraries of specific organs or tissues of grapevine were clearly under-represented in the 2003 unigene sets of TIGR or NCBI used for the design of commercial *Vitis* microarrays (see below). Furthermore, it was believed that sequence divergence would preclude the use of these chips for studies on non *vinifera* species. Because of this bias, custom cDNA microarrays are still produced in different laboratories regardless of financial and labor costs. Custom *Vitis* bud cDNA microarrays (9,216 inserts) were recently used to identify gene expression patterns associated with transition stages between endodormancy and ecodormancy in *V. riparia* overwintering buds as they accumulated chilling hours (Mathiason et al. 2009). This yielded 1,469 differentially expressed "microarray features". In the same way, custom dual label cDNAs microarrays including 2,712 inserts allowed to screen 121 transcripts representing 29 unique gene sequences differentially expressed between healthy young leaves of the *V. vinifera* cultivars Regent and Trincadeira (Figueiredo et al. 2008). However, in all previous approaches, the clones spotted on the arrays were randomly selected on non-normalized libraries. Important information, such as the number and sequences of independent genes probed by the array, is frequently lacking. This precludes further data mining, and the comparison of gene expression data with that obtained with other probe sets or microarrays platforms seems hardly feasible in practice. Moreover, due to their elevated T_m, the PCR products obtained through a generic couple of primers annealing in the vectors arms may hybridize to different genes inside multigenic families. This impairs the sensibility of cDNA microarray and may confuse the expression pattern of different isogenes.

The first high-throughput expression profiling results obtained with, respectively cDNA and oligo arrays were published quasi simultaneously on grapevine; and the dual label protocol with Cy3 and Cy5 fluorescent dyes was used on both systems (Terrier et al. 2005; Waters et al. 2005). Bioinformatic analysis of ESTs from seven libraries and candidate genes publicly available in 2002 helped to assemble a *Vitis* unigene set including 3,175 sequences. In order to prevent blocking of the sequence reaction by the polyT and to get rid of the 3′ untranslated region, and to improve BLASTX scores and functional annotation, sequencing was anchored on the 5′ ends of the clones in most EST programs in the early 2000. In spite of these difficulties, Terrier et al. (2005) deliberately favored 3′ sequencing of berry cDNA libraries to facilitate the design of isogen specific probes (Terrier et al. 2001). The orientation of the sequences was checked by identifying the polyA tail, and BLASTX comparison to protein sequences from public databases. Unoriented sequences were considered in both orientations. Fifty-mer oligonucleotides were designed preferentially in the 3′ non-coding regions close to the stop codon, taking into account Tm compatibility and specificity. Oligonucleotides design, synthesis, and spotting were made by

MWG Biotech, each 5′ amino-modified oligonucleotide being spotted in duplicate onto the arrays. These first oligoarrays were almost exclusively used on berries. Together with the EST libraries from which they were derived, they allowed the characterization of transcript level changes associated with the berry developmental program (Terrier et al. 2005; Glissant et al. 2008), flesh and skin color mutations (Ageorges et al. 2006) and the fleshless mutation (Fernandez et al. 2007). Expression profiles based on spot fluorescence was highly correlated with quantitative PCR data for all spots investigated.

11.2.2 14K Microarrays

Functional genomic resources for *Vitis vinifera* have proliferated mainly in the form of large, publicly available expressed sequence tag databases. The ca 100,000 EST and complete cDNA sequences available at NCBI in 2003 allowed to compute two different unigene sets. The TIGR unigene set and the NCBI unigene set including 19,451 and 14,000 sequences allowed the design of two 14k probe sets, respectively, Qiagen Operon 70-mers slides and Affymetrix chips.

11.2.2.1 Qiagen Operon Oligo Arrays

A 70-mer oligo set for investigating gene expression in grapevine is available from Operon (*http://omad.operon.com/download/storage/vitis_vinifera_V1.0.2_datasheet.pdf*). The *Vitis vinifera* Array-Ready Oligo Set (AROS) V1.0 became commercially available in 2003. This AROS includes a total of 14,56,270-mers to be spotted by the user. Data on sequences and putative annotations are freely downloadable at *http://omad.operon.com/download/index.php*. The 70-mers in the *Vitis vinifera* Array-Ready Oligo Set (AROS) V1.0 were designed from transcript sequences obtained from TIGR Grape Gene Index (VvGI), release 3.0, from August 18, 2003, including 19,451 unique sequences (*http://compbio.dfci.harvard.edu/tgi/gi/vvgi/release_notes.html*) The Grape gene Index has been transferred to Dana-Fraber Cancer Institute (DFCI, *htpp://compbio.dfci.harcard.edu/tgi/cgi-bin/tgi/gimain.pl?gudb=grape*). The latest release 6 has been put on lineon July 30, 2008.

The VvGI includes consensus sequences (TCs) assembled from ESTs and non-redundant transcripts (mRNAs and conceptual transcripts derived from genomic sequences) that overlap for at least 40 bases with at least 94% sequence identity, plus singletons, that do not satisfy the overlapping constraints. The assembly is performed by the TGICL program with a first clustering step by a slightly modified MegaBlast program, before a final assembly using the well known CAP3 program (see *http://compbio.dfci.harvard.edu/tgi/gifaq.html*). The initial purpose of such a assembly program is to generate as much complete open reading frames as possible, from EST, but the whole procedure can yield artefacts, such as chimeric constructs. Presently, this rather complex strategy yields an unigene set of as much as 78,976 sequences (VvGI release 6, 30 July 2008) in a three-fold excess with respect to the unigene Vvi build 9 (18 Sept 2008) at NCBI (23710 sequences). For many reasons, the Vvi unigene set of NCBI is far more stringent: multiple fragments of a similar gene are rejected since unigene must be anchored at the 3′ end, conceptual transcripts obtained from genome sequencing are not considered in the input data, and the unigene fasta file is just composed of the longest sequence, instead of uncurated assemblies.

No particular effort was made in order to favor annealing in the 3′ unstranslated region, as this was the case for MWG 3 k slides. The specificity of the oligos in the *V. vinifera* Array-Ready Oligo Set (AROS) V1.0 exclusively relies on sequence data available in the VvGI release 3.0: each oligo selected has less than or equal to 70% identity to all other transcripts and cannot have greater than 20 contiguous bases common to any non-self transcript. Obviously, both constraints are then dependent on the size of the sequence database used for the BLAST. Microarrays spotted with the *V. vinifera* Array-Ready Oligo Set (AROS) V1.0 were used to characterize the effect of ethylene on berry transcriptome (Chervin et al. 2008), change in gene expression upon phytoplasma infection of midribs (Rotter et al. 2008) or finding putative targets of the transcription factors VvMybA1, (Cutanda-Perez et al. 2009) VvMybPA1 and VvMyb PA2 ectopically expressed in grapevine hairy roots (Terrier et al. 2009). Twenty-six slides including 52 hybridizations were published with this platform.

11.2.2.2 Affymetrix

The gene chip *V. vinifera* genome array released by Affymetrix (*http://www.affymetrix.com/index.affx*) consists of 16,436 probe sets, which can interrogate 12,908 Genbank accessions of *Vitis vinifera* and 1,547 of other *Vitis* species or hybrids. Each probe set is composed of 16 oligonucleotides (25-mer). Oligonucleotide probes are spotted on quartz chips by photolithography. Ten papers including data from 314 hybridizations were published using this platform. Affymetrix chips are used to measure absolute values of gene expression—a single sample is hybridized.

11.2.2.3 Combimatrix Whole Genome Microarray Chip

Whole genome microarrays based on the sequenced grapevine genome (Jaillon et al. 2007) and DFCI Grape Gene Index release 6 were made in Verona, Italy with a Combimatrix CustomArrayTM Synthesizer (*http://*

ddlab.sci.univr.it/FunctionalGenomics). Researchers can customize the probe oligonucleotide design for up to 90,000 features. Probe length can be up to 50-mer, as successfully proposed on the early 3K MWG slides. A first version of the microarray was made containing three replicates of 24,622 grapevine oligonucleotide probes. A newer design is in work as for the latest update of the grapevine genome (M. Delledonne, pers. comm.). Chips are used for one-channel analysis, a single sample is hybridized. They can be washed and be reused for three times. Absolute values are obtained and any slide can be compared to any other slide, provided they have the same internal standards. One should notice that the *Vitis vinifera* Array-Ready Oligo Set (AROS) V1.0 also included probes for internal standards, but the dye swap protocol was always preferred to "absolute" measurements. No data has still been published with the Combimatrix arrays.

11.2.3 Oligonucleotide Annotation

Microarray designs for all but the Combimatrix whole genome are based largely on EST data. As predicted transcript sequences have been updated regularly, probe annotation and even probe association have to be updated. For this, single oligonucleotide probes have to be mapped to either newer versions of the DFCI Grape Gene Index, the grapevine genome or other available sequences. Considering only perfect match mappings of the VvAROS V1.0 oligonucleotid probe sequences against the predicted mRNA of the grapevine genome (Jaillon et al. 2007), only 4,915 can be associated, among which 166 maps to multiple predicted mRNA. Decreasing the mapping constraints yields a higher number of associated genes, but with multiple mapping, and thus with loss of specificity.

For example, in the Affymetrix gene chip, 330 probes correspond to 21 different probe sets assumed to correspond each to an aquaporin gene. Among these probes, only 174 perfectly match (25/25 nucleotides) with 12 predicted mRNA of the grapevine genome. One to 42 probes match with each putative aquaporin MRNA (Table 11-1).

A more definitive unigene set consistent with complete mRNA, EST, and genomic data still needs to be assembled, in order to facilitate the reannotation of past microarray probes, and the design of future so called exhaustive oligoarrays. The DFCI grape unigene set presently includes 78976 sequences (VvGI release 6, 30 June 2008), three fold more than unigene Vvi build 9 (18 Sep 2008) at NCBI (23,710 sequences), and two fold more than the realistic estimate of 30400 conceptual transcripts from grape genome (Vitis_vinifera_mRNA_V1.fa, downloadable at *http://www.genoscope.cns.fr/spip/Vitis-vinifera-whole-genome.html*).

Table 11-1 Number of Affymetrix probes matching with a selected set of different putative aquaporin genes.

Grapevine 8x genome identification	Number of matching probes
GSVIVT00029248001	42
GSVIVT00013854001	8
GSVIVT00000446001	8
GSVIVT00034350001	31
GSVIVT00026881001	1
GSVIVT00033619001	2
GSVIVT00022146001	12
GSVIVT00023192001	9
GSVIVT00024536001	24
GSVIVT00036133001	16
GSVIVT00024394001	16
GSVIVT00026882001	5

11.2.4 Data Pre-processing and Identification of Differentially Expressed Genes

Bioinformatics analyses are a crucial part in microarray studies. Data pre-processing such as background correction and spot intensities normalization within and/or between microarrays are necessary to reduce technical biases. Determination of expression values and identification of differentially expressed genes depends strongly on it (Rotter et al. 2008). There are several commercial tools available to do this, but the use of open source tools as R/Bioconductor (Gentleman et al. 2004, *http://www.bioconductor.org*) is certainly an advantage to understand data processing and to be able to reproduce it. Bioconductor is an open source project providing tools to analyze and understand genomic data.

11.3 Berry Development and Ripening

Due to its economical importance (about 58 million tonnes of both table and wine grapes are produced worldwide each year) and its complexity, berry ripening is the process that was studied the first, and most extensively by transcriptomic approaches. Global approaches based on microarray analysis using tomato as a model had yielded a great deal of information about the ripening of fleshy fruits (Giovannoni 2001; White 2002). However, unlike grape berry, tomato is a climacteric fruit and represents only a subset of the full range of angiosperm fruit types.

Grape berry development is a dynamic process that involves a complex series of molecular genetic and biochemical changes divided into three major phases (Deluc et al. 2007; Conde et al. 2007). During the first few weeks of development (phase I), rapid cell division occurs and existing pericarp cells expand. Tartaric acid accumulates during the initial stages of berry development while malic acid is accumulated in the flesh cells at the end of the first growth phase. Tannins, and hydroxycinnamates, which are precursors of volatile phenols, also accumulate during phase I (Romeyer et al. 1983). The second major phase is a lag phase during which cell expansion ceases and sugars begin to accumulate. *Véraison* (the onset of ripening) marks the beginning of the third major phase during which berries undergo a second period of sigmoidal growth due to additional mesocarp cell expansion. This period is characterized by a massive increase of several compounds, the major ones being glucose and fructose accumulated in the vacuoles of flesh cells. Anthocyanin synthesis rapidly increases after *véraison* and then stabilizes. Thus, at veraison, berries switch from a status where they were small, hard and acidic with little sugar to a status where they are larger, softer, sweet, less acidic, and strongly flavored and colored.

The structural genes involved in ripening, berry softening, sugar accumulation and pigment synthesis, and the regulatory genes that may control these processes have been extensively studied over the last 15 years. The majority of early work in the molecular biology of grape berry development were first confined to the level of individual genes or small groups of genes with the exception of four studies which attempted to ascertain genome-wide transcript level changes, using the techniques of EST analysis (Ablett et al. 2000; Terrier et al. 2001) and differential screening (Davies and Robinson 2000) and cDNA-AFLP (Venter et al. 2001). The first wide scale transcriptomic studies of berry ripening were published in 2005 (Terrier et al. 2005; Waters et al. 2005).

Terrier et al. (2005) achieved high-throughput identification of differentially expressed genes in Shiraz berries by the use of 50-mer oligoarrays bearing a set of 3,200 unigenes from *Vitis vinifera* to compare berry transcriptome at nine developmental stages. Comparisons of the green and ripening stages showed that 781 transcripts (25%) were significantly activated or repressed between both developmental periods. Change in average expression ranged from 1.3 to 132 and exceeded two-fold variation for about 400 transcripts. Analysis of transcript profiles revealed that most activations were triggered simultaneously with softening, occurring within only 24 hours for an individual berry, just before any change in coloration or water, sugar, and acid content can be detected. Many induced genes belong to unknown functional categories. However, several transcripts that were already known to be induced during ripening were confirmed in this study. These included grape ripening induced proteins (GRIPs) which are putatively involved in cell wall structure (GRIP3/4, 13 and 28) and in stress response (GRIP 22 and 61) (Davies and Robinson 2000), and defence related transcripts like thaumatins, PR-4 or endochitinase (Robinson et al. 1997; Davies and Robinson 2000; Tattersall et al. 2007) as well as UFGT

(UDP glucose-flavonoid 3-*O*-glucosyl transferase) implicated in grape berry skin coloration (Boss et al. 1996). The expression of 660 transcripts strongly decreased along ripening. They included genes related to photosynthetic activity (photoreaction centers I and II, chlorophyll a/b precursor, etc.), which is expected since the berries switch from a photosynthetically active status (green stage) to a sink status when they import sugars from the mature leaves. Two new partial putative hexose transporter-like genes were induced during Shiraz berry development. Their similar expression pattern was strictly parallel with sugar accumulation and declined with over-ripeness, when sugar import stopped. Likewise, a glutathione S-transferase exhibited the same expression profile as anthocyanin accumulation. Focusing on isogenes potentially significant in development regulation (hormonal control of transcription factor) revealed a possible role for several hormones (cytokinin, gibberellin, or jasmonic acid). The analysis of transcription factors revealed the induction of RAP2 and WRKY genes at *véraison*, which suggests increasing biotic and abiotic stress conditions during ripening.

Using the same microarray design as Terrier et al. (2005), Glissant et al. (2008) studied gene expression along eight stages of development in Chardonnay and Shiraz berries and addressed specifically the molecular basis of berry softening. Sixty three transcripts corresponding to genes involved in cell wall metabolism were studied in detail. A complex picture emerged since within a given gene family, different members exhibit quite distinct patterns of expression along berry development. This is the case for glucanases, α-mannosidases, xyloglucan endotransglycosylase (XET), pectin methyl esterases, prolin-rich proteins and GRIPs. The same observation could be made for members of the PIPs (plasma membrane intrinsic proteins) and TIPs (tonoplast intrinsic proteins) which mediate transmembrane water transport and control the water content of cells and tissues.

Fouquet et al. (2008) designed specific probes based on the 30 untranslated regions of each cDNA for the preparation of cDNA macroarray filters. Macroarray data pointed out to a global decrease of aquaporin gene expression during berry ripening.

The specific expression pattern found for different members within a gene family stress that detailed studies must be undertaken to identify the transcription factors, which control the expression of specific sets of genes. This can be done by several techniques among which transcriptomic analysis may also provide an initial screen for the identification of such factors, through the demonstration of co-expression patterns. For example, in a study by Glissant et al. (2008), the expression profile of a zinc finger protein was strongly parallel to that of a polygalacturonase like protein and the expression profiles of two pectin acetylesterase isoforms were close to that of a zinc finger protein (B-box type).

Waters et al. (2005) constructed microarray slides from 4,608 PCR-amplified cDNA clones derived from a cDNA library made from whole Shiraz ripening berries. They studied the expression profile of the genes that code for these cDNAs over the course of berry skin development. Berry skins are of particular interest, since the skin is metabolically active during development and ripening and may have an endocrinal function (Kanellis and Roubelakis-Angelakis 1993). Furthermore, the skin is the site of anthocyanin biosynthesis, a key component of wine quality. Statistical analysis revealed four sets of genes with distinctive and similar expression profiles over the course of berry development. The first set was composed of genes that had maximal transcript level in flowers, followed by a steady decrease in expression encoded proteins involved in photosynthesis. The second set was dominated by genes involved in flavonoid biosynthesis. The data showed co-ordinate regulation of the genes coding for 4-coumarate-CoA ligase, chalcone synthase, chalcone isomerase, flavonone hydroxylase, anthocyanidin reductase and cytochrome b5. The expression of these genes peaked week 2 post-flowering. The third set of cDNAs exhibited maximal expression week 5 post-flowering, midway between flowering and véraison, a period of rapid berry growth. This set of cDNAs was dominated by genes which code for structural cell wall proteins. The fourth set of genes was dramatically up-regulated at *véraison* and remained up-regulated until 13 weeks post-flowering. This set mostly contained genes with unknown function.

Using the Affymetrix *Vitis* GeneChip®, Pilati et al. (2007) investigated gene expression in berries sampled before and after véraison in three growing years, in order to investigate seasonal influences on gene expression and berry ripening. They identified a core set of 1,477 genes, which were similarly modulated in all seasons. They were able to separate ripening specific isoforms within gene families and to identify ripening related genes, which also appeared strongly regulated by the seasonal weather conditions. Annotation by Gene Ontology vocabulary revealed that functional categories corresponding to cell wall organization and biogenesis, carbohydrate and secondary metabolisms and stress response were specifically induced during the ripening phase, while photosynthesis was strongly repressed. Altogether, these data confirmed those previously obtained by Terrier et al. (2005) and Waters et al. (2005). Pilati et al. (2007) also pinpointed to the possible control of ripening by hormones, but instead of cytokinin, gibberellins and jasmonic acid (Terrier et al. 2005), auxin, ethylene and light emerged as the main stimuli influencing berry development in their data. In addition, an oxidative burst, previously not detected in grapevine, characterized by rapid accumulation of H_2O_2 starting from *véraison* and by the modulation of many ROS scavenging enzymes, was observed (Pilati et al. 2007).

Deluc et al. (2007) studied gene expression in Cabernet Sauvignon berries at seven stages of berry development with the Affymetrix GeneChip® *Vitis*, and in parallel they profiled selected metabolites. The aim was to understand the effect of transcriptional regulatory processes on specific metabolite production. Over the course of berry development, 28% of transcripts, (4,151 unigenes), had pronounced (≥ 2-fold) differences in mRNA expression. The subset of 4,151 unigenes was split into 20 well-correlated expression profiles. mRNA expression patterns of transcription factors, abscisic acid (ABA) biosynthesis, and calcium signaling genes identified candidate factors likely to participate in the progression of key developmental events such as véraison and genes possibly associated with processes like auxin partitioning in berry cells, aroma compound production, and pathway regulation and sequestration of flavonoid compounds. Finally, analysis of sugar metabolism gene expression patterns indicated the existence of an alternative pathway for glucose and triose phosphate production that is invoked from *véraison* to mature berries.

Using the Affymetrix *Vitis* chip, Grimplet et al. (2007) studied gene expression in three major tissue types (skin, pulp and seed) composing the berries (Cabernet Sauvignon), along development and after water stress. The skin provides volatile and non volatile aroma and color compounds, which assist in seed dispersal by avian and mammalian vectors. The pulp contributes organic acids and sugars, and the seeds provide condensed tannins, all of which are important for the formation of organoleptic characteristics of wine. Berry tissues expressed approximately 76% of genes represented on the Affymetrix chip. More than 28% of genes showed pronounced (2-fold or greater) differences in transcript level in one or more of the three major tissue types. The largest difference in tissue-specific expression was observed between the seed and pulp/skin. Exocarp tissue, which is involved in pathogen defense and pigment production, was characterized by the abundance of transcripts related to flavonoid biosynthesis, pathogen resistance, and cell wall modification. The mesocarp, which is considered a nutritive tissue, exhibited a high abundance of transcripts related to cell wall function and transport processes. Seeds, which supply resources for embryo development, expressed mostly genes encoding phenylpropanoid biosynthetic enzymes, seed storage proteins, and late embryogenesis abundant proteins. Water-deficit stress affected the transcript abundance of 13% of the genes with differential expression patterns occurring mainly in the pulp and skin. The genes affected by water stress encode storage proteins and novel genes (no-hit functional assignments). In the skin of berries from water-deficit stressed plants, transcripts related to general phenypropanoid and ethylene metabolism, pathogenesis-related responses, energy, and interaction with the environment were over-represented.

One important parameter for the final quality of the wine is the harvest date, which is still determined from macroscopic and relatively empirical parameters. Furthermore, berries on a given cluster may not be at the same ripening stage. More precise markers of ripening are thus needed, and transcriptomic approaches may address this issue by identifying genes whose expression changes (up- or down-regulation) would correlate with the desired ripening stage. Lund et al. (2008) analyzed gene expression in individual Cabernet Sauvignon grape berries sampled from two clusters at 50% ripening initiation using both the affymetrix gene chip, and RT-PCR experiments for those genes that were not represented on the chip. Global transcriptional profiles did not statistically differ when sampling was done from clusters on plant 1 or 2, or between different positions (proximal or distal to the shoot) within each cluster. Global transcriptional profiles could be significantly associated with berry pigmentation, which was thus considered as a good indicator of ripening. Special attention was also given to signal transduction functions in order to make progress in the regulatory network modeling of ripening initiation. Three receptor-like kinases and 17 transcription factors were up-regulated during the two earliest ripening transitions (green hard to green soft, and green soft to pink soft). A 9-cis-epoxycarotenoid gene (*VvNCED2*) and a putative ortholog to an abscisic acid receptor, VvGCR2, were well correlated with ripening initiation. Six transcripts corresponding to receptor like kinases and 15 transcripts corresponding to transcription factors were down-regulated during the same transitions. Guillaumie et al. (pers. comm.) used the 14 K Qiagen/Operon microarray to identify genes that may be markers of ripening in Chardonnay berries. Transcriptomic analysis were run for two consecutive years with berries collected at three developmental stages around technological maturity, and a comparison was made between berries that were sorted or not by density. Among the 14,000 probes spotted on the slide, only a few genes that were consistently up-regulated or down-regulated in the different series analyzed were identified, among which several genes encoded for enzymes involved in the metabolism of aroma compounds.

The use of transcriptomic approaches on natural mutants is a powerful technique to identify the genes directly or indirectly involved in the mutant phenotype. Likewise, transcriptomics are also used to analyze gene expression in transgenic plants overexpressing transcription factors for which the molecular targets are not yet all identified, as this will be detailed below in the paragraph on flavonoid synthesis.

In order to identify genes putatively involved in the early development of grapevine fruit, Fernandez et al. (2007) used the *fleshless* berry mutant (natural mutant derived from Ugni Blanc variety) exhibiting very small fruits, due to a lack of pericarp development. The 14 K Qiagen/Operon microarray helped to identify 53 and 50 genes that were down- and up-regulated, respectively, in the mutant. Furthermore, suppression subtractive hybridization experiments between the mutant and the wild type (WT), allowed the identification of new transcripts differentially expressed during the first stages of pericarp development. The mutation promotes the expression of several genes related to ripening and/or to stress and impairs the expression of several regulatory genes. Among those, five genes encoding proteins previously reported to be associated with, or involved in, developmental processes in other species (a specific tissue protein 2, ATHB13, a BURP domain protein, PISTILLATA, and YABBY2), were identified. Their expression in the pericarp was confirmed, specific spatial and/or temporal patterns were detected and differences were observed between the WT and the mutant during fruit development. Expression of these genes was affected during young fruit development in the mutant, suggesting that they may play a role in grape berry morphogenesis.

11.3.1 Biosynthesis of Flavonoids

The biosynthesis of flavonoid compounds, including anthocyanins and proanthocyanidins has been studied with special attention in some transcriptomic studies. Anthocyanins are responsible for the red, blue, and purple color of plant tissues. They play an important role in plant reproduction, by attracting pollinators and seed dispersers, and in protection against photo-oxidative stress (Winkel-Shirley 2002). The skin of grape berry accumulates large amounts of anthocyanins, which contribute to the sensory attributes of wine. Considerable attention has also been paid recently to the health benefits of anthocyanins, since epidemiological investigations have indicated that moderate consumption of red wine is associated with a lower risk of cardiovascular disease due to the presence of flavonoids (Hou 2003).

Anthocyanin synthesis and red/white color determination in grape berries are controlled by numerous transcription factors belonging to the Myb, bHLH and WD40 families. The expression of UFGT is also controlled by a myb-factor. In order to further explore co-ordinated gene expression patterns associated with color determination, Ageorges et al. (2006) developed a dual screening strategy using both a substractive hybridization (SSH) library construction and 3 K-oligo-arrays- profiling. Two subtractive libraries (SSH) were obtained by comparing two red-skinned cultivars with either red (Lacryma) or white pulp (Gamay). Microarray profiling was performed in several red/white samples to identify isogenes consistently expressed in red tissues whatever the source of variation (genotype, development, tissue). A set of nine isogenes were consistently associated with the colored berry tissues. Four of these isogenes were related to the flavonoid metabolism pathway (chalcone synthase (isogene *CHS3*), glutathione S-transferase (*GST*), *UFGT*, caffeoyl methyl transferase (*CaOMT*)), two of them (*GST* and *CaOMT*) being reported and characterized for the first time in grape berries. These results suggest the involvement of upstream (*CHS3*) and downstream (*GST*, *CaoMT*) isogenes of the anthocyanins pathway, in addition to UFGT for the determination of grape berry color.

The global atmospheric warming trend may affect grape berry ripening in the future. Jones et al. (2005) suggested that in regions producing high quality grapes on the margins of their climatic limits, future climate change would exceed a threshold, which would make it difficult to reach the balanced berry composition needed at harvest to produce given wine styles. In particular, anthocyanin accumulation is inhibited at high temperatures in red and black grapes. It is, therefore, important to investigate the molecular basis of the environmental effects on flavonoid metabolism. Mori et al. (2007) studied the effects of high temperature on anthocyanin composition and the transcriptome responses in Cabernet Sauvignon grape berries. Although high temperature (maximum 35°C) reduced the total anthocyanin content to less than half of that in the control berries (maximum 25°C), Affymetrix Vitis GeneChip analysis indicated that 405 genes were differentially transcribed (2-fold threshold). However, the anthocyanin biosynthetic genes were not strongly down-regulated at high temperature, with the exception of *DFR* (di-hydroflavonol reductase) and F3 H2 (flavanone 3 hydroxylase) and this was confirmed by quantitative real time-PCR. To test the possibility that high temperature increases anthocyanin degradation in grape skin, stable isotope-labeled tracer experiments were carried out by incubating softened green berries with aqueous L-[1-^{13}C]phenylalanine and the changes in ^{13}C-labeled anthocyanins were studied under different temperatures. In the berries cultured at 35°C, the content of total ^{13}C-labeled anthocyanins that were produced before exposure to high temperature was markedly reduced as compared with those cultured at 15°C and 25°C. Thus, the decrease in anthocyanin accumulation at high temperature results from anthocyanin degradation as well as the transcriptional repression of a few anthocyanin biosynthetic genes.

VvMYBA1 is a transcription factor thought to control anthocyanin biosynthesis in grape because it controls the expression of UFGT, but other targets are not excluded. In order to identify the whole set of downstream regulated genes, Cutanda-Perez et al. (2009) targeted constitutive ectopic expression of *VlmybA1-2*, a *Vitis labruscana* allele of *VvMybA1*, into grapevine hairy roots and plants. The ectopic expression of *VlmybA1-2* triggered *de novo* production and storage of anthocyanins in all transgenic vegetative organs but did not interfere with proanthocyanidin (PA) biosynthesis. A transcriptomic analysis using the Qiagen/operon 14K microarray revealed that the ectopic expression of *VlmybA1-2* activated only few genes, most of which are involved in both PA and anthocyanin biosynthesis, while the expression of *BAN* and *LAR* (two specific genes of the PA biosynthesis pathway) was unaffected. Among these, four genes were up-regulated as much as *VlmybA1-2* itself. They included *UFGT*, an isogen of *GST*, an O-methyltransferase, as well as a candidate gene putatively involved in the vacuolar anthocyanin transport (*antho-MATE*). These results suggest that MybA1 activates the last steps of anthocyanin synthesis and transport through the regulation of a narrow, specific range of genes regulated as a cluster.

In addition to anthocyanins, grapevine flavonoids also include proanthocyanidins, which contribute to plant defense mechanisms against biotic stress and also play a critical role in organoleptic properties of wine. *VvMybPA1* was identified as the first transcription factor involved in the regulation of the proanthocyanidin pathway during seed development in grapevine (Bogs et al. 2007). In contrast, *VvMybPA2* (Terrier et al. 2009), is mainly expressed in the exocarp of young berries and in the leaves. This transcription factor shows very high protein sequence homology with other plant Myb factors, which regulate flavonoid biosynthesis. Ectopic expression of either *VvMybPA1* or *VvMybPA2* in grapevine hairy roots induced a 5-fold increase in the proanthocyanidin content. Transcriptomic analyses of transformed hairy roots were run with the 14K Qiagen/Operon Only 2 to 3% of the genes spotted on the microarray exhibited significant expression changes induced by *VvMybPA1* or *VvMybPA2* overexpression. Both *VvMybPA1* and *VvMybPA2* significantly activated enzymes of the flavonoid pathway, including ANR (anthocyanidin reductase) and LAR1 (leucoanthocyanidin reductase 1), the specific terminal steps in the biosynthesis of epicatechin and catechin, respectively, but not LAR2. The functional annotation of the genes whose expression was altered also revealed putative new actors of the proanthocyanidin pathway, such as glucosyltransferases and transporters (Terrier et al. 2009).

Chervin et al. (2008) screened the gene pool affected by ethylene at the ripening inception in Cabernet Sauvignon berries with the 14K Qiagen/Operon microarray. Only 73 genes were significantly modulated by a 24-hour ethylene treatment performed 8 weeks after flowering. Ethylene application at véraison increased berry diameter due to sap uptake and cell wall modifications, which enable cell elongation. In parallel, the expression of two groups of genes were affected: aquaporins, which mediate transmembrane water transport, and genes controlling cell wall metabolism (polygalacturonases, xyloglucan endotransglucosylases (*XTH*), pectin methyl esterases, cellulose synthases and expansins). Q-PCR experiments were conducted to study the expression of selected genes in the skin, flesh and seeds. One XTH and one aquaporin gene were identified as good candidates to explain the ethylene-induced berry expansion.

The genetic enhancement of the expression of flavanoid (nutraceutical) compounds in two major North American grape species *Vitis rotundifolia* ("Noble") and *V. aestivalis* ("Cynthiana") was also studied with the 14K Qiagen/Operon microarray for two developmental stages, véraison and physiological maturity (Samuelian et al. 2009). The expression profiles of 28 genes involved in critical steps of the flavonoid biosynthesis and accumulation or strongly differentially expressed during the process of ripening were validated in the two cultivars by real-time PCR. Less than half of the transcripts (43%) had significantly similar expression patterns in "Noble" and "Cynthiana". Among them anthocyanidin synthase and anthocyanidin reductase, responsible for color stability in grapes, were significantly more expressed in "Cynthiana" than in "Noble" (Samuelian et al. 2009).

11.4 Abiotic Stress

Grapes are grown in semiarid environments where there may experience drought and salinity. Global warming is expected to increase problems of water supply during summer time, especially in southern areas. The molecular basis of adaptation to water stress and salinity therefore deserves detailed investigations. Cramer et al. (2007) used the 14K Affymetrix gene chip in order to study the shared and divergent transcriptomic responses of Cabernet Sauvignon vines to a gradually applied and long-term (16 days) water-deficit stress and equivalent salinity stress. Water deficit caused more rapid and greater inhibition of shoot growth than did salinity at equivalent stem water potentials. Water deficit resulted in an early increase of the RuBisCo activase transcripts (day 4), whereas this increase occurred much later in salt-stressed plants (day 12). Water deficit and salinity affected different functional categories. As water deficit progressed, a greater number of affected transcripts were involved in metabolism, transport, and the biogenesis of cellular components than did salinity. In contrast, salinity affected a higher percentage of transcripts involved in transcription, protein synthesis, and protein fate than did water deficit. Metabolite profiling revealed that concentrations of glucose, malate, and proline were higher in water-deficit treated plants compared to salinized plants. The metabolite differences were linked to differences in transcripts related to energy metabolism and nitrogen assimilation, particularly photosynthesis, gluconeogenesis, and photorespiration. Water deficit-treated plants appear to have a higher demand than salinized plants to adjust osmotically, detoxify free radicals (reactive oxygen species), and cope with photoinhibition.

In contrast with the long term stresses imposed in the experiments described above, Tattersall et al. (2007) exposed Cabernet Sauvignon grapevines to various sudden treatments including chilling (5°C), water deficit (polyethyene glycol, PEG), and an iso-osmotic salinity (120 mM NaCl and 12 mM $CaCl_2$) for 1, 4, 8, and 24 hours. Microarray analysis of transcript abundance in shoot tips detected no significant differences in transcript abundance between salinity and PEG before 24 hours. Chilling stress relates to changes in membrane structure, and transcript abundance patterns were predicted to reflect this. Forty-three percent of transcripts affected by stress vs. control for 1 through 8 hours were affected only by chilling. The functional categories most affected by stress included metabolism, protein metabolism, and signal transduction. Osmotic stress affected more protein synthesis and cell cycle transcripts, whereas chilling affected more calcium signaling transcripts. Many hormones (ABA, ethylene, and jasmonate) and transcription factor transcripts were also affected by the stress applied. The concentrations and transporter transcripts of several anions increased with time, including nitrate, sulfate, and phosphate. A larger and more complex response in the acclimation process of a gradual long-term stress was observed compared to a sudden stress.

Berries used for some sweet dessert wines (e.g., Recioto) and dry fortified wines (e.g., Amarone) produced in Italy undergo a phase of post-harvest dehydration (withering), which can last up to 3 months, during which metabolism is significantly modified and the sugar content increases (Kays 1997). In post-harvest berries, water loss induces cell wall enzyme activity, increased respiration and ethylene production, and causes the loss of volatiles and changes in polyphenol levels (Zamboni et al. 2008). Air drying also induces fruit softening. During post-harvest drying, berry cells undergo an initial water stress response at 10–12% weight loss, characterized by the accumulation of ABA, proline, and lipoxygenase. A second dramatic change in metabolism occurs at > 19% weight loss, characterized by the accumulation of proline and an increase in alcohol dehydrogenase activity. This two-step metabolism leads initially to the formation of C6 compounds, ethanol and acetaldehyde, which subsequently decrease due to the formation of ethyl acetate (volatile acidity) (Costantini et al. 2006).

Zamboni et al. (2008) carried out a detailed transcriptomic analysis of post-harvest grape berries by AFLP-transcriptional profiling. AFLP-TP identified 699 withering-specific genes, 167 and 86 of which were unique to off-plant and on-plant withering, respectively. Off-plant withering induced a stronger dehydration stress response resulting in the high level expression of genes involved in stress protection mechanisms, such as dehydrin and osmolite accumulation. Genes involved in hexose metabolism and transport, cell wall composition, and secondary metabolism (particularly the phenolic and terpene compound pathways) were similarly regulated in both processes.

It may be interesting to compare the changes in gene expression induced by this post-harvest dehydration of the berries with that of berries collected from water stressed plants.

11.5 Biotic Stress

Grapevine species (*Vitis* sp.) are prone to several diseases, fungi being the major pathogens compromising its cultivation and economic profit around the world. During the 19th century, pathogens such as powdery mildew (*Uncinula necator*) and downy mildew (*Plasmopara viticola* (Berk. et Curt.) were introduced in Europe from North America and became major risk factors for viticulture. Since *V. vinifera* does not carry any resistance to mildew fungi, multiple fungicide applications per growing season became indispensable for traditional cultivars. One way to overcome these problems is to develop resistant cultivars producing high quality fruits. Understanding the molecular basis for resistance to fungus infection is a prerequisite to build up strategies, which will improve resistance of sensitive grapevine cultivars. There are few reports directly based on microarray technology and related to the study of grapevine/pathogen interactions.

Figueiredo et al. (2008) analyzed transcript and metabolic profiles of the *Vitis vinifera* cultivars Regent and Trincadeira (resistant and susceptible to fungi, respectively) by cDNA microarray, quantitative real-time PCR, and nuclear magnetic resonance spectroscopy. Transcript and metabolite profiles differed between both cultivars and this may be associated with the innate resistance of Regent towards the mildews. Several transcripts related to stress and defence, namely a subtilisin-like protease, phenylalanine ammonia lyase, S-adenosylmethionine synthase, WD-repeat protein like, and J2P, were up-regulated in Regent suggesting an intrinsic resistance capability of this cultivar. Inositol and caffeic acid, which are known to confer resistance to fungi, accumulated in the resistant cultivar. This combination of transcriptomic and metabolite analysis may give some clues to discriminate among resistant and susceptible grapevine cultivars.

Using the Affymetrix chip, Fung et al. (2007) compared the transcriptome of young leaves of *V. vinifera* L. and *Vitis aestivalis* Michx. Approximately 12% of the genes were differentially expressed in the two grapevine species ($P < 0.001$). Over 200 probe sets were identified, which consistently detected transcripts in one grapevine species, but not in the other. No broad functional category in which transcript abundance was overall different in any one species could be identified. Among the genes expressed only in *V. aestivalis* leaves, a class IV chitinase which had been previously shown to have a flower- and fruit-specific expression in *V. vinifera* was identified. Among the transcripts which were differentially expressed ($P < 0.001$) in both species, there were genes encoding key enzymes in flavonoid, monolignol, and proanthocyanidin biosynthesis. Interestingly, statistical analysis suggested that sequence divergence between the predominantly *V. vinifera*-derived GeneChip probes and the *V. aestivalis* cRNA did not confound the hybridization data and that the reliability of the microarray results was similar in the two grapevine species.

Fung et al. (2008) also studied the evolution of grapevine transcriptome up to 48 hours post inoculation with powdery mildew in *V. vinifera* (susceptible) and *V. aestivalis* (resistant to powdery mildew). Only three genes responsive to powdery mildew were found in *V. aestivalis*, compared to 625 in *V. vinifera*, which indicated that the resistance of *V. aestivalis* was not associated with overall reprogramming of the transcriptome. Yet, attack by powdery mildew induces defense-oriented reprogramming of the transcriptome in *V. vinifera*, but this was not sufficient to prevent fungus invasion.

The oomycete *Plasmopara viticola*, which causes downy mildew, is strictly biotrophic, and thus entirely depends on living host cells for its survival. The molecular basis of compatibility and disease development in this system is poorly understood. Polesani et al. (2008) carried out a large-scale cDNA-AFLP analysis to identify grapevine and *P. viticola* genes associated with the infection process. They used artificially infected leaves of the susceptible cultivar Riesling at the oil spot stage, water-treated leaves and pure sporangia as controls. Selective amplifications with 128 primer combinations allowed the visualization of about 7,000 transcript-derived fragments (TDFs) in infected leaves, among which 1,196 (17%) were differentially expressed. Among 984 sequenced fragments, 804 were homologous to grapevine transcripts and 96, which were homologous to sequences in *Phytophthora* spp. Databases, were attributed to *P. viticola*. There were 82 orphan TDFs. Many grapevine genes spanning almost all functional categories were down-regulated during infection, especially genes involved in photosynthesis. Grapevine genes homologous to known resistance genes also tended to be repressed, as were several resistance gene analogs and carbonic anhydrase (recently implicated in pathogen resistance). In contrast, genes encoding cytoskeletal components, enzymes of the phenylpropanoid and beta-oxidation pathways, and pathogenesis related proteins were primarily up-regulated during infection. This type of approach may help elucidate the molecular basis of the infection process and identify genes and chemicals that could help to inhibit the pathogen.

Viral diseases affect grapevine cultures without inducing any resistance response. These plants thus develop systemic diseases and are chronically infected. Using the 14K Affymetrix chip, Espinoza et al. (2007b) published the first study of global gene profiling in grapevine in response to viral infections using DNA microarray. They carried out this analysis with the red wine cultivars Carménère and Cabernet-Sauvignon naturally infected with

GLRaV-3. The genes that are induced or repressed in viral infected leaves cover a wide spectrum of biological functions, including translation and protein targeting, metabolism, transport, and cell defense. Numerous genes related to the membrane and endomembrane systems were induced, and most of them are associated with transport, indicating that viral infection has a major impact on this function. In contrast, chloroplastic genes were mostly repressed.

The outcome of microarray analysis in two-channel expression data depends on preprocessing tools that are used (Nettleton 2006). Rotter et al. (2008) investigated the influence of the selection of preprocessing methods (background correction, normalization, and duplicate spots correlation calculation) on the discovery of differentially expressed genes. They analyzed the grapevine transcriptome from healthy and phytoplasma infected midribs with the 14K Qiagen/Operon microarray. The list of differentially expressed genes and the expression values of selected genes depended significantly on the preprocessing approach applied. Rotter et al. (2008) thus proposed to combine two normalization methods which are theoretically distinct from one another on the same dataset. Then, the intersection of results was used to reach a more accurate estimation of the genes that were *de facto* differentially expressed. Fifty-five genes were present in all of the data analysis combinations performed for the healthy and phytoplasma infected midrib samples. They could be grouped into 10 functional categories with transport, stress, miscellaneous enzymes being affected most by the experimental conditions.

11.6 What is next?

Although whole genome transcriptomics gives an overview on gene expression in certain cell or tissue types, its biological interpretation can be rather tedious. MapMan is a software tool that displays large datasets (e.g., gene expression data) onto diagrams of metabolic pathways or other processes and thus enables easier interpretation of results. To enable easier visualization grapevine transcriptomic data, Rotter et al. (2009) set an annotation of Unigene (version 5) to MapMan ontology and created new pictorial representations in the grape mapping file focusing on three selected pathways: carotenoid pathway, terpenoid pathway and flavonoid pathway. These new tools were validated with Affymetrix microarrays grape véraison experiment where new aspects of process regulation were discovered, and with Operon microarray data related to interactions of grape plantlets with the fungal pathogen *Eutypa lata*.

Microarray-based transcriptomic approaches are already challenged by high-throughput sequencing technologies, and this trend will probably become sharper with technological progress and decreasing prices, though even these techniques need an in-depth bioinformatical analysis. The release of the grape genome allows the use of 5′SOLiD sequencing, a powerful massive parallel signature sequencing (MPSS) technology for transcription profiling on a genome-wide scale. Concerning grape research, this approach offers several potential advantages over microarrays. First, it avoids the problem of poor hybridization, which might be encountered when using arrays for *Vitis* species or genus that are not genetically close to the ones, which were used to design the array probes. This technology allows the visualization of mRNAs present at a low copy level per cell (1–30 copies per cell). Furthermore, quantitative changes in low abundance mRNAs can readily be measured. The high sensitivity of the MPSS technologies is the key to quantifying the behavior of some low-abundance transcription factors and signal transduction factors that are key players in gene regulatory networks. Thus, this approach is very suitable to analyze developmental processes or responses to abiotic and biotic stresses. Iolandino et al. (2008) performed mRNA profiling using MPSS and combined it with available EST data. The sequence depth of MPSS allowed them to capture and quantify almost all the transcripts at a specific stage in the development of the grape berry. A total of 2,635,293 17-base and 2,259,286 20-base signatures were obtained, representing at least 30,737 and 26,878 distinct sequences. The average normalized abundance per signature was ~ 49 TPM (transcripts per million). Comparisons of the MPSS signatures with available *Vitis* species' ESTs and a unigene set demonstrated that 6,430 distinct contigs and 2,190 singletons have a perfect match to at least one MPSS signature. Among the matched sequences, ESTs were identified from tissues other than berries or from berries at different developmental stages. Additional MPSS signatures not matching to known grape ESTs can extend our knowledge of the *V. vinifera* transcriptome, during the unique stage in development that immediately precedes the onset of ripening. Importantly, the MPSS dataset also revealed evidence of antisense expression not previously reported in grapes but comparable to that reported in other plant species.

Whatever the approach used to obtain a precise and complete picture of the transcriptome (transcriptomics, see above), EST profiling (da Silva et al. 2005; Moser et al. 2005) or cDNA AFLP (Polesani et al. 2008), a complete and robust annotation of the genome is needed, the data should be available in public repository with a standard format, and must be organized according to metabolic maps and/or gene ontology category. This obviously requires a coordinate effort from the grapevine research community, which may be catalyzed by the International Grape Genome Program (IGGP). The IGGP (*http://www.vitaceae.org/*) was formed in 2001 to promote international collaboration and to develop resources for the grape research community. One advance since the beginning of this program has been the placement of data in public databases such as NCBI. The IGGP strongly recommends submitting all the public expression data to public databases such as NCBI Gene Expression Omnibus (GEO) (*http://www.ncbi.nlm.nih.gov/geo/index.cgi*). The NCBI taxonomy web address for *V. vinifera*, the main species used for wine production, summarizes the data that have been deposited. At the time of writing, an EST query for Vitis yields the following figures: *V. vinifera* (353,748), *V. shuttleworthii* (10,704), *Vitis* hybrid cultivar (6,548), *V. arizonica* x *V. rupestris* (5,421), *V. aestivalis* (2,101). The gene Index Project (*http://compbio.dfci.harvard.edu/tgi/cgi-bin/tgi/gimain.pl?gudb=grape*) presently lists 347,879 ESTs and 33,638 Tcs.

The build up of the metabolome of the grape berry, and its effect on wine ageing and sensory properties are still poorly known. Significant progress in the understanding of the build up of berry quality may be expected from integrative systems biology approaches. Wine quality still largely depends on the content of berry at harvest, which itself depends on complex interactions between the genome of the rootstock, the genome of the scion, the climate and viticultural practices. In view of remarkable technical advances in transcriptomics and metabolomics, integrated analysis by data-mining informatics is an excellent tool for prediction and identification of gene function, particularly for genes involved in complicated metabolic pathways (Saito et al. 2008). Sequence annotation of the *Arabidopsis* genome indicates that many genes are assigned to families controlling secondary metabolism, and this is even more true for grapevine. These genes control the synthesis, modification, degradation and/or transport of particular metabolites, but for a vast majority of genes, the function remains to be determined. The availability of *Arabidopsis* public transcriptome datasets containing data of > 1,000 microarrays reinforces the potential for prediction of gene function by transcriptome coexpression analysis. Integrated analyses of the transcriptome with the metabolome, correlating the expression pattern of all genes with the accumulation pattern of all metabolites help in deciphering the functions of genes that do not exhibit apparent phenotypic changes in mutant or overexpressing lines. Recent progress in metabolomics technology has increased the feasibility of this strategy. According to the so-called "guilt-by-association" principle, it is conceivable that a set of genes involved in a biological process are co-regulated and thus coexpressed under the control of a shared regulatory system. Therefore, if an unknown gene is coexpressed with known genes in a particular metabolic pathway, this unknown gene is probably involved in the pathway. This principle, has been successfully applied to functional characterization of a wide range of genes in non-plant organisms, such as yeast, humans and other mammals, based on the assumption of functional relationships between genes that are coexpressed, and it has also been successfully used in *Arabidopsis*, only because large sets of transcriptomic and metabolomic analysis are available (Saito et al. 2008). Development of similar approaches for grapevine requires the availability of large data sets shared between the community.

Several outputs may be expected in terms of application of transcriptional profiling in viticulture and breeding. The transcriptomic data allow the identification of transcripts playing a major role in a wide range of experimental conditions (for example healthy/infected plants; watered/watered stressed plants, markers of ripening). The same can be done in order to study the effects of any given viticultural practice on fruit quality or disease resistance. These transcripts may be ascribed to candidate genes, which are responsible for disease tolerance/sensitivity, or water stress resistance, or ripening in the examples given. This information can be strengthened and/or complemented by the results of other approaches (EST profiling, physical and genetic maps, RT-PCR). When candidate genes have been definitively identified, one may study their allelic variation within a natural population and correlate it with phenotypic traits. The exact function of the gene, if unknown, may be studied by a wide range of functional approaches combining cell localization, heterologous or homologous expression by transgenesis, RNA interference, the latter being associated with an in-depth phenotypical and metabolic analysis. Selection markers for this allele or for this gene can be designed and considerably increase the efficiency of the selection process after a cross between two varieties for example. For the markers of ripening, it may also be envisaged to produce antibodies against the protein encoded by the gene, and to design immunological tools able to estimate the amount of protein in the field. A set of two or three genes whose expression is positively correlated with the ripening stage, and a set of two or three genes whose expression is negatively correlated with the ripening stage would give a reliable differential analysis. An alternative strategy would be to characterize the products of the activity of the enzymes, if the markers encode active enzymes, and to monitor the amount of this product along ripening.

References

Ablett E, Lee S, Henry R (2000) New insights into leaf and berry physiology from analysis of 5,000 Vitis ESTs. In: K Roubelakis-Angelakis (ed) 6th Int Symp on Grapevine Physiology and Biotechnology, Heraklion, Greece, p 96.

Adam-Blondon AF, Bernole A, Faes G, Lamoureux D, Pateyron S, Grando MS, Caboche M, Velasco R, Chalhoub B (2005) Construction and characterization of BAC libraries from major grapevine cultivars. Theor Appl Genet 110: 1363–1371.

Ageorges A, Fernandez L, Vialet S, Merdinoglu D, Terrier N, Romieu C (2006) Four specific isogenes of the anthocyanin metabolic pathway are systematically co-expressed with the red colour of grape berries. Plant Sci 170: 372–383.

Bogs J, Jaffé FW, Takos AM, Walker AR, Robinson SP (2007) The grapevine transcription factor VvMYBPA1 regulates proanthocyanidin synthesis during fruit development. Plant Physiol 143: 1347–1361.

Boss PK, Davies C, Robinson SP (1996) Analysis of the expression of anthocyanin pathway genes in developing *Vitis vinifera* L. cv Shiraz grape berries and the implication for pathway regulation. Plant Physiol 111: 1059–1066.
Chervin C, Tira-Umphon A, Terrier N, Zouine M, Severac D, Roustan JP (2008) Stimulation of the grape berry expansion by ethylene and effects on related gene transcripts, over the ripening phase. Physiol Plant 134: 534–546.
Conde C, Silva P, Fontes N, Dias ACP, Tavares RM, Sousa MJ, Agasse A, Delrot S, Geros H (2007) Biochemical changes throughout grape berry development and fruit and wine quality. Food 1: 1–22.
Costantini V, Bellincontro A, De Santis D, Botondi R, Mencarelli F (2006) Metabolic changes of Malvasia grapes for wine production during post-harvest drying. J Agri Food Chem 54: 3334–3340.
Cramer G, Ergül A, Grimplet J, Tillett R, Tattersall E, Bohlman M, Vincent D, Sonderegger J, Evans J, Osborne C, Quilici D, Schalauch KA, Schooley DA, Cushman JC (2007) Water and salinity stress in grapevines: early and late changes in transcript and metabolite profiles. Funct Integr Genom 7: 111–134.
Cutanda-Perez, MC, Ageorges A, Gomez C, Vialet S, Terrier N, Romieu C, Torregrosa L (2009) Ectopic expression of VlmybA1 in grapevine activates a narrow set of genes involved in anthocyanin synthesis and transport. Plant Mol Biol 69: 633–648.
da Silva FG, Iandolino A, Al-Kayal F, Bohlmann MC, Cushman MA, Lim H, Ergul A, Figueroa R, Kabuloglu EK, Osborne C, Rowe J, Tattersall E, Leslie A, Xu J, Baek J, Cramer GR, Cushman JC, Cook DR (200) Characterizing the grape transcriptome. Analysis of expressed sequence tags from multiple Vitis species and development of a compendium of gene expression during berry development. Plant Physiol 139: 574–597.
Davies C, Robinson SP (2000) Differential screening indicates a dramatic change in mRNA profiles during grape berry ripening Cloning and characterization of cDNAs encoding putative cell wall and stress response proteins. Plant Physiol 122: 803–812.
Deluc LG, Grimplet J, Wheatley MD, Tillett RL, Quilici DR, Osborne C, Schooley CA, Schlauch KA, Cushman JC, and Cramer GR (2007) Transcriptomic and metabolite analyses of Cabernet Sauvignon grape berry development. BMC Genom 8: 429.
Doddapaneni H, Lin H, Walker MA, Yao J, Civerolo EL (2008) VitisExpDB: A database resource for grape functional genomics. BMC Plant Biol 8: 23.
Espinoza C, Medina C, Somerville S, Arce-Johnson P (2007a) Senescence-associated genes induced during compatible viral interactions with grapevine and Arabidopsis. J Exp Bot 58: 3197–3212.
Espinoza C, Vega A, Medina C, Schlauch K, Cramer G, and Arce-Johnson P (2007b) Gene expression associated with compatible viral diseases in grapevine cultivars Funct Integr Genom 7: 95–110.
Fernandez L, Torregrosa L, Terrier N, Sreekantan L, Grimplet J, Davies C, Thomas M, Romieu C, Ageorges A (2007) Identification of genes associated with flesh morphogenesis during grapevine fruit development. Plant Mol Biol 63: 307–323.
Figueiredo A, Fortes A, Ferreira S, Sebastiana M, Choi Y, Sousa L, Acioli-Santos B, Pessoa F, Verpoorte R, Pais M (2008) Transcriptional and metabolic profiling of grape (*Vitis vinifera* L.) leaves unravel possible innate resistance against pathogenic fungi. J Exp Bot 59: 3371–3381.
Fouquet R, Léon C, Ollat N, Barrieu F (2008) Identification of grapevine aquaporins and expression analysis in developing berries. Plant Cell Rep 27: 1541–1550.
Franks T, Powell K, Choimes S, Marsh E, Iocco P, Sinclair B, Ford C, van Heeswijck R (2006) Consequences of transferring three Sorghum genes for secondary metabolite (cyanogenic glucoside) biosynthesis to grapevine hairy roots. Transgen Res 15: 181–195.
Fung R, Qiu W, Su Y, Schachtman D, Huppert K, Fekete C, Kovács L (2007) Gene expression variation in grapevine species *Vitis vinifera* L. and *Vitis aestivalis* Michx. Genet Resour Crop Evol 54: 1541–1553.

Fung RW, Gonzalo M, Fekete C, Kovacs LG, He Y, Marsh E, McIntyre LM, Schachtman DP, Qiu W (2008) Powdery mildew induces defense-oriented reprogramming of the transcriptome in a susceptible but not in a resistant grapevine. Plant Physiol 146: 236–249.

Gentleman RC, Carey VJ, Bates DM, Bolstad B, Dettling M, Dudoit S, Ellis B, Gautier L, Ge Y, Gentry J, Hornik K, Hothorn T, Huber W, Iacus S, Irizarry R, Leisch F, Li C, Maechler M, Rossini AJ, Sawitzki G, Smith C, Smyth G, Tierney L, Yang JY, Zhang J (2004) Bioconductor: open software development for computational biology and bioinformatics. Genome Biol 5: R80.

Giovannoni J (2001) Molecular biology of fruit maturation and ripening. Annu Rev Plant Physiol Plant Mol Biol 52: 725–749.

Glissant D, Dédaldéchamp F, Delrot S (2008) Transcriptomic analysis of grape berry softening during ripening. J Int Sci Vigne Vin 42: 1–13.

Grimplet J, Deluc LG, Tillett RL, Wheatley MD, Schlauch KA, Cramer GR, Cushman JC (2007) Tissue-specific mRNA expression profiling in grape berry tissues. BMC Genom 8: 187.

Hou DX (2003) Potential mechanisms of cancer chemoprevention by anthocyanins. Curr Mol Med 3: 149–159.

Hughes TR, Marton MJ, Jones AR, Roberts CJ, Stoughton R, Armour CD, Bennett HA, Coffey E, Dai H, He YD, Kidd MJ, King AM, Meyer MR, Slade D, Lum PY, Stepaniants SB, Shoemaker DD, Gachotte D, Chakraburtty K, Simon J, Bard M, Friend SH (2000) Functional discovery via a compendium of expression profiles. Cell 102: 109–126.

Iandolino A, Nobuta K, da Silva FG, Cook DR, Meyers BC (2008) Comparative expression profiling in grape (*Vitis vinifera*) berries derived from frequency analysis of ESTs and MPSS signatures. BMC Plant Biol 8: 53.

Jaillon O, Aury JM, Noel B, Policriti A, Clepet C, Casagrande A, Choisne N, Aubourg S, Vitulo N, Jubin C, Vezzi A, Legeai F, Hugueney P, Dasilva C, Horner D, Mica E, Jublot D, Poulain J, Bruyère C, Billault A, Segurens B, Gouyvenoux M, Ugarte E, Cattonaro F, Anthouard V, Vico V, Del Fabbro C, Alaux M, Di Gaspero G, Dumas V, Felice N, Paillard S, Juman I, Moroldo M, Scalabrin S, Canaguier A, Le Clainche I, Malacrida G, Durand E, Pesole G, Laucou V, Chatelet P, Merdinoglu D, Delledonne M, Pezzotti M, Lecharny A, Scarpelli C, Artiguenave F, Pè E, Valle G, Morgante M, Caboche M, Adam-Blondon AF, Weissenbach J, Quétier F, Wincker P (2007) The grapevine genome sequence suggests ancestral hexaploidization in major angiosperm phyla. Nature 449: 463–467.

Jones GV, White MA, Cooper OR, Storchmann K (2005) Climate change and global wine quality. Climate change 73: 319–343.

Kanellis AK, Roubelakis-Angelakis KA (1993) Grape. In: G Seymour, J Taylor, G Tucker (eds) Biochemistry of Fruit Ripening. Chapman & Hall, London, UK, pp 189–234.

Kays SJ (1997) Stress in harvested products In: SJ Kays (ed) Post-harvest Physiology in Perishable Plant Products. Athens, Giorgia, pp 335–408.

Keilin T, Pang X, Venkateswari J, Halaly T, Crane O, Keren A, Ogrodovitch A, Ophir R, Volpin H, Galbraith D, Or E (2007) Digital expression profiling of a grape-bud EST collection leads to new insight into molecular events during grape-bud dormancy release. Plant Sci 173: 446–457.

Lin H, Doddapaneni H, Takahashi Y, Walker MA (2007) Comparative analysis of ESTs involved in grape responses to *Xylella fastidiosa* infection. BMC Plant Biol 7: 8.

Lisso J, Steinhauser D, Altmann T, Kopka K, Mussig C (2005) Identification of brassinosteroid-related genes by means of transcript co-response analyses. Nucl Acids Res 33: 2685–2696.

Lund S, Peng F, Nayar T, Reid K, Schlosser J (2008) Gene expression analyses in individual grape (*Vitis vinifera* L.) berries during ripening initiation reveal that pigmentation intensity is a valid indicator of developmental staging within the cluster. Plant Mol Biol 68: 301–315.

Marguerit E, Boury C, Manicki A, Donnart M, Butterlin G, Némorin A, Wiedemann-Merdinoglu S, Merdinoglu D, Ollat N, Decroocq S (2009) Genetic dissection of sex determinism, inflorescence morphology and downy mildew resistance in grapevine. Theor Appl Genet 118: 1261–1278.

Mathiason K, He D, Grimplet J, Venkateswari J, Galbraith D, Or E, Fennell A (2009) Transcript profiling in *Vitis riparia* during chilling requirement fulfillment reveals coordination of gene expression patterns with optimized bud break. Funct Integr Genom 9: 81–96.

Meyers BC, Tej SS, Vu TH, Haudenschild CD, Agrawal V, Edberg SB, Ghazal H, Decola S (2004) The use of MPSS for whole-genome transcriptional analysis in Arabidopsis. Genome Res 14: 1641–1653.

Mori K, Goto-Yamamoto N, Kitayama M, Hashizume K (2007) Loss of anthocyanins in red-wine grape under high temperature. J Exp Bot 58: 1935–1945.

Moser C, Segala C, Fontana P, Salakhudtinov I, Gatto P, Pindo M, Zyprian E, Toepfer R, Grando MS, Velasco R (2005) Comparative analysis of expressed sequence tags from different organs of *Vitis vinifera* L. Funct Integr Genom 5: 208–217.

Nettleton D (2006) A discussion of statistical methods for design and analysis of microarray experiments for plant scientists. Plant Cell 18: 2112–2121.

Persson S, Wei H, Milne J, Page GP, Somerville CR (2005) Identification of genes required for cellulose synthesis by regression analysis of public microarray data sets. Proc Natl Acad Sci USA 102: 8633–8638.

Pilati S, Perazzolli M, Malossini A, Cestaro A, Demattè L, Fontana P, Dal Ri A, Viola R, Velasco R, Moser C (2007) Genome-wide transcriptional analysis of grapevine berry ripening reveals a set of genes similarly modulated during three seasons and the occurrence of an oxidative burst at veraison. BMC Genom 8: 428.

Polesani M, Desario F, Ferrarini A, Zamboni A, Pezzotti M, Kortekamp A, Polverari A (2008) cDNA-AFLP analysis of plant and pathogen genes expressed in grapevine infected with *Plasmopara viticola*. BMC Genom 9: 142.

Rensink WA, Buell CR (2005) Microarray expression profiling resources for plant genomics. Trends Plant Sci 10: 603–609.

Robinson SP, Jacobs AK, Dry IB (1997) A class IV chitinase is highly expressed in grape berries during ripening. Plant Physiol 114: 771–778.

Romeyer FM, Macheix JJ, Goiffon JP, Reminiac CC, Sapis JC (1983) The browning capacity of grapes 3. Changes and importance of hydroxycinnamic acid-tartaric acid esters during development and maturation of the fruit J Agri Food Chem 31: 346–349.

Rotter A, Hren M, Baebler S, Blejec A, Gruden K (2008) Finding differentially expressed genes in two-channel DNA microarray datasets: how to increase reliability of data preprocessing. J Integr Biol 12: 171–182.

Rotter A, Camps C, Lohse M, Kappel C, Pilati S, Hren M, Stitt M, Coutos-Thevenot P, Moser C, Usadel B, Delrot S, Gruden K (2009) Gene expression profiling in susceptible interaction of grapevine with its fungal pathogen *Eutypa lata*: Extending MapMan ontology for grapevine. BMC Plant Biol 9: 104.

Samuelian SK, Camps C, Kappel C, Simova EP, Delrot S, Colova V (2009) Differential screening of overexpressed genes involved in flavonoid biosynthesis in North American native grapes: 'Noble' *muscadinia* var. and 'Cynthiana' *aestivalis* var. Plant Sci 177: 211–221.

Saito K, Hirai MY, Yonekura-Sakakibara K (2008) Decoding genes with co-expression networks and metabolomics-'majority report by precogs'. Trends Plant Sci 13: 36–43.

Steinhauser D, Usadel B, Luedemann A, Thimm O, Kopka J (2004) CSB:DB: A comprehensive systems-biology database. Bioinformatics 20: 3647–3651.

Stuart JM, Segal E, Koller D, Kim KK (2003) A gene co-expression network for global discovery of conserved genetic modules. Science 302: 249–255.

Tattersall E, Grimplet J, DeLuc L, Wheatley M, Vincent D, Osborne C, Ergül A, Lomen E, Blank R, Schlauch K, Cushman J, Cramer G (2007) Transcript abundance profiles reveal larger and more complex responses of grapevine to chilling compared to osmotic and salinity stress. Funct Integr Genom 7: 317–333.

Terrier N, Ageorges A, Abbal P, Romieu C (2001) Generation of ESTs from grape berry at various developmental stages. J Plant Physiol 158: 1575–1583.

Terrier N, Glissant D, Grimplet J, Barrieu F, Abbal P, Couture C, Ageorges A, Atanassova R, Léon C, Renaudin JP, Dédaldéchamp F, Romieu C, Delrot S, Hamdi S (2005) Isogene specific oligo arrays reveal multifaceted changes in gene expression during grape berry (*Vitis vinifera* L.) development. Planta 222: 832–847.

Terrier N, Torregrosa L, Ageorges A, Vialet S, Verriès C, Cheynier V, Romieu C (2009) Ectopic expression of VvMybPA2 promotes proanthocyanidin biosynthesis in grapevine and suggests additional targets in the pathway. Plant Physiol 149: 1028–1041.

This P, Lacombe T, Thomas MR (2006) Historical origins and genetic diversity of wine grapes. Trends Genet 22: 511–519.

Venter M, Burger AL and Botha FC (2001) Molecular analysis of fruit ripening : the identification of differentially expressed sequences in *Vitis* vinifera using cDNA-AFLP technology. *Vitis* 40: 191–196.

Waters DLE, Holton TA, Ablett EM, Lee LS, Henry RJ (2005) cDNA microarray analysis of developing grape (*Vitis vinifera* cv. Shiraz) berry skin. Funct Integr Genom 5: 40–58.

White PJ (2002) Recent advances in fruit development and ripening: an overview. J Exp Bot 53: 1995–2000.

Winkel-Shirley B (2002) Biosynthesis of flavonoids and effects of stress. Curr Opin Plant Biol 5: 218–223.

Zamboni A, Minoia L, Ferrarini A, Tornielli GB, Zago E, Delledonne M, Pezzotti M (2008) Molecular analysis of post-harvest withering in grape by AFLP transcriptional profiling. J Exp Bot 59: 4145– 4159.

12

Functional Genomics: Proteomics and Metabolomics

Steven T. Lund[1,*] and *Grant R. Cramer*[2]

ABSTRACT

Expression profiling is a powerful experimental approach used by molecular biologists to correlate changes in abundances of molecules to biological processes of interest. Positive or negative correlations can be used to develop hypotheses regarding gene functions and build cellular signaling and metabolic network models. Although massive parallel RNA profiling can now be performed in a straightforward manner, changes in transcript abundances do not routinely correlate well with corresponding protein levels, nor do they necessarily reflect accurately changes in metabolism in time and space. It is increasingly recognized that proteomic and metabolomic expression profiling should be conducted in tandem with RNA expression profiling to most accurately assess molecular changes underlying biological processes of interest. An increasing number of proteomic and targeted metabolite studies have been applied successfully in recent years to characterize aspects of grape berry ripening and responses to abiotic and biotic stresses; however, technical aspects inherent of associated analytical techniques significantly limit the detection of most of the proteome or metabolome in any tissue. As these technical hurdles are overcome, it is anticipated that proteomic and metabolomic profiling will advance our understanding of grapevine molecular biology and its interaction with the natural and human-made environment to improve vine management and berry quality along each season.

Keywords: abiotic and biotic stress, berry ripening, expression profiling, mass spectrometry, metabolomics, proteomics

[1]241- 2205 East Mall, Wine Research Centre, Faculty of Land and Food Systems, University of British Columbia, Vancouver, BC, Canada, V6T 1Z4.
[2]1664 North Virginia Street, Department of Biochemistry and Molecular Biology, University of Nevada, Reno, NV 89557, USA .
*Corresponding author: *steven.lund@ubc.ca*

12.1 Introduction

Beyond the RNA world (see Chapters 10 and 11) lay proteins, small peptides, and thousands of non-proteinaceous plant natural products. These molecules have diverse roles in plant biology including cellular structure, signaling, transport, and metabolism, as well as pest defense and attraction of pollinators and foragers. Expression patterns of proteins and fluxes of plant natural products in biological systems may not be reflected accurately in transcriptional accumulation patterns for corresponding genes. In order to gain a more global picture of regulatory and metabolic networks in grapevine, it is important to consider proteomic and metabolomic approaches to complement transcriptomic analyses.

High-throughput quantitation of proteins and plant natural products is complicated by the relatively large dynamic ranges of concentrations of these molecules present in cells, tissues, and organs coupled with the current lack of capture technologies such as monoclonal antibody arrays for sensitive, comprehensive detection. In order to simplify analyses of complex biological samples, separation techniques using two-dimensional gel electrophoresis (proteins) and chromatography (peptides and plant natural products) are employed. Quantitation and identification are then carried out on the fractionated samples using techniques such as time-of-flight mass spectrometry (TOF/TOF MS), triple quadrupole (MS/MS) and ion trap mass spectrometry (ITMS), gas chromatography mass spectrometry (GC-MS), nuclear magnetic resonance (NMR) spectrometry, or Fourier transform ion cyclotron resonance (FT-ICR) mass spectrometry. We next examine different permutations of uses of these experimental approaches within the context of proteomics and metabolomics research, including examples from the literature.

12.2 Proteomics

High-throughput analyses of the protein complement of the genome or, more simply, proteomics, is a relatively new and rapidly advancing field. Still, the most widely used separative technique in proteomics today was first published nearly 20 years before the term "proteomics" was coined by Marc Wilkins in 1994. O'Farrell (1975) demonstrated a two-dimensional gel-based electrophoretic method for separating proteins in complex biological samples by their isoelectric points and then their sizes. Protein samples from two or more biological samples are each subjected to two-dimensional gel electrophoresis (2DGE). Separated proteins are visualized following staining of the gels. In modern laboratories, computer software is used to scan gel images for differences in staining patterns in order to select putative differentially expressed proteins between samples. Candidate spots are excised from the gels and subjected to enzymatic digestion, typically using trypsin, and then the cleaved peptides are purified prior to matrix-assisted laser desorption/ionization (MALDI). The peptide preparations are added to a crystalline matrix such as sinapic acid and individual preparations are spotted onto distinct locations on a MALDI plate. A laser is used to ionize each cleaved protein sample by firing at specific addresses on the plate. The ions fly into the mass analyzer and the velocity that a given peptide migrates through it is dependent upon its mass to charge ratio (m/z). In this manner, m/z can be calculated for each tryptic peptide per protein sample based upon the time it takes to fly (i.e., time-of-flight, TOF) through the mass spectrometer and reach the detector. For each given protein, a pattern or "fingerprint" of peptide masses is calculated and databased for each protein (i.e., 2DGE spot) analyzed. Alternately, peptide preparations in solution can be fractioned by nanoscale liquid chromatography, commonly ion exchange followed by reverse-phase. Then, in an automated manner, fractions are subjected sequentially to electrospray ionization (ESI) and applied to a mass spectrometer. ESI does not require spotting to a plate and digested protein samples are analyzed en masse rather than physically one at a time. Protein abundances can be quantified as either absolute values within biological samples or relative values among biological samples (Ong and Mann 2005). Relative quantitation in 2D gels is accomplished through densitometry, the calculation of sizes of replicated spots between gels that exhibit the same migration pattern, which raises both precision and accuracy issues, especially considering cases of multiple protein occurrence in the same spot. Spots can appear pure by ordinary analysis with MALDI-TOF/TOF, but multiple proteins have been identified in approximately 75% of the spots sampled with more sensitive LC-ESI-MS/MS techniques (Yang et al. 2007; GR Cramer, unpublished data in prep.). It is recommended, then, that only spots verified to be pure should be included in expression analyses and reported in publications.

Two major advances in the 1990s improved the identification of proteins in complex samples. These were the introduction of mass spectrometry to replace the laborious Edman degradation technique and the concomitant increase in gene and predicted protein sequence databases. The peptide mass fingerprinting (PMF) technique was first introduced for use with a single mass spectrometer. PMF entails the use of software to find best matches between m/z patterns of cleaved peptides identified by mass spectrometry for a given sample and a database of hypothetical m/z values for tryptic peptides predicted from gene sequence databases. Unless a fully sequenced and annotated genome is available providing full length amino acid sequences for the entire proteome, protein annotations via PMF can be hampered by truncated peptides predicted, for example, from an expressed sequence tag (EST) database. A further improvement to protein identification was the implementation of tandem MS (triple quad or TOF/TOF) which provides *de novo* sequencing of peptides in the second MS. Peptides selected in the first MS are sent to a "collision cell" whereupon they are fragmented via collision with a gas. The amino acid sequence of the peptide is resolved by software from resulting "daughter ions" which have masses that are consistent with known molecular weights of dipeptides, tripeptides, and so on. The resulting amino acid sequence is then used to search predicted protein databases using software such as Mascot (Perkins et al. 1999) to find similar sequences with high statistical confidence. As with PMF, similarity searches are hampered when sequence databases do not contain full length protein predictions, plus sequence diversity between the query sequences and databased sequences can also reduce the ability to determine statistically significant annotations.

2DGE has been used to characterize differential protein expression underlying diverse aspects of grapevine biology. The berry is critically important for consumption as fruit, as well as several value-added products including juices, wines, and spirits. Previous proteomics studies have focused on grapevine berries with the common goal of better understanding protein-level control of ripening and compositional chemistry. A 2DGE approach was used to study protein accumulation along ripening in the pericarp (Giribaldi et al. 2007), whereas the mesocarp (Sarry et al. 2004), exocarp (Deytieux et al. 2007; Negri et al. 2008a) and the mesocarp, exocarp and seed (Grimplet et al. 2009) of the grape berry were analyzed separately in order to characterize tissue compartmentalization of protein expression. Sarry et al. (2004) identified proteins important for sucrose and malate metabolism, which had previously been identified mainly at the transcript level only. Prevalent proteins detected in the mesocarp included a hexose transporter, apoplastic and vacuolar invertases, and enzymes involved in malate oxidation (gluconeogenesis). Evidence was provided in support of the apoplastic invertase, VvIVR1, being a key enzyme regulating sucrose unloading from the phloem and driving, in part, the uptake and accumulation of glucose and fructose at ripening initiation (véraison). In the exocarp, Deytieux et al. (2007) and Negri et al. (2008a) each determined that the accumulation of anthocyanin anabolic enzymes was temporally and spatially correlated with the accumulation pattern for anthocyanin pigments along véraison. As in the mesocarp (Sarry et al. 2004), Deytieux et al. (2007) and Negri et al. (2008a) each detected enzymes involved in primary carbon and nitrogen metabolism in the exocarp. In addition to corroborating findings of the other authors cited here, Giribaldi et al. (2007, 2010), in their extensive 2DGE study along seven berry developmental stages, identified components of abscisic acid (ABA) and ethylene biosynthesis, which have been previously demonstrated to play major and minor roles, respectively, in regulating véraison. Grimplet et al. (2009) assayed 1,047 proteins in the pericarp with 90 identified proteins differing between the exocarp and the mesocarp. There were 169 proteins of the total 695 proteins surveyed in the seed that were distinct from the pericarp. Proteins with differential abundances between the tissues included proteins involved in phenylpropanoid metabolism, defense, ROS scavenging, abiotic stress, and seed storage. Although several hundred proteins were newly identified in grapevine in the studies cited here, a collective trend in these data was the detection of relatively abundant proteins with similarity to previously described housekeeping-type proteins in other plant species, rather than discovery of low abundance proteins. Further, from the published proteome data for grapevine berries, it is difficult to identify mesocarp- versus exocarp-specific protein accumulation along ripening beyond the anthocyanin biosynthetic pathway. Nonetheless, these studies constitute pioneering work in grape berry proteomics that have provided important insights to protein networks underlying sucrose and malate metabolism at véraison, hormone biosynthesis, signal transduction, phenylpropanoid metabolism, ripening, abiotic and biotic stress, as well as proteins important for cell wall architecture and extensibility, such as xyloglucan endotransglycosylase, during rapid cell expansion in the pericarp.

Subcellular compartmentalization in the grapevine pericarp has been analyzed via the isolation of cell wall (Negri et al. 2008b) or membrane-localized (Zhang et al. 2008) total proteins. Total protein preparations are made difficult in grapevine berries by the relatively high sugar, phenolic, and organic acid content, as compared with other plant tissues. Negri et al. (2008b) tested three methods for the specific isolation of cell wall proteins in the pericarp and concluded that a phenol-based method yielded the highest number of proteins as well as the largest number of spots in 2DGE. These data are consistent with a previous study (Vincent et al. 2006), which determined that phenol-based extraction of berries gave larger protein yields with higher spot resolution on 2D gels. Comparison of 2D gels of cell wall-enriched proteins against cytoplasm proteins revealed that approximately one third of the over 800 spots were unique to one fraction or the other. Negri et al. (2008b) identified several proteins in the cell wall-enriched subcellular fraction that are likely involved in pathogen defense, including acidic glucanases, chitinases, and nectarin IV-like proteins. Like the cell wall's specialized roles in cell enlargement and pathogen defense, the membrane fraction is rich in hormone receptors, other signaling components such as arabinogalactan proteins, and transporters that can be important for regulating growth and compositional chemistry along ripening. Zhang et al. (2008) acknowledged the technical limitations of membrane protein solubilization in conjunction with the 2DGE approach, but the authors described extraction procedures designed to help protect membrane proteins from oxidation or polymerization with phenolic flavonoid compounds.

Sixty-two membrane proteins, which Zhang et al. (2008) classified by the number of predicted transmembrane domains, were identified with high confidence from three stages of berry development (II, véraison, and III). The annotated membrane proteins from the green berry stage (II) were mainly abundant housekeeping proteins involved in energy status, such as ATP synthase and a ribose importing ABC transporter, signaling, including three Ras-like GTPases, and cellular structure and biogenesis. Few proteins were identified by Zhang et al. (2008) as being differentially expressed along development, but accumulation of xyloglucan endotransglycosylase was clearly correlated with véraison and the onset of ripening, consistent with the finding of another group (Deytieux et al. 2007).

Healthy grapevines are required for the production of high quality fruit. 2DGE has been implemented to characterize proteome changes in response to environmental stresses, including water stress and high salinity (Vincent et al. 2007; Jellouli et al. 2008), herbicide treatment (Castro et al. 2005), and constitutive pathogen defense (Agüero et al. 2008; Basha et al. 2010). Vincent et al. (2007) used contrasting physiological responses to water stress and salinity in Chardonnay in comparison to Cabernet Sauvignon. A new genetic model for comparative stress proteomics in grapevine was carefully validated using physiological parameters. The authors determined that the shoot growth of water-deficit and salt-stressed Chardonnay vines was less inhibited by negative water potentials in comparison to water-deficit and salt-stressed Cabernet Sauvignon vines, indicating that Chardonnay is more stress-tolerant. Significant differences in protein accumulation were detected via 2DGE, which were determined via analysis of variance to be most strongly associated with the genotype rather than the water deficit or salt stress treatments within genotypes. For example, the more stress-sensitive cultivar, Cabernet Sauvignon, expressed higher levels of proteolytic enzymes than the more stress-tolerant Chardonnay. Few of the identified proteins were determined to be responsive to the stress treatments; however, this finding is likely a result of the detection sensitivity limitations of the 2DGE approach. Nevertheless, the inhibition of growth was highly correlated with the abundance of proteins involved in photosynthesis and protein synthesis. Jellouli et al. (2008) investigated salt stress proteomics in a Tunisian *V. vinifera* cultivar, Razegui, since approximately one third of the Tunisian grape industry is impacted by high soil salinity. Using a phenol-based protein extraction procedure followed by 2DGE, these authors also determined few differentially expressed proteins in response to a similar stepwise salt stress treatment to that used by Vincent et al. (2007). One salt stress-responsive protein isolated from leaves and stems was determined via Edman degradation analysis to be pathogenesis-related (PR) protein 10, which has several reported functions including ribonuclease activity. More sensitive techniques to 2DGE will need to be employed in order to better understand signaling and metabolic protein networks that specifically mediate water and salt stress responses in different grapevine organs.

Agüero et al. (2008) used 2DGE to identify secreted proteins in the xylem of Chardonnay plants. Seven proteins with putative anti-microbial or other plant defense roles were identified. These proteins included a serine protease, cell wall modifying enzymes such as a xylosidase, and NtPRp27, which was also identified by Vincent et al. (2007) in water-stressed shoot tips of Cabernet Sauvignon and Chardonnay. Empirical evidence that the signal peptides on the N-termini of the proteins identified by Agüero et al. (2008) directed the proteins to the apoplast and, ultimately, the xylem, provides a basis for engineering of xylem-secreted proteins which could be used to improve grapevine defense against important xylem-localized pathogens such as *Xylella fastidiosa*. Basha et al. (2010) directly studied proteome responses to *X. fastidiosa* infection using 2DGE. A comparative approach was undertaken by analyzing differential proteome responses in the xylem sap of susceptible *V. vinifera* cultivars with those in resistant cultivars of *V. rotundifolia*. The authors determined that infected *V. rotundifolia* accumulated distinct defense-related (e.g., PR-2, peroxidase) and other proteins of unknown function(s) in xylem sap in comparison with the xylem sap proteome in infected *V. vinifera*. These differential protein expression data, as well as metabolite profiling data showing elevated carbohydrate and amino acid levels in xylem sap of *V. vinifera* in comparison with *V. rotundifolia*, may help explain why *X. fastidiosa* is able to colonize *V. vinifera* more effectively.

Protein expression responses to a potential abiotic stress, the herbicide "flumioxazin", were examined by Castro et al. (2005). Flumioxazin is a relatively new pre-emergence herbicide used in some vineyards for controlling weeds and unwanted grasses but the effects on non-target dicots such as *Vitis* spp. are not well understood (Saladin et al. 2003). Flumioxazin is thought to act on unwanted species by blocking chlorophyll and heme biosynthesis. Castro et al. (2005) used micropropagated Chardonnay plantlets and determined via 2DGE that a limited number of proteins were induced in response to flumioxazin application in the growth media. Induced proteins were identified via TOF/TOF MS and included a phosphoglycerate mutase and a UTP-glucose pyrophosphorylase likely involved in carbon metabolism, an orcinol *O*-methyltransferase, and eight proteins representing two isoforms of PR-10. Castro et al. (2005) claimed that repeated identification of the large subunit of ribulose-1,5-bisphosphate carboxylase/oxygenase (RuBisCO) provided evidence of chlorophyll degradation resulting from the flumioxazin treatment, but this is unclear given that complexed proteins are reduced in preparation for 2DGE. Further work with vineyard-grown vines will be needed to clarify and potentially "fingerprint" the response to commercial applications of flumioxazin and other commonly used herbicides. This information could prove useful in determining potential unwanted side effects of commercial herbicide applications in the vineyard or monitoring herbicide drift and contamination in organic operations.

The most accurate and replicable means of determining absolute and relative quantities of proteins in a given sample(s) is through isotopic labeling or "tagging" of individual proteins prior to separation followed by tandem mass spectrometry. The technical advantage of tagging is that peptides can be directly quantified via MS proportional to the amount of tag detected in a given biological sample for a given peptide sequence. Several methods for tagging of peptides have been published and can be classified via the positioning and attachment chemistry of the tag (Yan and Chen 2005). A groundbreaking advance in peptide tagging reported by Gygi et al. (1999) is called "isotope-coded affinity tags (ICAT)". A more recent version of the ICAT technique uses biotinylated reagents incorporating ^{12}C or ^{13}C that are covalently attached to sulfhydryl-containing amino acids (Yi et al. 2005). Two biological samples are labeled with ^{12}C ("light") or ^{13}C ("heavy") ICAT reagents, the samples are mixed together, digested with trypsin, tagged peptides are isolated using affinity chromatography directed against the biotin moiety, and then tagged peptides are subjected to nanoLC-MS/MS. The small mass difference between the light and heavy tags allows quantitation of each peptide in each sample prior to ionization and then *de novo* sequencing is carried out via the second MS. Although ICAT applications to plant proteomics have been reported for other plant species, no studies have been published for grapevine to date using this technique.

It is important to understand ICAT and its limitations in light of a more recent and technically superior peptide tagging technique, iTRAQ (isobaric tagging for relative and absolute quantitation). ICAT is limited by employing only two tags (light and heavy) and requiring an affinity purification step; thus, only two samples can be analyzed in each experiment and, invariably, information is lost because those peptides lacking a cysteine or threonine are not recovered. iTRAQ uses four or eight isobaric tags that are attached to free amine groups. No affinity purification step is required and amine linkage means that virtually every peptide is labeled and can potentially be analyzed via LC-MS/MS. The tags are isobaric since the attachment chemistries for each tag are balanced such that there is no net mass difference amongst the four or eight tags (Pierce et al. 2008). Following co-elution and selection of peptides in the first MS, the different tag masses are dissociated from the peptides in the collision cell and then the quantity of each is calculated from the second MS. Unlike in ICAT, both quantitation and *de novo* sequencing are done in the second MS step using the iTRAQ technique. Few reports exist for iTRAQ applications in plant proteomics due to the recent development of this technique. In grapevine, Lücker et al. (2009) used iTRAQ to investigate relative changes in protein expression along véraison in grapevine berries. Proteins identified as increasing along véraison included enzymes required for brassinosteroid and abscisic acid hormone biosynthesis as well as anthocyanin flavonoid and general terpenoid metabolism. Lücker et al. (2009) also reported on software the authors created to predict open reading frames from grapevine EST data. The software was designed to specifically define tryptic peptides for each protein while avoiding truncated peptides that can arise due to the non-full-length nature of EST data. Using this database construction approach, the authors demonstrated an improvement in protein annotation from MS/MS data in comparison to searches employing the commonly used MSDB protein database (Perkins, Imperial College School of Medicine, London UK). Marsh et al. (2010) used iTRAQ to investigate proteome changes in Cabernet Sauvignon leaves in response to challenge with the casual agent of powdery mildew disease, *Erysiphe necator*. Elevated levels of sterol biosynthesis, oxygen radical-scavenging, and pathogenesis related (particularly, PR-10) proteins were detected in response to pathogen infection, whereas photosynthetic protein abundances were varied over time.

12.3 Metabolomics

Plant metabolites can be divided into primary and secondary metabolites. Primary carbon metabolism involves basic biochemical processes necessary for life, which include photosynthesis, protein synthesis, and respiration. The primary metabolites include amino acids, nucleotides, sugars, organic acids, hormones, pigments, signaling molecules such as inositol triphosphate, and acyl lipids. Secondary carbon metabolism involves natural products that do not generally have direct roles in the basic processes in which primary metabolites participate. Secondary metabolites or "plant natural products" are not absolutely necessary for daily survival but do contribute to long-term health and reproductive fitness.

Different types or variants of plant natural products can be unique to a plant species; however, they can be divided into three general classes: Terpenoids, phenolics, and alkaloids. It has been estimated that there are more than 200,000 different metabolites in the plant kingdom (Weckwerth 2003; Dettmer et al. 2007). It is not clear how many different metabolites there might be in a single plant, but it is a minimum of several thousand. There are currently 28,955 metabolites listed in KNApSAcK (*http://kanaya.naist.jp/KNApSAcK//*) and 20,306 metabolites in the Madison Metabolomics Consortium Database (*http://mmcd.nmrfam.wisc.edu/*), both of which contain metabolites from a wide array of organisms. In plant-specific metabolite databases, there are presently 2,314 metabolites in the Plant Metabolic Network database (*http://www.plantcyc.org/*) and 1,956 metabolites in AraCyc (*http://www.arabidopsis.org/biocyc/index.jsp*). The total number of grape metabolites has been estimated from the current build of metabolic maps for grapevine to be about 1,600 (J Grimplet, pers. comm.), but the actual number is expected to be much larger.

The goal of metabolomics is to identify and quantify the entire metabolome, which is defined as all chemical compounds present in a biological sample; however, the detection and identification of *all* metabolites has similar problems to protein detection and identification. No extraction protocol can isolate all metabolites in a single sample (Dettmer et al. 2007), nor is there a detection method sensitive enough to cover the entire quantitative scale of metabolites in a particular sample, which has been estimated to be seven orders of magnitude (Fiehn et al. 2008). Thus, multiple extraction protocols and detection methods are required for different classes of metabolites. Metabolite profiling refers to a quantitative metabolite analysis of a subset of the metabolome; in this sense, all metabolomics methods to date have been limited to metabolite profiling. Metabolite fingerprinting uses a global approach to classify samples (chromatograms) based upon metabolite patterns (fingerprints). This approach is used to distinguish samples or treatment effects.

Primary metabolites in plants are commonly extracted with a ternary solvent mixture of both lipophilic and hydrophilic organic solvents with water and measured by GC-MS (Broeckling et al. 2005; Lisec et al. 2006; Fiehn et al. 2007). Cramer et al. (2007), Deluc et al. (2007), and Fenoll et al. (2009) have measured between 120 and 180 metabolites in a single grapevine sample (depending on the tissue) using similar methods. Some have reported single chromatograms from plant samples having 300 primary metabolites or more (Fiehn et al. 2008). The most important primary metabolites for grape and wine quality are sugars (glucose, fructose) and organic acids (malate, tartrate). Secondary metabolites are more difficult to measure. Terpenoids are quite volatile and have been measured by two-dimensional GC-MS with 56 different terpenoids detected in a single sample (Rocha et al. 2007). Terpenoids can be stored as glycoside derivatives in grapevine berry vacuoles, so acid or enzyme cleavage techniques are used to volatilize the compounds prior to analyses via GC-MS. Phenolics are represented by approximately 8,000 different compounds (Robbins 2003). They are usually extracted with hot water, methanol, ethanol, acetone, or ethyl acetate. Ordinary low molecular weight, simple phenolics are adequately measured with reverse-phase high performance liquid chromatography (HPLC or LC) coupled with diode array detection (DAD) or MS (Robbins 2003; Cohen et al. 2008; Jeffery et al. 2008; Matsuda et al. 2009); however, this represents only a small proportion of the total phenolics; high molecular weight phenolic compounds require more specialized methods such as ESI-TOF MS and MALDI-TOF MS (Fulcrand et al. 2008). There are approximately 12,000 different alkaloid compounds in plants (Ziegler and Facchini 2008). Alkaloids are low molecular weight compounds with a nitrogen atom in the heterocyclic ring. Although orthologous *Vitis* genes are identified in the KEGG pathways for alkaloid biosynthesis I and II and for caffeine metabolism, no evidence for naturally occurring alkaloids in *Vitis* spp. could be found in the literature, possibly due to the specialized and/or sometimes toxic nature of the compounds, requiring specialized storage mechanisms *in planta*. Others researchers have used nuclear magnetic resonance spectroscopy (NMR) to study primary and secondary metabolite profiles in grapevine (Pereira et al. 2006; Figueiredo et al. 2008; Ali et al. 2009; Son et al. 2009). While NMR is a very powerful approach that is useful for metabolite fingerprinting, its sensitivity is lower when compared to GC- and LC-MS, often resulting in a smaller set of metabolites detected in a single sample (Weckwerth 2003; Dettmer et al. 2007).

The advantages of using more sophisticated techniques such as HPLC and GC coupled with MS or NMR is that the identification and quantification of compounds is more accurate. A single peak off of a GC or HPLC chromatogram may contain several compounds in it or may be obscured by the sudden increase of another peak due to changes in some abundant compound. A good example is the accumulation of glucose and fructose during berry development that can obscure lower abundance compounds like some organic and amino acids. In early berry developmental stages, the sugar abundance is quite low and it is relatively easy to detect citrate and lysine; however, their peaks become obscured by the glucose and fructose peaks after véraison because they have similar retention times. With MS and deconvolution software (e.g., AMDIS; *www.amdis.net/*), these compounds can be accurately separated and quantified.

Automated identification has many potential errors. Some of the problems occur due to shifts in the retention time in the chromatograms or the very similar or identical mass fingerprints of closely related compounds like sugars. The isomers, glucose, mannose, and galactose, or the sucrose-containing raffinose and stachyose sugars are good examples. Generally, they are distinguished by differences in their retention times. This requires the analyses of pure standards with individual laboratories' particular instruments and conditions, making it necessary to develop an "in-house" metabolite library. If there is a shift in the retention times on the chromatogram, the automated process may mis-identify the compound in question. For this and other reasons, the use of fatty acid methyl esters (FAMEs) as retention time standards have been advocated in order to calculate the more accurate retention index for a particular compound (Lisec et al. 2006). Alternately, a number of peak alignment programs (e.g., SpectConnect; spectconnect.mit.edu) or MetAlign (*www.metalign.wur.nl/UK/*) have been developed to match peaks and reduce noise and error.

At the very beginning of a study, it may be very useful to measure the metabolome with an "unbiased" approach, i.e., try to measure everything and determine what is changing. New discoveries are sure to come from this approach, but at what expense? It can be very costly and time consuming to make all of these types of measurements at different concentration ranges to attempt to measure the entire metabolome. If one has additional information and knowledge, it may be useful to take a more targeted metabolite profiling approach. For example, one could undertake a global analysis of the transcriptome with whole genome microarrays, analyze the data, determine the major metabolic pathways that have been affected, and then follow that up with targeted metabolite profiling for those metabolic pathways. This is the approach taken by Figueiredo et al. (2008) and Deluc et al. (2009) for their analyses of innate resistance to pathogen infection in leaves and the effects of water deficit on berry metabolism, respectively. Rather than try to measure everything in the leaf or berry, the authors focused on metabolic pathways known to be important for berry quality components from previous studies and used the transcriptomic data to validate and identify new pathways that were affected, as well. This approach used hypothesis-driven and discovery-based methods to identify important effects of water deficit on berry flavors and quality characteristics.

12.4 Outlook

Plant proteomics and metabolomics have advanced significantly in the most recent decade, owing mainly to improved extraction techniques and augmented detection sensitivity of the analytical equipment employed. Accurate quantitation of proteins and plant natural products is now possible both on an absolute and a relative scale, which was essentially unattainable for high-throughput protein analyses in complex biological samples until recently. Further, identification of both proteins and non-proteinaceous compounds has improved due to the availability of sequenced genomes and growing MS and NMR spectra libraries for plant natural products. For proteomics, the recent availability of a 12-fold coverage reference genome sequence from Pinot Noir (The French-Italian Public Consortium for Grapevine Genome Characterization 2007) will accelerate protein annotation. Lücker et al. (2009) compared high confidence annotation of pericarp proteins using databases either constructed to favor Cabernet Sauvignon-derived sequence data or a non-cultivar-biased database from all available *Vitis* spp. ESTs. Based on the finding of these authors, a predicted peptide database based upon the Pinot Noir genome sequence data should be valid for use with all *V. vinifera* cultivars since Lücker et al. (2009) did not find a significant advantage in using a cultivar-biased database.

From the publications reviewed in this chapter, it can be concluded that we have only just begun to identify global protein and metabolite profiles from grapevine tissues and build regulatory and metabolic networks associated with different aspects of development and/or defense. We are probably able to identify, at most, 10 to 20% of all proteins and metabolites in any grapevine tissue using leading edge techniques. A shotgun (LC-MS/MS) analytical approach was reported to improve detection sensitivity by Lücker et al. (2009) who worked with pericarp; however, significant limitations to detecting the majority of the proteome remain to be solved. One disadvantage a shotgun approach has in comparison to 2DGE is that there is no pre-selection or screening of proteins before samples are applied to the mass spectrometer. This invariably leads to over-sampling of abundant proteins such as, for example, H+–ATPase and thaumatin in grapevine pericarp samples (Lücker et al. 2009). Zolotarjova et al. (2007) developed an affinity column for human proteomics applications which effectively removed 14 abundant proteins from plasma samples, which was confirmed via ELISA analyses. Such an approach reduced the dynamic range of detection for the mass spectrometer and allowed the identification of low abundance proteins known to be present at the ng/mL level in human plasma. A similar approach has been taken with plants to remove RuBisCO using immunodepletion columns (Cellar et al. 2008). It would be beneficial to the grapevine community if additional antigen-specific antibodies could be developed in a similar manner against abundant proteins in tissues of interest such as the pericarp, seed, and xylem, in order to reveal interesting, low-abundance protein expression patterns associated with biological processes of interest. Further, the availability of more sensitive analytical equipment such as the reported 10-fold greater sensitivity attributed to FT-ICR MS over the triple quad technology will undoubtedly enable greater detection of low abundance proteins in complex grapevine samples.

As detection sensitivity and data replication issues are resolved and more of the grapevine proteome and metabolome are revealed in a statistically robust manner, these data are expected to become increasingly useful for improving grapevine genotypes and viticultural practices. Molecular markers have become increasingly useful in recent years for genetic mapping and breeding of superior traits in table grapes and, to a lesser degree, wine grapes, as molecular maps for grapevine became more saturated (Salmaso et al. 2008 and references therein). An emerging molecular mapping technique is the use of expression profiles as so-called "e-QTLs" (expression quantitative trait loci). e-QTLs are being used in plants for mapping complex traits controlled by several genes with some having only minor influences on phenotypes of interest (Kirst et al. 2005;

Rowe et al. 2008; see also, Chapter 7). e-QTL studies have thus far been limited to transcriptional profiling data with the goal of identifying loci that affect variations in gene expression that are statistically associated with quantitative expression of a trait. Using this approach, a mapping population of genotypes is screened for gene expression profiles in which each transcript is considered as a heritable phenotype. These data can be used to help construct regulatory and/or metabolic networks that contribute quantitatively to expression of a trait. Theoretically, similar trait mapping approaches could be taken more directly using proteome profiles in grapevine, as has been reported for other organisms (Stylianou et al. 2008). Note that a trait of interest can be related to metabolomics, such as compositional chemistry in grapevine berries. A complementary mapping strategy to eQTLs is to exploit the diverse genetic variation in *V. vinifera* via an association mapping approach. Significant associations between phenotypes of interest and random or sequence-targeted variation at the DNA level are defined in a genetically divergent population of related individuals (Aranzana et al. 2005). This information contributes to an increased understanding of regulatory and metabolic components underlying the expression of traits of interest. Similar to eQTLs, association mapping can be coupled to metabolomic data when these are considered as a quantitative trait.

With the realistic prospect of genetically modified (GM) grapevines becoming a commercially viable strategy for improving wine and table grape genotypes, substantial equivalence determination will become imperative for regulatory purposes, i.e., demonstrating whether non-targeted changes in molecular expression are caused by insertion of a foreign gene construct into a genome. Proteome and metabolome fingerprinting in grapevine will become increasingly important in this regard. For example, protein and metabolite fingerprinting have been employed to study the effects of manipulating alcohol dehydrogenase (VvADH2) expression in grapevine (Tesniere et al. 2006; Sauvage et al. 2007). Proteome and metabolome fingerprints will also be further developed as biomarkers for genotype identification, stress diagnoses, and predictive modeling of berry quality parameters along the season. Such an approach has been used to begin to distinguish cultivars based upon their anthocyanin (Mattivi et al. 2006) and resveratrol abundance (Gatto et al. 2008). In addition, certain compounds or profiles may be used to identify the health status of the vine (Figueiredo et al. 2008) or they may be used to determine markers for berry tissues (Grimplet et al. 2009), berry development and optimum ripeness (Lund and Bohlmann 2006; Deluc et al. 2007; Pilati et al. 2007), and berry withering (Zamboni et al. 2008).

References

Agüero CB, Thorne ET, Ibáñez AM, Gubler WD, Dandekar AM (2008) Xylem sap proteins from *Vitis vinifera* L. Chardonnay. Am J Enol Vitic 59: 306–311.
Ali K, Maltese F, Zyprian E, Rex M, Choi YH, Verpoorte R (2009) NMR metabolic fingerprinting based identification of grapevine metabolites associated with Downy Mildew resistance. J Agri Food Chem 57: 9599–9606.
Aranzana MJ, Kim S, Zhao K, Bakker E, Horton M, Jakob K, Lister C, Molitor J, Shindo C, Tang C, Toomajian C, Traw B, Zheng H, Bergelson J, Dean C, Marjoram P, Nordborg M (2005) Genome-wide association mapping in Arabidopsis identifies previously known flowering time and pathogen resistance genes. PLoS Genetics 1: 531–539.
Basha SM, Mazhar H, Vasanthaiah HKN (2010) Proteomics approach to identify unique xylem sap proteins in Pierce's Disease-tolerant *Vitis* species. Appl Biochem Biotechnol 160: 932–944.
Broeckling CD, Huhman DV, Farag MA, Smith JT, May GD, Mendes P, Dixon RA, Sumner LW (2005) Metabolic profiling of *Medicago truncatula* cell cultures reveals the effects of biotic and abiotic elicitors on metabolism. J Exp Bot 56: 323–336.
Castro AJ, Carapito C, Zorn N, Magné C, Leize E, Van Dorsselaer A, Clément C (2005) Proteomic analysis of grapevine (*Vitis vinifera* L.) tissues subjected to herbicide stress. J Exp Bot 56: 2783–2795.
Cellar NA, Kuppannan K, Langhorst ML, Ni W, Xu P, Young SA (2008) Cross species applicability of abundant protein depletion columns for ribulose-1,5-bisphosphate carboxylase/oxygenase. J Chromat B Anal Tech Biomed Life Sci 861: 29–39.
Cohen SD, Tarara JM, Kennedy JA (2008) Assessing the impact of temperature on grape phenolic metabolism. Anal Chim Acta 621: 57–67.
Cramer GR, Ergul A, Grimplet J, Tillett RL, Tattersall EA, Bohlman MC, Vincent D, Sonderegger J, Evans J, Osborne C, Quilici D, Schlauch KA, Schooley DA, Cushman JC (2007) Water and salinity stress in grapevines: early and late changes in transcript and metabolite profiles. Funct Integr Genom 7: 111–134.
Deluc LG, Grimplet J, Wheatley MD, Tillett RL, Quilici DR, Osborne C, Schooley DA, Schlauch KA, Cushman JC, Cramer GR (2007) Transcriptomic and metabolite analyses of Cabernet Sauvignon grape berry development. BMC Genom 8: 429.
Deluc LG, Quilici DR, Decendit A, Grimplet J, Wheatley MD, Schlauch KA, Mérillon JM, Cushman JC, Cramer GR (2009) Water deficit alters differentially metabolic pathways affecting important flavor and quality traits in grape berries of Cabernet Sauvignon and Chardonnay. BMC Genom 10: 212.
Dettmer K, Aronov PA, Hammock BD (2007) Mass spectrometry-based metabolomics. Mass Spect Rev 26: 51–78.
Deytieux C, Geny L, Lapaillerie D, Claverol S, Bonneu M, Doneche B (2007) Proteome analysis of grape skins during ripening. J Exp Bot 58: 1851–1862.
Fenoll J, Manso A, Hellín P, Ruiz L, Flores P (2009) Changes in the aromatic composition of the *Vitis vinifera* grape Muscat Hamburg during ripening. Food Chem 114: 420–428.
Fiehn O, Robertson D, Griffin J, van der Werf M, Nikolau B, Morrison N, Sumner L, Goodacre R, Hardy N, Taylor C, Fostel J, Kristal B, Kaddurah-Daouk R, Mendes P, van Ommen B, Lindon J, Sansone S-A (2007) The metabolomics standards initiative (MSI). Metabolomics 3: 175–178.
Fiehn O, Wohlgemuth G, Scholz M, Kind T, Lee do Y, Lu Y, Moon S, Nikolau B (2008) Quality control for plant metabolomics: Reporting MSI-compliant studies. Plant J 53: 691–704.
Figueiredo A, Fortes AM, Ferreira S, Sebastiana M, Choi YH, Sousa L, Acioli-Santos B, Pessoa F, Verpoorte R, Pais MS (2008). Transcriptional and metabolic profiling of grape (*Vitis vinifera* L.) leaves unravel possible innate resistance against pathogenic fungi. J Exp Bot 59: 3371–3381.

Fulcrand H, Mane C, Preys S, Mazerolles G, Bouchut C, Mazauric JP, Souquet JM, Meudec E, Li Y, Cole RB, Cheynier V (2008) Direct mass spectrometry approaches to characterize polyphenol composition of complex samples. Phytochemistry 69: 3131–3138.

Gatto P, Vrhovsek U, Muth J, Segala C, Romualdi C, Fontana P, Pruefer D, Stefanini M, Moser C, Mattivi F, Velasco R (2008) Ripening and genotype control stilbene accumulation in healthy grapes. J Agri Food Chem 56: 11773–11785.

Giribaldi M, Perugini I, Sauvage F-X, Schubert A (2007) Analysis of protein changes during grape berry ripening by 2-DE and MALDI-TOF. Proteomics 7: 3154–3170.

Giribaldi M, Gény L, Delrot S, Schubert A (2010) Proteomic analysis of the effects of ABA treatments on ripening *Vitis vinifera* berries. J Exp Bot 61: 2447–2458.

Grimplet J, Wheatley MD, Ben Jouira H, Deluc LG, Cramer GR, Cushman JC (2009) Proteomic and selected metabolite analysis of grape berry tissues under well watered and water-deficit stress conditions. Proteomics 9: 2503–2528.

Gygi SP, Rist B, Gerber SA, Turecek F, Gelb MH, Aebersold R (1999) Quantitative analysis of complex protein mixtures using isotope-coded affinity tags. Nat Biotechnol 17: 994–998.

Jeffery DW, Mercurio MD, Herderich MJ, Hayasaka Y, Smith PA (2008) Rapid isolation of red wine polymeric polyphenols by solid-phase extraction. J Agri Food Chem 56: 2571–2580.

Jellouli N, Ben Jouira H, Skouri H, Ghorbel A, Gourgouri A, Mliki A (2008) Proteomic analysis of Tunisian grapevine cultivar Razegui under salt stress. J Plant Physiol 165: 471–481.

Kirst M, Basten CJ, Myburg AA, Zeng Z-B, Sederoff RR (2005) Genetic architecture of transcript-level variation in differentiating xylem of a Eucalyptus hybrid. Genetics 169: 2295–2303.

Lisec J, Schauer N, Kopka J, Willmitzer L, Fernie AR (2006) Gas chromatography mass spectrometry-based metabolite profiling in plants. Nat Protoc 1: 387–396.

Lücker J, Laszczak M, Smith D, Lund ST (2009) Generation of a predicted protein database from EST data and application to iTRAQ analyses in grape (*Vitis vinifera* cv. Cabernet Sauvignon) berries at ripening initiation. BMC Genom 10: 50.

Lund ST, Bohlmann J (2006) The molecular basis for wine grape quality—a volatile subject. Science 311: 804–805.

Marsh E, Alvarez S, Hicks LM, Barbazuk WB, Qiu W, Kovacs L, Schachtman D (2010) Changes in protein abundance during powdery mildew infection of leaf tissues of Cabernet Sauvignon grapevine (*Vitis vinifera* L.). Proteomics 10: 2057–2064.

Matsuda F, Yonekura-Sakakibara K, Niida R, Kuromori T, Shinozaki K, Saito K (2009) MS/MS spectral tag-based annotation of non-targeted profile of plant secondary metabolites. Plant J 57: 555–577.

Mattivi F, Guzzon R, Vrhovsek U, Stefanini M, Velasco R (2006) Metabolite profiling of grape: flavonols and anthocyanins. J Agri Food Chem 54: 7692–7702.

Negri AS, Prinsi B, Rossoni M, Failla O, Scienza A, Cocucci M, Espen L (2008a) Proteome changes in the skin of the grape cultivar Barbera among different stages of ripening. BMC Genom 9: 378.

Negri AS, Prinsi B, Scienza A, Morgutti S, Cocucci M, Espen L (2008b) Analysis of grape berry cell wall proteome: A comparative evaluation of extraction methods. J Plant Physiol 165: 1379–1389.

O'Farrell PH (1975) High resolution two-dimensional electrophoresis of proteins. J Biol Chem 250(10): 4007–4021.

Ong S-E, Mann M (2005) Mass spectrometry-based proteomics turns quantitative. Nat Chem Biol 1: 252–262.

Pereira GE, Gaudillere JP, Pieri P, Hilbert G, Maucourt M, Deborde C, Moing A, Rolin D (2006) Microclimate influence on mineral and metabolic profiles of grape berries. J Agri Food Chem 54: 6765–6775.

Perkins DN, Pappin DJC, Creasy DM, Cottrell JS (1999) Probability-based protein identification by searching sequence databases using mass spectrometry data. Electrophoresis 20: 3551–3567.

Pierce A, Unwin RD, Evans CA, Griffiths S, Carney L, Zhang L, Jaworska E, Lee C-F, Blinco D, Okoniewski MJ, Miller CJ, Bitton DA, Spooncer E, Whetton AD (2008) Eight-channel iTRAQ enables comparison of the activity of 6 leukaemogenic tyrosine kinases. Mol Cell Prot 7: 853–863.

Pilati S, Perazzolli M, Malossini A, Cestaro A, Dematte L, Fontana P, Dal Ri A, Viola R, Velasco R, Moser C (2007) Genome-wide transcriptional analysis of grapevine berry ripening reveals a set of genes similarly modulated during three seasons and the occurrence of an oxidative burst at veraison. BMC Genom 8: 428.

Robbins RJ (2003) Phenolic acids in foods: an overview of analytical methodology. J Agri Food Chem 51: 2866–2887.

Rocha SM, Coelho E, Zrostlikova J, Delgadillo I, Coimbra MA (2007) Comprehensive two-dimensional gas chromatography with time-of-flight mass spectrometry of monoterpenoids as a powerful tool for grape origin traceability. J Chrom A 1161: 292–299.

Rowe HC, Hansen BG, Halkierb BA, Kliebenstein DJ (2008) Biochemical networks and epistasis shape the *Arabidopsis thaliana* metabolome. Plant Cell 20: 1199–1216.

Saladin G, Magné C, Clément C (2003) Impact of flumioxazin herbicide on growth and carbohydrate physiology in *Vitis vinifera* L. Plant Cell Rep 21: 821–827.

Salmaso M, Malacarne G, Troggio M, Faes G, Stefanini M, Grando MS, Velasco R (2008) A grapevine (*Vitis vinifera* L.) genetic map integrating the position of 139 expressed genes. Theor Appl Genet 116: 1129–1143.

Sarry J-E, Sommerer N, Sauvage F-X, Bergoin A, Rossignol M, Albagnac G, Romieu C (2004) Grape berry biochemistry revisited upon proteomic analysis of the mesocarp. Proteomics 4: 201–215.

Sauvage F-X, Pradal M, Chatelet P, Tesniere C (2007) Proteome changes in leaves from grapevine (*Vitis vinifera* L.) transformed for alcohol dehydrogenase activity. J Agri Food Chem 55: 2597–2603.

Son HS, Hwang GS, Kim KM, Ahn HJ, Park WM, Van Den Berg F, Hong YS, Lee CH (2009) Metabolomic studies on geographical grapes and their wines using 1H NMR analysis coupled with multivariate statistics. J Agri Food Chem 57: 1481–1490.

Stylianou IM, Affourtit JP, Shockley KR, Wilpan RY, Abdi FA, Bhardwaj S, Rollins J, Churchill GA, Paigen B (2008) Applying gene expression, proteomics and single-nucleotide polymorphism analysis for complex trait gene identification. Genetics 178: 1795–1805.

Tesniere C, Torregrosa L, Pradal M, Souquet J-M, Gilles C, Dos Santos K, Chatelet P, Gunata Z (2006) Effects of genetic manipulation of alcohol dehydrogenase levels on the response to stress and the synthesis of secondary metabolites in grapevine leaves. J Exp Bot 57: 91–99.

Vincent D, Wheatley MD, Cramer GR (2006) Optimization of protein extraction and solubilization for mature grape berry clusters. Electrophoresis 27: 1853–1865.

Vincent D, Ergül A, Bohlman MC, Tattersall EAR, Tillett RL, Wheatley MD, Woolsey R, Quilici DR, Joets J, Schlauch K, Schooley DA, Cushman JC, Cramer GR (2007) Proteomic analysis reveals differences between *Vitis vinifera* L. cv. Chardonnay and cv. Cabernet Sauvignon and their responses to water deficit and salinity. J Exp Bot 58: 1873–1892.

Weckwerth W (2003) Metabolomics in systems biology. Annu Rev Plant Biol 54: 669–689.

Yan W, Chen SS (2005) Mass spectrometry-based quantitative proteomic profiling. Brief Funct Genom Proteom 4: 27–38.

Yang Y, Thannhauser TW, Li L, Zhang S (2007) Development of an integrated approach for evaluation of 2D gel image analysis: Impact of multiple proteins in single spots on comparative proteomics in conventional 2D gel/MALDI workflow. Electrophoresis 28: 2080–2094.

Yi EC, Li X-J, Cooke K, Lee H, Raught B, Page A, Aneliunas V, Hieter P, Goodlett DR, Aebersold R (2005) Increased quantitative proteome coverage with $^{13}C/^{12}C$-based, acid-cleavable isotope-coded affinity tag reagent and modified data acquisition scheme. Proteomics 5: 380–387.

Zamboni A, Minoia L, Ferrarini A, Tornielli GB, Zago E, Delledonne M, Pezzotti M (2008) Molecular analysis of post-harvest withering in grape by AFLP transcriptional profiling. J Exp Bot 59: 4145–4159.

Zhang J, Ma H, Feng J, Zeng L, Wang Z, Chen S (2008) Grape berry plasma membrane proteome analysis and its differential expression during ripening. J Exp Bot 59: 2979–2990.

Ziegler J, Facchini PJ (2008) Alkaloid biosynthesis: Metabolism and trafficking. Annu Rev Plant Biol 59: 735–769.

Zolotarjova N, Mrozinski P, Chen H, Martosella J (2007) Combination of affinity depletion of abundant proteins and reversed-phase fractionation in proteomic analysis of human plasma/serum. J Chrom A, 1189: 332–338.

13

Bioinformatics Tools in Grapevine Genomics

Jérôme Grimplet,[1,5,*] *Julie Dickerson,*[2]
Anne-Françoise Adam-Blondon[3] and *Grant Cramer*[4]

ABSTRACT

With the release of the grapevine genome sequence and with the increasing affordability of high throughput analysis tools, an ever increasing wealth of grapevine bioinformatics resources have been developed over the last decade in highly diverse biological fields. The grapevine community can now access extensive databases containing information relative to molecular markers, genetic maps, the genome sequence and its annotation, gene expression, proteins and metabolites. This chapter presents how to access these resources and provides some review of their contents and use. The next challenge is the further development of systems biology for grapevine; regarding this aspect, the different tools currently available for the integration of different types of data are discussed.

Keywords: Database, maps, STS, EST, gene annotation, gene ontology, system biology

[1]South Dakota State University, Horticulture, Forestry, Landscape and Parks Department, Brookings, SD, 57006, USA.
[2]Electrical and Computer Engineering, Iowa State University, Ames, IA, 50011-3060, USA.
[3]UMR INRA UEVE ERL CNRS, Génomique Végétale (URGV), 2 rue Gaston Crémieux, BP5708, 91 057 Evry cedex, France.
[4]1664 North Virginia Street, Department of Biochemistry and Molecular Biology, University of Nevada, Reno, NY 89557, USA.
[5]Instituto de Ciencias de la Vid y del Vino (CSIC,UR, Gobierno de La Rioja), C/Madre de Dios 51, 26006 Logroño, Spain.
*Corresponding autor: *jerome.grimplet@icvv.es*

13.1 Introduction

The grapevine genomic data has been increasing very rapidly since the start of the International Grape Genome Program in 2001 (IGGP) (*www.vitaceae.org*). Starting from scratch at that time, the genome sequence has been published twice (Jaillon et al. 2007; Velasco et al. 2007), more than 360,000 expressed sequence tags (ESTs) are now available (*Vitis vinifera:* July 2010, *http://www.ncbi.nlm.nih.gov/nucest*), and large efforts in proteomics and metabolomics have been successful (Chapter 12). There is more than ever a need for common databases and bioinformatics tools to allow a wide and deep use of these important resources. The first policy of the IGGP was to advise that deposition of the resources should be placed in public databases like the National Center for Biotechnology Information (NCBI). This advice has been followed for the most part by the grapevine community. However, there is also a need for better coordination in many aspects and of an interface with NCBI. As the grapevine community is a rather small one when compared to the community working on *Arabidopsis*, it will be difficult to raise funds to end up with a highly centralized and productive system like TAIR (*http://www.arabidopsis.org*). Other more collaborative initiatives are proposed such as for the Sol project (Menda et al. 2008) and may be a good middle term system. Such a model could be followed in the future with the help of the French national repository for plant genomic data (URGI; *http://urgi.versailles.inra.fr/*).

13.2 Molecular Marker and Genetic Map Databases

The international consortium for sequencing the grape genome was strongly encouraged to deposit the marker sequences and primers in the STS (sequence tagged sites) section of the GenBank public database maintained by NCBI. Currently, there are 779 grapevine unique STS in UniSTS (*http://www.ncbi.nlm.nih.gov/genome/sts/unists_stats.html*). Most of these publically available markers are simple sequence repeat (SSR) markers, mainly developed from *Vitis vinifera*.

More than 500 markers, mainly simple sequence repeat (SSR) markers, have been mapped onto the 17 maps stored in UniSTS and in the 27 maps stored at URGI (*http://urgi.versailles.inra.fr/GnpMap*). These maps were first made available through the GnpMap database of the French national repository for plant genomic data maintained by URGI and then transferred to NCBI. URGI also provides access to the Cabernet Sauvignon physical map and its links to the grapevine genetic map and the grapevine reference genome sequence (Moroldo et al. 2008; *http://urgi.versailles.inra.fr/cmap/*) using the GMOD tool, CMap (*www.gmod.org*). A comprehensive view of the data stored at URGI is provided by the web page *http://urgi.versailles.inra.fr/index.php/urgi/Species/Vitis/*.

In the near future, large sets of single nucleotide polymorphisms (SNPs) should become available. A large resource of SNP markers was for instance developed for mapping at Instituto Agrario di San Michele All'Adige-Fundazione Edmund Mach (Italy, Troggio et al. 2007; Pindo et al. 2008; Vezzulli et al. 2008) and the maps and their links with a Pinot Noir physical map and genome sequence can be viewed at *http://genomics.research.iasma.it/iasma/*. The markers are however not yet deposited at NCBI. More recently, a project of SNP discovery using the new high-throughput sequencing technologies has been released (Myles et al. 2010). A grape Gramene database has been set up to help the visualization of the results when they are obtained (*http://www.gramene.org/Vitis_vinifera/*). Such projects are now possible thanks to the availability of the reference genome sequence.

13.3 Genome Databases and Gene Annotation

The published version (8×) of the grapevine reference genome annotated sequence (Jaillon et al. 2007) has been deposited at EMBL and can be queried through the GBrowse interface (*www.gmod.org*) provided by two members of the sequencing consortium, in France (*http://www.genoscope.cns.fr/Vitis*) and Italy (*http://www.appliedgenomics.org/vitis8x*). It has been decided by the international community to transfer the last version (12×) to URGI, which will maintain it for the future and provide regular annotation updates to NCBI. The 12x genome is now accessible at URGI (*http://urgi.versailles.inra.fr/cgi-bin/gbrowse/vitis_12x_pub/*) and EMBL websites. The annotation provided was generated automatically by integrating several layers of evidence (alignments with ESTs, with protein, *de novo* prediction and comparative genomics) using the software, GAZE, which was manually calibrated (Howe et al. 2002; Jaillon et al. 2007). In parallel, the genome assembly of another *Vitis vinifera* genotype (cv. Pinot Noir) and its automatic annotation is available at *http://genomics.research.iasma.it/iasma/* (Chapter 9 Section9.4).

The next major objective is to organize a coordinated community of manual annotators. This is not simple; for example, in the *Arabidopsis* community it has been done through the parallel, integrated projects of (i) communities of bioinformaticists and expert biologists that manually annotated gene families, both at the structural and the functional level in a highly coordinated way, with in-house tools and high standards (see for example, Aubourg et al. 2005) and (ii) the TAIR repository (*www.arabidopsis.org*) that regularly improves and updates the automatic annotation of the *Arabidopsis* genome sequence and also manually curates the results (Swarbreck et al. 2007). The goal is to progressively include and centralize the work achieved by experts like the one stored at the Center for Bioinformatics of Peking University (Grape Transcription Factor Database: *http://planttfdb.cbi.pku.edu.cn/index.php?sp=Vvi*) on transcription factors in a single genome browser and/or to NCBI TPA (Third Party Annotation). Several tools or databases can help such community surveys like PlantGDB, Gramene, FLAGdb++, and DFCI.

The *Vitis* section of PlantGDB (VvGDB; *http://www.plantgdb.org/VvGDB/*) encompasses primary sequence data obtained from GenBank, including the most current GSS (Genome Survey Sequences), bacterial artificial chromosome (BAC) assembly and annotation from the Genoscope, EST, cDNA sequences, and protein sequences derived from the annotated genes. The mapping of cDNA, EST and Affymetrix probesets sequences onto the genome as well as interpretations of the overall data are based on the PlantGDB spliced alignment results obtained with the GeneSeqer software (*http://deepc2.psi.iastate.edu/cgi-bin/gs.cgi*). The Grapevine (*Vitis vinifera*) Transcription Factor Database collected 867 predicted transcription factors (TFs) from wine grape transcripts using input sequences from the PlantGDB-assembled Unique Transcript-fragment derived from *Vitis_vinifera* mRNAs from October 2006 (based on GenBank release 155). Extensive annotations for those TFs have been made, including similarity searches against major databases (e.g., Uniprot, RefSeq, EMBL, TRANSFAC), InterPro domain information and EST expression information extracted from UniGene. Fifty-nine transcription factor families have been identified in grape.

Manual gene family functional annotation is greatly facilitated through the use of the databases FLAGdb++ (Samson et al. 2004; *http://urgv.evry.inra.fr/projects/FLAGdb++/HTML/index.shtml*), GRAMENE (*http://www.gramene.org*) and the DFCI Grape Gene Index (*http://compbio.dfci.harvard.edu/tgi/cgi-bin/tgi/gimain.pl?gudb=grape*). FLAGdb++ allows the query of gene families using keywords of Pfam domains (Finn et al. 2008), the retrieval of all their sequences in one query, their mapping (BLAST) on the *Arabidopsis* and rice genome. GRAMENE proposes whole genome alignments with the *Arabidopsis* and the poplar genome, phylogenetic trees for each gene, identification of orthologs and paralogs, prediction of alternative splicing, knowledge on genetic variation, mapping of different kind of domains and signatures. The DFCI Grape Gene Index integrates research data from international grapevine EST sequencing and gene research projects. The latest release of the Grape Gene Index (Release 7.0; 17 April, 2010) has been built from 352,730 ESTs and 25,497 ETs (mature transcripts corresponding to the genes from the heterozygous genome). The Grape Gene Index contains 80,778 unique sequences (34,154 TCs, 31,813 singleton ESTs and 14,811 singleton ETs). Several functional annotation tools are available at the DFCI Grape Gene Index: the prediction of alternative splice variants (5,306 clusters); the comparison of EST expression between different libraries or tissues for libraries with over 50 ESTs; the classification of tentative contigs (TCs) by Gene ontology (GO) vocabularies (16,633 TC sequences with Molecular Function, 16,781 TC sequences with Biological Process,

20,570 TC sequences with Cellular Component); and the list of all 70-mer oligo predictions for the TC and the association of TCs with metabolic and signaling pathways.

13.4 Gene Expression Databases

Gene expression database resources are available on grapevine, such as repositories for microarray experiments, which are not limited to grapevine (GEO, ArrayExpress, PlexDB), MPSS (massively parallel signature sequencing), small RNA, or EST databases (VitisExpDB).

Grape MPSS (*http://mpss.udel.edu/grape/*) is a database that collected sequences based upon MPSS. It is a limited database that allows one to estimate transcript abundance in different tissues (Iandolino et al. 2008). This deep sequencing approach has the power to measure very low abundance transcripts that ordinary microarrays are not able to measure adequately. A small RNA website (*http://smallrna.udel.edu/index.php*) has also been generated from the same research group. This site allows mapping of small RNAs against the plant genome of choice including grape. One can determine if your transcript of choice might be regulated by small RNAs.

VitisExpDB (Doddapaneni et al. 2008, *http://cropdisease.ars.usda.gov/vitis_at/main-page.htm*) is a relational database that houses annotated EST and gene expression data. The database contains about 300,000 ESTs downloaded from the NCBI website. Several tools are available such as BLAST and the GO annotation of the ESTs. Results from a custom microarray experiment on Pierce's disease infected tissues are also available. The expression through EST experiments that are mapped to an *Arabidopsis* gene can be viewed on 25 AraCyc pathways (Zhang et al. 2005).

There are currently three public repositories where microarray experiments for grapevine have been uploaded.

Gene Expression Omnibus (GEO; *www.ncbi.nlm.nih.gov/geo/*) is a public gene expression/molecular abundance repository supporting MIAME compliant data submissions, and a curated, online resource for gene expression data browsing, query and retrieval. Fourteen grapevine microarray experiments data (including published works from Peng et al. 2007; Fung et al. 2008; Rotter et al. 2008; Lund et al. 2008; Mathiason et al. 2009; Ophir et al. 2009; Mica et al. 2009; Albertazzi et al. 2009; Hren et al. 2009; Bellin et al. 2009; Sreekantan et al. 2010; Polesani et al. 2010) have been submitted to GEO. GEO also includes simple tools for exploration, visualization, and analysis of the data sets.

ArrayExpress (*http://www.ebi.ac.uk/microarray-as/ae/*) is supported by the European Bioinformatics Institute (EBI). Not only is it able to store datasets from different platforms (MIAME- and MINSEQE- compliant data), but has the capability of conducting large meta-analyses across different microarrays using different technologies. There are 13 grapevine microarray data sets that have been submitted to ArrayExpress (Pilati et al. 2007; Fung et al. 2008; D'Onofrio et al. 2009; Fernandez et al. 2010; Rotter et al. 2009; Hren et al. 2009; Camps et al. 2010; Polesani et al. 2010).

PLEXdb (Plant Expression Database, *http://www.plexdb.org/* Wise et al. 2007) is a unified public resource for gene expression for plants and plant pathogens. PLEXdb serves as a bridge to integrate new and rapidly expanding gene expression profile data sets with traditional structural genomics and phenotypic data. The integrated tools of PLEXdb allow investigators to use commonalities in plant biology for a comparative approach to functional genomics through the use of large-scale expression profiling data sets. In addition to the microarray experiments repository aspects, analysis tools are available, such as gene expression normalization and statistical comparison across experiments, graphical visualization of gene expression. PLEXdb is the largest repository for grapevine microarray experiments: it contains data for published works from Pilati et al. (2007), Deluc et al. (2007), Grimplet et al. (2007), Tattersall et al. (2007), Cramer et al. (2007), Espinoza et al. (2007), Fung et al. (2008), Lund et al. (2008), Albertazzi et al. (2009), Koyama et al. (2009) in addition to secured unpublished works and all experiments available in GEO. Experiments included studies on berry development, tissue expression profiles, abiotic stresses, viral disease and other biotic stresses. PLEXdb also provides a submission service to the GEO repository for submitted experiments.

13.5 Protein or Metabolite Databases

Proteomic and metabolite databases are less well developed than gene expression and genomic databases. We shall give a brief description of these databases. Currently there are no grape-specific databases in use, although these databases are under development in MetNet (see below).

PRIDE (*http://www.ebi.ac.uk/pride/*), which stands for Proteomics Identifications database, is an open source central repository for all proteomic information and data. This database provides a common data exchange format for deposited proteomic data and publications that can be accessed by other programs for analysis. For example, the program SkyPainter (*http://www.reactome.org/cgi-bin/skypainter2?DB=gk_current*) can be used to identify the statistical overabundance of a protein, its presence in specific pathways and types of reactions in a set of proteins of interest.

KNApSAcK (*http://kanaya.naist.jp/KNApSAcK/*) is a comprehensive species-metabolite database. The database provides information on the metabolite, species that the metabolite was found in, and the mass spectrum of the metabolite, which is very useful for the identification of the metabolites. It includes data from NMR, LC-MS and GC-MS technologies.

PRIMe: (*http://prime.psc.riken.jp/*) stands for Platform for RIKEN Metabolomics. It is a web-based program for metabolomics and transcriptomics. Metabolites are measured by multi-dimensional NMR spectroscopy, GC-MS, LC-MS, and CE-MS technologies and can be linked with KNApSAcK. Additional tools are available for analysis of metabolites, transcripts, and integrated data sets.

MetNet (*http://www.metnetdb.org/*) contains software and a database that allows the global analysis of transcripts, proteins, and metabolites. One can visualize and model metabolic and regulatory networks. At the moment, the database contains functions for grapevine, *E.coli*, *Arabidopsis* and soybean data.

The UniProt (*http://www.uniprot.org*) database contains 31,254 grapevine protein sequences, mostly putative proteins from the heterozygote genome sequencing project completed by 821 other sequences (July 1, 2010). The UniProt database contains 421 reviewed grapevine proteins with highly curated annotation. UniProt displays automatic GO annotation in addition to domains and protein families automatically identified from a variety of sources (e.g., INTERPRO, Gene3D, Pfam, PIR, PRINTS, SMART, TIGRFAMs, PROSITE, and ProtoNet).

The PlantMetabolomics.org database contains metabolite data for *Arabidopsis* from a consortium of laboratories and contains both fingerprinting and targeted metabolomics data (Bais et al. 2010). Detailed tutorials on sample production are provided along with analysis tools and visualizations of results. The metabolite data can be placed in pathways using metabolite mappings from the AraCyc and MetNet pathway databases.

13.6 Genetic Resources and Phenotypes

The grapevine is a perennial and highly heterozygous species. As such, grapevine genetic resources are conserved in vineyards and distributed as cuttings most of the time, although seeds and pollen exchanges can also occur through scientists. Two databases have been established in Europe aiming first at facilitating the management of the collections and second, at their characterization by collaborative networks (Table 13-1).

The current goal is to improve these databases in order to link genotypic variation to phenotypic variation. This requires a common ontology for the description of developmental stages, phenotypes and traits. In grapevine, there is a long history of international coordination that led to a list of 128 OIV descriptors (OIV 1983) aiming at a precise identification of the accessions. Other descriptors were added later, for traits description (e.g., resistance to diseases); the European database is still under improvement through a collaborative project (*http://www1.montpellier.inra.fr/grapegen06/*). There is now a need to go further, with the aim to contribute to the international Plant Ontology Consortium (PO; *http://www.plantontology.org/*).

Table 13-1 Public databases for grapevine genetic resources management and characterization.

Country	Network	Web Site	Main Features Publically Available
France	French Network for Grapevine Genetic Resources	*http://urgi.versailles.inra.fr/siregal/*	-Passport data -core-collections
Europe	European Network for Grapevine Genetic Resources Conservation and Characterization (GENRES #081)	*http://www.genres.de/eccdb/vitis/*	-Passport data -primary and secondary descriptors

13.7 Integration of Different Data and Systems Biology of the Grapevine

The sequencing of the grapevine genome raised numerous questions about the potential function of all the identified genes. The observed complexity of the genome led grapevine researchers to rethink their approach for studying gene function. The reductionist approach, that has been used for decades or even centuries is limited. Grapevine is a complex living organism with multiple levels of hierarchical control from the individual cell to the tissue, organ and whole plant. Every part must be fully integrated with its environment. Understanding the role of these individual parts and their higher levels of control requires the integration of information in order to understand complex processes such as growth, fruit production and reproduction. Systems biology is the study of how objects interact with each other in a system and can be used for understanding biological phenomena (Mesarovic 1968). In its current use, systems biology refers to the integrative study of the molecular parts of an organism known as transcripts, proteins and metabolites.

Tools to integrate (correlate) metabolite, protein and transcript profiles are seriously underdeveloped, but the availability will be greater in the near future.

For instance, the Tomato Functional Genomics database (*http://ted.bti.cornell.edu/*) integrates the tomato metabolite and transcript data using Pearson correlations to find links between metabolites and transcripts using a newly developed package, Plant MetGenMap (*http://bioinfo.bti.cornell.edu/cgi-bin/MetGenMAP/home.cgi*). Currently the program supports *Arabidopsis*, rice and tomato data, but will support data from other species upon request. A metabolite database is under development at EBI (European Bioinformatics Institute), the institute that has developed the ArrayExpress, UniProt and PRIDE databases (to name a few) with the long-term goal to integrate these databases with a systems biology approach. A statistical method to analyze or make correlations with two data sets (Bylesjo et al. 2007) was recently extended to three datasets (Bylesjo et al. 2009), which is a first step toward such integrative approaches.

In parallel to these efforts towards an integration of heterogeneous data, several tools have been developed that allow transcript, protein and metabolite "omics" data to be displayed on molecular pathways. Each of these mapping programs presents specific advantages that are summarized in Table 13-2.

Table 13-2 Comparison of mapping programs available for plants. T: Transcripts, P: Proteins, M: Metabolites. N/A: numerical values about pathways composition are not available.

Program	Number of Pathways	*Mapped Genes*	*Mapped Metabolites*	*Metabolic Pathways*	*Regulatory Pathways*	"Omics Visualization"
Arabidopsis data						
Mapman	64	N/A	N/A	yes	yes	T/M, embedded
PMN/Aracyc	423	5335	2720	yes	no	T or P/M, embedded
REACTOME	454	2646	N/A	yes	yes	T or P/M, via Cytoscape
KaPPA-View	153	2610	1427	yes	no	T/M, embedded
MetNet	334	N/A	N/A	yes	yes	T/P/M, via Cytoscape
Vitis data						
DFCI GI	N/A	N/A	N/A	yes	no	No
KEGG vitis	106	2,265	N/A	yes	yes	No
VitisNet/ MetNet	219	13,145	1,593	yes	yes	T/P/M, via Cytoscape

13.7.1 Mapping Programs for Plant Data

MapMan (*http://gabi.rzpd.de/projects/MapMan/*) displays large data sets onto pictorial diagrams that symbolically depict areas of biological function. The originality of MapMan is that genes are initially organized in blocks rather than as pathways. This allows genes to be tentatively assigned, even when their function is only approximately known and to present pathways not extensively described. Fifty-seven pathways are currently available. Twenty-seven mapping files allow specific microarray platforms to be mapped on the pathways and are downloadable on the MapMan website. The mapping files exist for *Arabidopsis*, rice, potato, barley, maize, wheat, tomato, *Medicago* and grapevine. MapMan also allows the visualization of metabolite abundance.

The Plant Metabolic Network (PMN, *http://plantcyc.org/*) is a collaborative project among databases and biochemists with a common goal to build a broad network of plant metabolic pathway databases. PlantCyc provides access to manually curated or reviewed information about shared and unique metabolic pathways present in over 340 plant species. PlantCyc-based databases include AraCyc (*http://www.arabidopsis.org/biocyc/index.jsp*) and 10 other plant "Cyc" pathway databases and have been built with the Pathway Tools software. The PlantCyc release 4.0 contains 762 pathways regrouping 11,058 enzymes and 2,966 compounds. AraCyc contains 423 pathways regrouping 5,506 enzymes and 2,719 metabolites. AraCyc includes an expression data visualization tool, Omics Viewer that paints data values from the user's data sets onto the Cellular Overview diagram. A VitisCyc database has been created from the 12 × sequence data and will be available at the PlantCyc website.

REACTOME (*http://www.reactome.org/*) is probably the most extensive pathway database available, it contains up to 5,000 proteins and 1,000 pathways for some animal species. Pathways for *Arabidopsis* and rice are available, but they have been electronically inferred from human gene annotation. REACTOME allows the exportation of the pathways into the biological pathways standard format such as SBML and BioPAX. REACTOME does not include a molecule abundance viewing tool, but data can be exported into Cytoscape (*www.cytoscape.org*), where the abundance can be uploaded, visualized and analyzed with plugins provided by a very active user community. These tools include OmicsViz (Xia and Dickerson 2008) and Genoscape (Clement-Ziza et al. 2009).

KaPPA-View4 (*http://kpv.kazusa.or.jp/kpv4/*) allows users to display quantitatively data sets of transcripts and/or metabolites on comprehensive plant metabolic pathway maps. In addition, *Arabidopsis*, rice, tomato and *Lotus japonicus* genes were assigned to the pathway maps. KaPPA-View contains 153 pathways regrouping 2610 genes and 1427 metabolites for *Arabidopsis*.

The MetNet database (MetNetDB; *http://metnetonline.org* Wurtele et al. 2007) contains integrative information on networks of metabolic and regulatory and interactions in *E. coli*, *Arabidopsis*, soybean and grapevine. The grapevine pathways are manually curated and described in Grimplet et al. (2009). The pathway information is based on input from biologists in their area of expertise. In addition to the MetNet-curated interactions, AraCyc-curated pathways and AGRIS-curated regulatory networks are provided. MetNet allows export of pathways to the Cytoscape format for visualization of "omics" data.

13.7.2 Mapping Efforts for Grapevine

The grapevine nucleotide sequences have been mapped onto the KEGG pathways at the DFCI Grape Gene Index and at KEGG itself (*http://www.*

genome.jp/kegg-bin/show_organism?org=dvvi). The DFCI Grape Gene Index presents the mapping of the contigs onto the pathways. Sequences with a GO annotation linked to an E.C. number have been assigned to the pathways where that E.C number appears. At KEGG, 2265 Genoscope gene sequences have been annotated with KAAS (KEGG Automatic Annotation Server) blast for ortholog assignment and mapped on 106 pathways. This database currently lacks the tools for the visualization of "omics" data.

To date, grapevine data integration has been limited to manual construction of maps and visualization (Cramer et al. 2007; Deluc et al. 2007; Deluc et al. 2009). However new approaches are emerging. The first approach from Zamboni et al. (2008) uses the O2PLS method (Bylesjö et al. 2007) to combine the expression of the genes obtained from microarray technology with the profiles of metabolites obtained from HPLC-MS analysis. This powerful method allows the detection of genes whose expression correlates with metabolite abundances. The second approach, called VitisNet, from Grimplet et al. (2009) is based on the construction of molecular networks that regroup genes, transcripts, proteins and metabolites within biological pathways. These "omics" data can then be uploaded onto the networks and abundance of molecules can be visualized through a user-defined color scheme with Cytoscape. The pathways are available at MetNet (*http://www.metnetonline.org/*) and at *http://vitis-dormancy.sdstate.org*. Two hundred and nineteen pathways are available that incorporate more than 13,000 genes; novel pathways involving hormone signaling, transport, and transcription factors that have not been elucidated in other species yet are included in this mapping effort.

13.8 Conclusions

There has been a rapid explosion of bioinformatics tools in the last few years. The most significant contribution to these efforts has been the sequencing of the grapevine genome. The next major task for grapevine biology is to annotate many of the unknown genes and elucidate their function. This will be done with the large toolset now available to grapevine biologists. Bioinformatics databases are essential for capturing this information and facilitating additional analyses, insights and hypotheses. Grapevine biology has never been more exciting and promises to continue that way for some time to come. It is expected that there will be a rapid acceleration in grapevine research with wide exploration of the many grapevine genetic resources that are available in the many grapevine repositories around the world. One can predict the discovery of many new genes that play important roles in development, flavor, and stress tolerance.

References

Albertazzi G, Caffagni A, Milc JA, Francia E, Roncaglia E, Ferrari F, Tagliafico E, Stefani E, Pecchioni N (2009) Gene expression in grapevine cultivars in response to Bois Noir phytoplasma infection. Plant Sci 176: 792–804.

Aubourg S, Brunaud V, Bruyère C, Cock M, Cooke R, Cottet A, Couloux A, Déhais P, Deléage G, Duclert A, Echeverria M, Eschbach A, Falconet D, Filippi G, Gaspin C, Geourjon C, Grienenberger JM, Houlné G, Jamet E, Lechauve F, Leleu O, Leroy P, Mache R, Meyer C, Nedjari H, Negrutiu I, Orsini V, Peyretaillade E, Pommier C, Raes J, Risler JL, Rivière S, Rombauts S, Rouzé P, Schneider M, Schwob P, Small I, Soumayet-Kampetenga G, Stankovski D, Toffano C, Tognolli M, Caboche M, Lecharny A (2005) Genefarm, structural and functional annotation of Arabidopsis gene and protein families by a network of experts. Nucl Acids Res 33: D641–D646.

Bais P, Moon SM, He K, Leitao R, Dreher K, Walk T, Sucaet Y, Barkan L, Wohlgemuth G, Roth MR, Wurtele ES, Dixon P, Fiehn O, Lange BM, Shulaev V, Sumner LW, Welti R, Nikolau BJ, Rhee SY, Dickerson JA (2010) PlantMetabolomics org: A Web Portal for Plant Metabolomics Experiments. Plant Physiol 152: 1807–16.

Bellin D, Ferrarini A, Chimento A, Kaiser O, Levenkova N, Bouffard P, Delledonne M (2009) Combining next-generation pyrosequencing with microarray for large scale expression analysis in non-model species. BMC Genom 10: 555.

Bylesjo M, Eriksson D, Kusano M, Moritz T, Trygg J (2007) Data integration in plant biology: the O2PLS method for combined modeling of transcript and metabolite data. Plant J 52: 1181–1191.

Bylesjo M, Nilsson R, Srivastava V, Gronlund A, Johansson AI, Jansson S, Karlsson J, Moritz T, Wingsle G, Trygg J (2009) Integrated analysis of transcript, protein and metabolite data to study lignin biosynthesis in hybrid aspen. J Proteome Res 8: 199–210.

Camps C, Kappel C, Lecomte P, Léon C, Gomès E, Coutos-Thévenot P, Delrot S (2010) A transcriptomic study of grapevine (*Vitis vinifera* cv. Cabernet-Sauvignon) interaction with the vascular ascomycete fungus Eutypa lata. J Exp Bot 61: 1719–37.

Clément-Ziza M, Malabat C, Weber C, Moszer I, Aittokallio T, Letondal C, Rousseau S (2009) Genoscape: a Cytoscape plug-in to automate the retrieval and integration of gene expression data and molecular networks. Bioinformatics 25: 2617–2618.

Cramer GR, Ergül A, Grimplet J, Tillett RL, Tattersall EA, Bohlman MC, Vincent D, Sonderegger J, Evans J, Osborne C, Quilici D, Schlauch KA, Schooley DA, Cushman JC (2007) Water and salinity stress in grapevines: early and late changes in transcript and metabolite profiles. Funct Integr Genom 7: 111–134.

Deluc LG, Grimplet J, Wheatley MD, Tillett RL, Quilici DR, Osborne C, Schooley DA, Schlauch KA, Cushman JC, Cramer GR (2007) Transcriptomic and metabolite analyses of cabernet sauvignon grape berry development. BMC Genom 8: 429.

Deluc LG, Quilici DR, Decendit A, Grimplet J, Wheatley MD, Schlauch KA, Mérillon JM, Cushman JC, Cramer GR (2009) Water deficit induces cultivar-specific effects in multiple metabolic pathways affecting important flavor and quality traits throughout grape berry ripening. BMC Genom 10: 212.

Doddapaneni H, Lin H, Walker AM, Yao J, Civerolo EL (2008) Vitisexpdb: A database resource for grape functional genomics. BMC Plant Biol 8: 23.

D'Onofrio C, Cox A, Davies C, Boss PK (2009) Induction of secondary metabolism in grape cell cultures by jasmonates. Funct Plant Biol 36: 323–338.

Espinoza C, Vega A, Medina C, Schlauch K, Cramer G, Arce-Johnson P (2007) Gene expression associated with compatible viral diseases in grapevine cultivars. Funct Integr Genom 7: 95–110.

Fernandez L, Torregrosa L, Segura V, Bouquet A, Martinez-Zapater JM (2010) Transposon-induced gene activation as a mechanism generating cluster shape somatic variation in grapevine. Plant J 61: 545–57.

Finn RD, Tate J, Mistry J, Coggill PC, Sammut JS, Hotz HR, Ceric G, Forslund K, Eddy SR, Sonnhammer EL, Bateman A (2008) The Pfam protein families database. Nucl Acids Res 36: D281–D288.

Fung RW, Gonzalo M, Fekete C, Kovacs LG, He Y, Marsh E, McIntyre LM, Schachtman DP, Qiu W (2008) Powdery mildew induces defense-oriented reprogramming of the transcriptome in a susceptible but not in a resistant grapevine. Plant physiol 146: 236–249.

Grimplet J, Deluc LG, Tillett RL, Wheatley MD, Schlauch KA, Cramer GR, Cushman JC (2007) Tissue-specific mRNA expression profiling in grape berry tissues. BMC Genom 8: 187.

Grimplet J, Cramer GR, Dickerson JA, Mathiason K, Van Hemert J, Fennell AY (2009) VitisNet: "Omics" integration through grapevine molecular networks. PLoS One 4: e8365.

Howe KL, Chothia T, Durbin R (2002) GAZE: a generic framework for the integration of gene-prediction data by dynamic programming. Genome Res 12: 1418–1427.

Hren M, Nikolić P, Rotter A, Blejec A, Terrier N, Ravnikar M, Dermastia M, Gruden K (2009) 'Bois noir' phytoplasma induces significant reprogramming of the leaf transcriptome in the field grown grapevine. BMC Genom 10: 460.

Iandolino A, Nobuta K, da Silva FG, Cook DR, Meyers BC (2008) Comparative expression profiling in grape (*Vitis vinifera*) berries derived from frequency analysis of ESTs and MPSS signatures. BMC Plant Biol 8: 53.

Jaillon O, Aury JM, Noel B, Policriti A, Clepet C, Casagrande A, Choisne N, Aubourg S, Vitulo N, Jubin C, Vezzi A, Legeai F, Hugueney P, Dasilva C, Horner D, Mica E, Jublot D, Poulain J, Bruyère C, Billault A, Segurens B, Gouyvenoux M, Ugarte E, Cattonaro F, Anthouard V, Vico V, Del Fabbro C, Alaux M, Di Gaspero G, Dumas V, Felice N, Paillard S, Juman I, Moroldo M, Scalabrin S, Canaguier A, Le Clainche I, Malacrida G, Durand E, Pesole G, Laucou V, Chatelet P, Merdinoglu D, Delledonne M, Pezzotti M, Lecharny A, Scarpelli C, Artiguenave F, Pè ME, Valle G, Morgante M, Caboche M, Adam-Blondon AF, Weissenbach J, Quétier F, Wincker P (2007) The grapevine genome sequence suggests ancestral hexaploidization in major angiosperm phyla. Nature 449: 463–467.

Koyama K, Sadamatsu K, Goto-Yamamoto N (2009) Abscisic acid stimulated ripening and gene expression in berry skins of the Cabernet Sauvignon grape. Funct Integr Genom Epub ahead of print.

Lund ST, Peng FY, Nayar T, Reid KE, Schlosser J (2008) Gene expression analyses in individual grape (*Vitis vinifera* L.) berries during ripening initiation reveal that pigmentation intensity is a valid indicator of developmental staging within the cluster. Plant Mol Biol 68: 301–15.

Mathiason K, He D, Grimplet J, Venkateswari J, Galbraith DW, Or E, Fennell A (2009) Transcript profiling in *Vitis riparia* during chilling requirement fulfillment reveals coordination of gene expression patterns with optimized bud break. Funct Integr Genom 9: 81–96.

Menda N, Buels RM, Tecle I, Mueller LA (2008) A community-based annotation framework for linking solanaceae genomes with phenomes. Plant Physiol. 147: 1788–1799.

Mesarovic MD (1968) Systems theory and biology—view of a theoretician. Syst Theor Biol 351: 59–87.

Mica E, Piccolo V, Delledonne M, Ferrarini A, Pezzotti M, Casati C, Del Fabbro C, Valle G, Policriti A, Morgante M, Pesole G, Pè ME, Horner DS (2009) High throughput approaches reveal splicing of primary microRNA transcripts and tissue specific expression of mature microRNAs in Vitis vinifera. BMC Genom 10: 558.

Moroldo M, Paillard S, Marconi R, Fabrice L, Canaguier A, Cruaud C, De Berardinis V, Guichard C, Brunaud V, Le Clainche I, Scalabrin S, Testolin R, Di Gaspero G, Morgante M, Adam-Blondon AF (2008) A physical map of the heterozygous grapevine 'cabernet sauvignon' allows mapping candidate genes for disease resistance. BMC Plant Biol 8: 66.

Myles S, Chia J-M, Hurwitz B, Simon C, Zhong GY, Buckler E, Ware D (2010) Rapid Genome Characterization of the Genus Vitis. PLoS One 5: e8219.

[OIV] Organisation Internationale de la Vigne et du Vin (1983) Code des caractères descriptifs des variétés et espèces de *Vitis*. OIV, 18 rue d'Aguesseau, Paris, France.

Ophir R, Pang X, Halaly T, Venkateswari J, Lavee S, Galbraith D, Or E (2009) Gene-expression profiling of grape bud response to two alternative dormancy-release stimuli expose possible links between impaired mitochondrial activity, hypoxia, ethylene-ABA interplay and cell enlargement. Plant Mol Biol 71: 403–23.

Peng FY, Reid KE, Liao N, Schlosser J, Lijavetzky D, Holt R, Martínez Zapater JM, Jones S, Marra M, Bohlmann J, Lund ST (2007) Generation of ESTs in *Vitis vinifera* wine grape (Cabernet Sauvignon) and table grape (Muscat Hamburg) and discovery of new candidate genes with potential roles in berry development. Gene 402: 40–50.

Pilati S, Perazzolli M, Malossini A, Cestaro A, Demattè L, Fontana P, Dal Ri A, Viola R, Velasco R, Moser C (2007) Genome-wide transcriptional analysis of grapevine berry ripening reveals a set of genes similarly modulated during three seasons and the occurrence of an oxidative burst at veraison. BMC Genom 8: 428.

Pindo M, Vezzulli S, Coppola G, Cartwright DA, Zharkikh A, Velasco R, Troggio M (2008) SNP high-throughput screening in grapevine using the SNPLEX™ genotyping system. BMC Plant Biol 8: 12.

Polesani M, Bortesi L, Ferrarini A, Zamboni A, Fasoli M, Zadra C, Lovato A, Pezzotti M, Delledonne M, Polverari A (2010) General and species-specific transcriptional responses to downy mildew infection in a susceptible (Vitis vinifera) and a resistant (V. riparia) grapevine species. BMC Genom 11: 117.

Rotter A, Hren M, Baebler S, Blejec A, Gruden K (2008) Finding differentially expressed genes in two-channel DNA microarray datasets: how to increase reliability of data preprocessing. Omics 12: 171–182.

Rotter A, Camps C, Lohse M, Kappel C, Pilati S, Hren M, Stitt M, Coutos-Thévenot P, Moser C, Usadel B, Delrot S, Gruden K (2009) Gene expression profiling in susceptible interaction of grapevine with its fungal pathogen Eutypa lata: extending MapMan ontology for grapevine. BMC Plant Biol 9: 104.

Samson F, Brunaud V, Duchêne S, De Oliveira Y, Caboche M, Lecharny A, Aubourg S (2004) Flagdb++: a database for the functional analysis of the *Arabidopsis* genome. Nucl Acids Res 32: D347–D350.

Sreekantan L, Mathiason K, Grimplet J, Schlauch K, Dickerson JA, Fennell AY (2010) Differential floral development and gene expression in grapevines during long and short photoperiods suggests a role for floral genes in dormancy transitioning. Plant Mol Biol 73: 191–205.

Swarbreck D, Wilks C, Lamesch P, Berardini TZ, Garcia-Hernandez M, Foerster H, Li D, Meyer T, Muller R, Ploetz L, Radenbaugh A, Singh S, Swing V, Tissier C, Zhang P, Huala E (2007) The *Arabidopsis* information resource (TAIR): gene structure and function annotation. Nucl. Acids Res 36: D1009–D1014.

Tattersall EA, Grimplet J, DeLuc L, Wheatley MD, Vincent D, Osborne C, Ergül A, Lomen E, Blank RR, Schlauch KA, Cushman JC, Cramer GR (2007) Transcript abundance profiles reveal larger and more complex responses of grapevine to chilling compared to osmotic and salinity stress. Funct Integr Genom 7: 317–333.

Troggio M, Malacarne G, Coppola G, Segala C, Cartwright DA, Pindo M, Stefanini M, Mank R, Moroldo M, Morgante M, Grando MS, Velasco R (2007) A dense single-nucleotide polymorphism-based genetic linkage map of grapevine (*Vitis vinifera* l.) anchoring pinot noir bacterial artificial chromosome contigs. Genetics 176: 2637–2650.

Velasco R, Zharkikh A, Troggio M, Cartwright DA, Cestaro A, Pruss D, Pindo M, Fitzgerald LM, Vezzulli S, Reid J, Malacarne G, Iliev D, Coppola G, Wardell B, Micheletti D, Macalma T, Facci M, Mitchell JT, Perazzolli M, Eldredge G, Gatto P, Oyzerski R, Moretto M, Gutin N, Stefanini M, Chen Y, Segala C, Davenport C, Demattè L, Mraz A, Battilana J, Stormo K, Costa F, Tao Q, Si-Ammour A, Harkins T, Lackey A, Perbost C, Taillon B, Stella A, Solovyev V, Fawcett JA, Sterck L, Vandepoele K, Grando SM, Toppo S, Moser C, Lanchbury J, Bogden R, Skolnick M, Sgaramella V, Bhatnagar SK, Fontana P, Gutin A, Van de Peer Y, Salamini F, Viola R (2007) A high quality draft consensus sequence of the genome of a heterozygous grapevine variety. PLoS ONE 2: e1326.

Vezzulli S, Micheletti D, Riaz S, Pindo M, Viola R, This P, Walker MA, Troggio M, Velasco R (2008) A SNP transferability survey within the genus *Vitis*. BMC Plant Biol 8: 128.
Wise RP, Caldo RA, Hong L, Shen L, Cannon EK, Dickerson JA (2007) BarleyBase/PLEXdb: A Unified Expression Profiling Database for Plants and Plant Pathogens. In: D Edwards (ed) Plant Bioinformatics—Methods and Protocols. Methods in Molecular Biology, vol 406 Humana Press, Totowa, NJ, USA, pp 347–363.
Wurtele ES, Li L, Berleant D, Cook D, Dickerson JA, Ding J, Hofmann H, Lawrence M, Lee EK, Li J, Mentzen W, Miller L, Nikolau BJ, Ransom N, Wang Y (2007) MetNet: Systems Biology Software for *Arabidopsis*. In: BJ Nikolau, ES Wurtele (eds) Concepts in Plant Metabolomics Springer Dordrecht, The Netherlands.
Xia T, Dickerson JA (2008) OmicsViz: Cytoscape plug-in for visualizing omics data across species. Bioinformatics 24: 2557–2558.
Zhang P, Foerster H, Tissier C, Mueller L, Paley S, Karp P, Rhee S (2005) MetaCyc and AraCyc Metabolic Pathway Databases for Plant Research. Plant Physiol 138: 27–37.
Zamboni AL, Minoia L, Ferrarini A, Chimento A, Di Carli M, Desiderio A, Benvenuto E, Guzzo F, Toffali K, Delledonne M, Pezzotti M (2008) Systems biology of berry ripening and postharvest withering processes. 8th Int Symp on Grapevine Physiology and Biotechnology. Adelaide, Australia, 23–28 Nov 2008.

14

Future Prospects

A-F Adam-Blondon,[1,*] *M Boulay*[2] and *J-M Martinez-Zapater*[3]

ABSTRACT

Viticulture is a very valuable crop in terms of income and still attracts new producers. However, in order to stay this way, growers will have to meet several challenges with the help of scientists. First, the use of pesticides will have to be greatly reduced through the development of global strategies using fewer chemicals with other methods such as biological control, genetic resistance and new agronomical practices. The design of such integrated systems requires in-depth knowledge at several levels on how the grapevine functions in its biotic and abiotic environment. This will be achieved using metagenomics strategies to generate inventories of the Grapevine-Associated Micro-Flora, as well as transcriptomic, proteomic and metabolomic analyses of both grapevine and pests and pathogens during their interaction. Modern genome based approaches will also increase the efficiency in the breeding of new varieties combining high quality and durable resistance. Second, viticulture will have to adapt to changing growing conditions due to global climate changes and to better and more rapidly meet consumer demands. This will require the ability to model the function of grapevine in the environment based on a combination of genomics, genetic and ecophysiologic information as a way to quickly target the relevant trait and the relevant strategy to address (breeding, training practices, enology practices, etc.).

Keywords: Pathogen and pest resistance, agrochemicals, global warming, grapevine breeding, genomics

[1]UMR INRA UEVE ERL CNRS Génomique Végétale, 2 rue Gaston Crémieux, BP5708, 91 057 Evry cedex, France.
[2]60 rue de Vaux, 77000 Livry sur Seine, France, ex. Responsable du Développement Viticole Moët & Chandon 20 Avenue de Champagne 51200 Epernay, France.
[3]Instituto de Ciencias de la Vid y del Vino (CSIC, Universidad de La Rioja, Gobierno de La Rioja), C/Madre de Dios 51, 26006 Logroño, Spain.
*Corresponding author: *adam@evry.inra.fr*

14.1 Evolution of Viticulture Around the World

The analysis of viticulture, wine market and consumption around the world presented in Chapter 1 and summarized in Fig. 1-1 in terms of trends, shows that Europe has a traditional and strong viticulture and as such still maintains the largest vineyard areas, together with the largest wine exports and imports. European consumers still account for most of the worldwide consumption (Fig. 14-1). However, countries like North America and China will perhaps compete for this last point in the future. The only exception in the grapevine-related market is for fresh/dried berries consumption, where Europe lies just behind Asia (*http://faostat.fao.org/*; data not shown). During the last 60 years however, the vineyard areas have decreased roughly by one-third across Europe with a reorientation towards wine quality: low quality varieties have been replaced by high-quality ones and low quality *terroirs* have been abandoned. After the genetic diversity bottleneck caused in the European viticulture by the introduction of Phylloxera and fungal

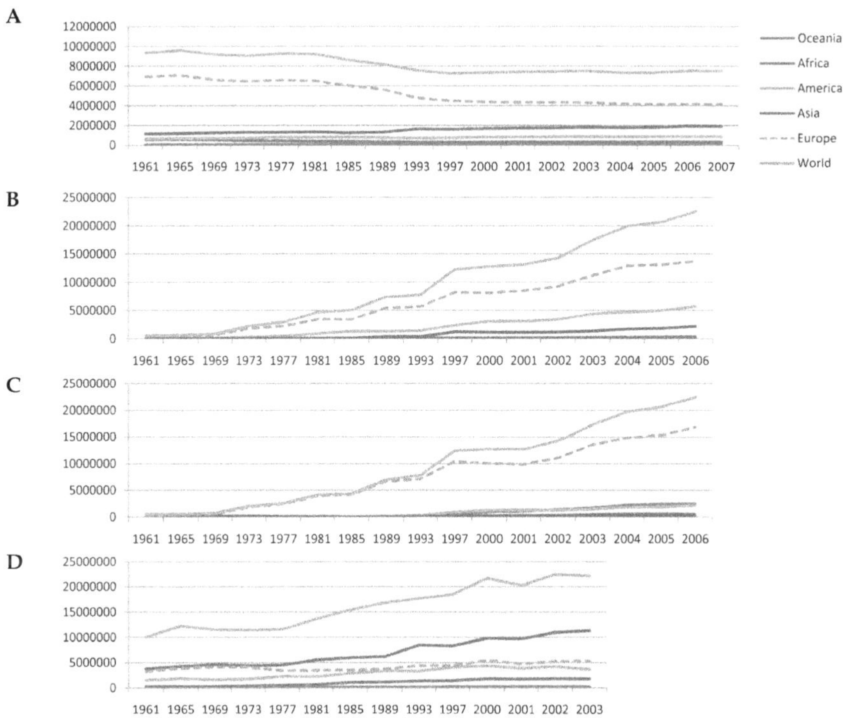

Figure 14-1 World statistics for the wine sector (from FAOSTAT; *http://faostat.fao.org/*) (A) Surface harvested (ha) (B) Wine imports (1000 $). (C) Wine exports in (1000$) (D) Wine consumption (t).

Color image of this figure appears in the color plate section at the end of the book.

pathogens in the second half of the 19th century, this market selection of high quality wines resulted in further reduction of cultivar diversity across the vineyards (Bouquet and Boursiquot 1999). In parallel, as pointed out in Chapter 1, drastic changes in consumer habits were also recently observed, mainly in France, Italy and Spain, with a shift towards less wine consumption but of higher quality and a progressive disappearance of regular wine consumers that, as a whole, resulted in a global reduction of wine consumption (Fig. 14-1d).

The worldwide vineyard area has remained roughly stable despite its reduction in Europe thanks to a sharp increase of plantations in America, Asia and Australia (Fig. 14-1a). This expansion has recently been particularly impressive in China (Fig. 14-2a), with Chinese wine production and importations (Fig. 14-2b) being completely devoted to supply the large increase of internal consumption (Fig. 14-2d). This is quite different from the situation in most of the other producing countries that were expanding their exports timely with the increase of production (e.g., Australia: Fig. 14-2c), progressively reducing the position of Europe as the major wine exporter (Fig. 14-1c). All around the world, viticulture is still a very valuable crop with a total US $22,423 million value for worldwide wine exports and US $ 4,601 million value for worldwide grape and grape juices exports in 2006 (*http://faostat.fao.org/*). It may remain the same, provided viticulture copes with several important challenges in the near future which could be facilitated by major advances in scientific knowledge.

14.2 Viticulture Will Have to Reduce the Use of Pesticides and Perhaps to Face the Expansion of New Diseases due Either to Such Reduction or to Climatic Changes

The first challenge is to drastically reduce the use of pesticides while maintaining the quality of the production. Until now, except for Phylloxera pest control, this was mainly achieved through the use of chemicals. However, it is now widely recognized that this approach is not sustainable as some pesticide active molecules can have effects on the environment and on human health. Their use has now become much more restricted in Europe (*http://www.europarl.europa.eu/news/expert/briefing_page/45150-012-01-03-20081217BRI45149-12-01-2009-2009/default_p001c004_en.htm*), which may lead to a technical dead-end for current practices. Although it will probably not be possible to protect the vineyard without using any chemicals, their use will continue to be reduced. In fact, insecticides have been intensively reduced by the use of biological control such as Phytoseids or pheromones, as well as herbicides by inter row under grassing. For fungicides, reductions are obtained thanks to the increased use of disease forecast models as well as prophylaxis techniques. Following these already initiated patterns, new

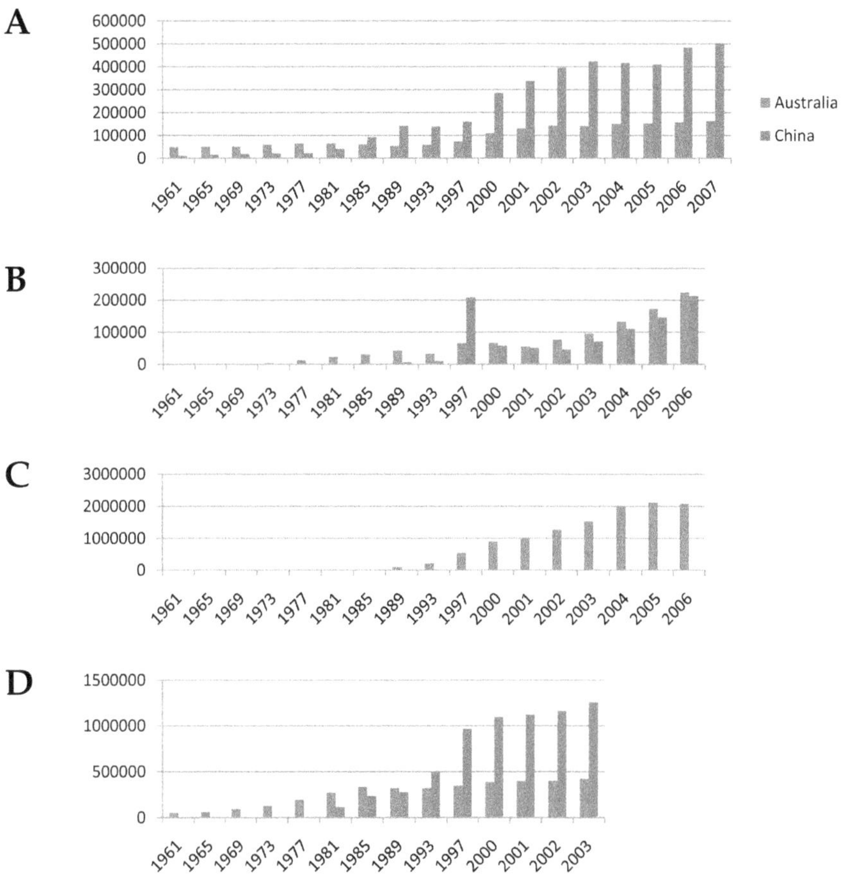

Figure 14-2 Statistics for the wine sector in Australia and China (from FAOSTAT; *http://faostat.fao.org/*) (A) Surface harvested (ha) (B) Wine imports (1,000 $) (C) Wine exports (1,000$) (D) Wine consumption (t).

grapevine management technologies will probably include the use of fewer chemicals (more environmentally friendly) in a global strategy together with methods for biological control, genetic resistance and new agronomical practices for a more sustainable viticulture. The design of such integrated systems requires in-depth knowledge at several levels on how the grapevine functions in its biotic and abiotic environment.

14.2.1 Inventory of the Grapevine-Associated Micro-Flora

The first layer of knowledge, which concerns the complete description of the microorganisms that live in vineyard soils, but also on leaves, trunks or berries as epiphytes or in perennial organs such as roots, or trunks as endophytes microorganisms is now-a-days a near complete black box. Yet, some of these microbial populations are known to have the potential to favor plant growth and adaptation to abiotic stresses, but also to prime defense responses against pathogens in plants (Ait Barka et al. 2002, 2006; Conrath et al. 2006; Van Wees et al. 2008), which may lead to the development of new strategies for pests and pathogens control (McMahan et al. 2001; Schmidt et al. 2001; Mohamed et al. 2007). However, the use of a living organism as a pesticide is always more complex than the use of a chemical and the activation by elicitors of the defense pathways normally switched on by these beneficial microbes (Zimmerli et al. 2000) does not give a straight forward solution for field conditions either. These pathways may be antagonistic to other pathways also important for plant adaptation and in the molecular "dialog" that represents the host-parasite relationship, the parasite being at different stages of development under field conditions, may always have time in advance to the host defence. Therefore, it is important to better understand the vineyard ecological system in order to develop new agronomical practices taking into account beneficial microbial populations (Beckers and Conrath 2007). The first step is to record all the microorganisms present in the vineyards in/on grapevines or in its immediate environment, to study the evolution of such communities in different conditions (healthy/non healthy vine, under abiotic stresses, etc.) and the physiological reactions of grapevine and of individual microbes in interaction. This will require metagenomic approaches that are currently under massive development for instance to better understand human diseases or to find solutions to polluted environments (Daniel 2005; Ventura et al. 2009), but are still in their infancy with regards to agriculture (Jiao et al. 2006; Rabosto et al. 2006; Van Elsas et al. 2008).

14.2.2 Inventory of Grapevine Interactions with its Main Pests and Pathogens with a "Both Sides" Point of View

Another important layer of knowledge concerns the molecular mechanisms underlying the interaction of grapevine with its major pests and pathogens. Many laboratories are currently studying grapevine transcriptome changes under pest stress as a first step (Espinoza et al. 2007; Lin et al. 2007; Roper et al. 2007; Figueiredo et al. 2008; Fung et al. 2008; Polesani et al. 2008; Polesani et al. 2010) but it would also be very important to study gene regulation in pathogens in the same conditions to fully understand the molecular dialog between the two interacting organisms (e.g., Lambais et al. 2000; Choquer et al. 2007) as evoked above. For this purpose, it will be important to develop genomic tools for the main grapevine pests and especially those responsible for a lot of treatments in the vineyard (e.g., *Erysiphe necator, Plasmopara viticola, Botrytis cinerea*) and those for which no control solution is available yet (the virus pathogen GFLV transmitted by the dagger type nematode *Xiphinema index,* the bacterial pathogen *Xylella fastidiosa,* as well as some other viral diseases like *GLRaV*, leaf roll viruses, etc.). A few of those pathogens have their genome sequenced [e.g., *X. fastidiosa* (Van Sluys et al 2003), *B. cinerea* (Fillinger et al. 2007) and GFLV (Naraghi-Arani et al. 2001)] and two others, responsible for the stolbur and the flavescence dorée are under way (Cimerman et al. 2006; Malembic-Maher et al. 2008). Unfortunately, so far comprehensive collections of express sequence tags (ESTs) have only been produced for *B. cinerea* (Viaud et al. 2005).

14.2.3 Inventory of the Genetic Diversity for Resistance to Diseases in the Vitis Genus and Grapevine Improvement

Last but not the least, as in other crops, it is important to introduce resistant varieties in viticulture. These varieties will have to produce high quality wine given the current trends of the market and the resistance will have to be durable (plantations have an average duration of 30 years or even more for high quality red wines!). Regarding resistance durability, it will be important to characterize the natural pathogen populations, their genetic diversity, their epidemiology, but also the selective pressures exerted by their host and the management system (McDonald and Linde 2002). From the plant side, it will be very important to combine in the resistant varieties different resistance mechanisms against a given pest (pyramiding resistance; see Chapter 7 Section 7.3), which supports the interest to characterize genetic resources in the genus *Vitis* and to study the genetic basis of their pathogen resistance mechanisms. The increase of grapevine research in China that will most probably come with the increase of vineyard areas will probably bring new such resources from the Asian pool, yet poorly known. The design of high quality varieties requires enough generations of backcrossing in order to recover the *Vitis vinifera* best quality level while retaining the resistance traits. Many groups are currently working at the development of molecular markers to facilitate and accelerate such programs (see Chapters 5 and 7) and some interesting varieties have been produced and are already planted (see Chapter 1 Section 1.3.2). They will have to prove their worth during the coming years, both in terms of resistance to diseases and of organoleptic qualities of their issued wines, before being considered into high quality production systems. However, if some variety proves in the future to be of such high interest, it is likely that it will succeed, especially if a technical dead-end is reached with regards to disease control. It was demonstrated in the past when the Phylloxera epidemic drew to the general use of rootstocks from either American *Vitis* species or hybrids between them and *Vitis vinifera,* greatly reshaping the plant production chain (grafted plants versus own roots).

14.2.4 Is There a Place for Genetically Modified Grapevine in the Future?

On the other hand, depending on social acceptance of genetically modified grapevine, improving grapevine resistance to disease by gene transfer is also possible. This is particularly true for resistance to viruses or phytoplasms, where no major resistance sources are known in the *Vitis* genus and different transgene-based strategies are developed all over the world for resistance to GFLV or GLRaV (see for review, Laimer et al. 2009). Another possibility raised by such an approach is a quicker development of high quality varieties transformed with *Vitis*-derived resistance genes (see Chapter 8). However, the introduction of genetically modified organism (GMO) in viticulture is still a challenging issue and will necessarily have to be co-developed within the society (Laimer et al. 2009). In this respect, an interesting and partly successful initiative has been set up in France (Joly and Rip 2007).

14.3 Viticulture Will Have to Adapt to New Climatic Conditions in Many Areas

Another challenge, global climatic change, may lead to changes in the cultivars planted. The projections for 2100 of the Intergovernmental Panel on Climatic Change (IPCC; *http://www.ipcc.ch*) include an increase of the average temperature, a change in the rain regime especially in the Mediterranean areas with less rain in summer combined with a water demand increase by plants and a doubling of atmospheric CO_2 content. While the change in carbon dioxide concentration perhaps will have no or few effects on wine quality as demonstrated recently on a Portuguese red cultivar (Gonçalves et al. 2009), temperature increases will have. An elevation of summer temperatures and a consequent shortening of the grapevine growing period have already been observed during the last 10 to 15 years and have recorded consequences in viticulture production as reviewed by Jones (2007). For instance in France, vintages are now taking place generally one month earlier than 15 years ago. Furthermore, harvested berries show an increase of their sugar content, a reduction of their acidity and lower phenolic maturity. The berry ripening phase often occurs now during the hottest periods, which is not favorable for an optimal aromatic production. Finally, the increasing water deficit in soils in spring and summer impairs nitrogen assimilation by grapevine resulting in less free amino acids content in the berries, which in turn may lead to problems in must fermentation and wine ageing. An indirect effect of climatic change may also be the further spreading of pests that were currently stopped by cold winters (e.g., vector of Pierce's disease in Europe).

There is, therefore, a need to better understand the interaction of grapevine with its physical environment and its consequences on the process of plant production and berry content. This will need integrated system approaches combining more powerful phenotypic analyses including transcriptomic, proteomic and metabolomic studies in different plant tissues and under different environmental conditions (Castellarin et al. 2007a and b; Cramer et al. 2007; Deluc et al. 2007; Mori et al. 2007; Pilati et al. 2007; Vincent et al. 2007; Iandolino et al. 2008; Lund et al. 2008; Zhang et al. 2008; Bellin et al. 2009; Zenoni et al. 2010; see Chapters 11 and 12). In addition, this analysis should also integrate the existent genetic variation represented by the different grapevine cultivars as well as their differential interactions with the environment. This information will be the basis to develop better tools for the assessment of quality potential before harvest, to provide better criteria for the selection of new cultivars or to screen genetic resources for varieties that were abandoned because of their low potential under former climatic conditions but that could now be quite acceptable. Yet, the classification of the grapevine varieties according to their optimal climate growing conditions proposed by Jones (2007) could be extended to other genetic resources using appropriate screening procedures (Duchêne et al 2010). Currently, the new available tools derived from genomic information are generating many opportunities in different wine producing countries to develop research programs together with the private sector, addressing these types of viticulture issues.

14.4 Viticulture Will Have to Adapt Better and Faster to Consumer Demands

Traditionally, the main consumers were located in countries having a wine culture. This means that they were more or less educated about wine history or wine culture and were willing and able to choose their wines among a large diversity of products from wine regions associating different cultivars, soils and viticulture practices. More recently, some cultivars have been pushed forward and used throughout the world, sometimes in association with new enological practices (e.g., use of oak chips) with the aim to obtain a more standard product, better adapted to a worldwide demand. This demand is evolving and diversifying for example towards wines that are more fruity and fresh on one hand or wines that could be drunk on occasions where they must compete with other beverages like spirits. In order to answer such diversified demands, the enologists have a long tradition of improvement of the wine making process (development of new yeast strains, new extraction methods for aromatic compounds, better control of fermentation, etc.). However, a better understanding of the whole integrated process from berry composition and ripening to fermentation and enological technologies can allow new approaches in the design of a desirable type of wine called "reverse engineering" (H. Ojeda, pers. comm.). Under this approach the design of a new wine would define appropriate cultivar characteristics, viticulture practices and enological techniques.

Even, if the evolution of the berry composition in water, sugar, acids, phenolics and terpens is well known (for review see Ollat et al. 2002; Conde et al. 2007), there is a need for a better understanding of the regulation of metabolic pathways involved, of their crosstalks and of the molecular basis of the genetic variation shown by grapevine cultivars. Furthermore, it is important to understand the genotype environment interactions on berry composition that will help develop more precise diagnostic tools to monitor vintages.

As pointed above, genome derived technologies allowed for a more thoroughly phenotypic description of berry evolution at gene expression, protein synthesis and metabolic composition and several transcriptome, proteome and metabolome analyses of developing berries have been recently published, drawing a first rough sketch of the pattern of co-regulated genes (see Chapters 10, 11 and 12 for reviews). These preliminary data will be further completed with future analyses using either more representative genome microarrays or open expression analyses methods derived from the new high throughput sequencing techniques now that the genome sequence is available. Additional studies will also have to consider different genotypes as well as different growing conditions to develop a complete picture of berry development and the ripening process. However, the identification of the key genes responsible for these processes will only be possible through a combination of system biology approaches including statistical and mathematical modelling based on these "omics" data and of genetic approaches. The latter relies on the study of natural or induced genetic variation. Quantitative trait loci (QTL) detection has been a quite useful approach in the identification of genetic loci responsible for quality traits in grapes as well as resistance traits (see Chapters 6 and 7), although only the gene(s) behind one of these major loci is as yet close to being identified (see Chapter 8). The study of natural somatic variants (reviewed in Chapter 3) has been successful in the identification of specific genes involved in reproductive development and berry color (e.g., Boss and Thomas 2002; Kobayashi et al. 2004; Fernández et al. 2010). However, if one wishes to generalize such mutant studies, one caveat that has to be overcome is the production of large mutant collections in grapevine, such as those which have been developed for tomato (Menda et al. 2004) or pea (Dalmais et al. 2008) and that are very useful in the assignment of gene biological function. Limitations relate in this case to the need of space to maintain the genetic resources and the length of generation time required for segregation analyses. The difficulties involving these genetic analyses and the resources required has led the grapevine community to start the development of tools and knowledge for application of association genetics strategies (see for instance Barnaud et al. 2009; Fournier-Level et al. 2009; Myles et al. 2010 and Chapter 4 for review), as it is been done in human or animals genetics for similar reasons (Karlsson et al. 2007; Slatkin 2008; McCarthy et al. 2008). In addition, based on the finding of a natural somatic mutant derived from Pinot Meunier (Boss and Thomas 2002), short cycling grapevine genotypes suitable either for transformation or for genetic studies were recently successfully developed (Chaib et al. 2010). Such lines will be of tremendous interest as they are dwarf to semi-dwarf, needing therefore less space, and as they are produce fruits within 6 months (Chaib et al. 2010).

Within agricultural production viticulture and wine production results from a series of complex interactions between the plant, physical environment and human environment. The wine as the final product summarizes this complexity and displays a large diversity representing the different ways and the different tools and environments that have been used in human history for its production. The complexity of interactions required to generate a high quality product have made viticulture and enology to be considered almost as an art. Understanding the grapevine genome as well as the genomes of other organisms such as yeast and *Oenococcus oeni* bacteria that participate in the process of wine making can help us to more deeply understand the complexities of berry development and ripening and wine production. The possibilities to manage large amounts of data enable us now to address the analysis of more complex processes and this can help explain scientifically all kinds of interactions taking place in the vineyard and in the fermentation vat. It is expected that this knowledge will help in the design of new strategies to cope with some of the described challenges in viticulture and perhaps other unforeseen ones. It will be difficult to make better wines than the current ones but the goal is to maintain this high quality level under the frame of sustainability within a changing environment.

References

Ait Barka E, Gognies S, Nowak J, Audran JC, Belarbi A (2002) Inhibitory effect of endophyte bacteria on *Botrytis cinerea* and its influence to promote the grapevine growth. Biol Control 24: 35–142.

Ait Barka E, Nowak J, Clément C (2006) Enhancement of chilling resistance of inoculated grapevine plantlets with a plant growth-promoting rhizobacterium, *Burkholderia phytofirmans* strain PsJN. Appl Environ Microbiol 72: 7246–7252.

Barnaud A, Laucou V, This P, Lacombe T, Doligez A (2009) Linkage disequilibrium in wild French grapevine, *Vitis vinifera* L. subsp. *Silvestris*. Heredity 104: 431–437.

Battilana J, Costantini L, Emanuelli F, Sevini F, Segala C, Moser C, Velasco R, Versini G, Grando MS (2009) The 1-deoxy-D: -xylulose 5-phosphate synthase gene co-localizes with a major QTL affecting monoterpene content in grapevine. Theor Appl Genet 118: 653–669.

Beckers GJ, Conrath U (2007) Priming for stress resistance: from the lab to the field. Curr Opin Plant Biol 10: 425–431.

Bellin D, Ferrarini A, Chimento A, Kaiser O, Levenkova N, Bouffard P, Delledonne M (2009) Combining next-generation pyrosequencing with microarray for large scale expression analysis in non-model species. BMC Genom 10: 555.

Boss PK, Thomas MR (2002) Association of dwarfism and floral induction with a grape 'green revolution' mutation. Nature 416: 847–850.

Bouquet A, Boursiquot J-M (1999) Safeguarding old varieties and developing recently obtained new cultivars a joint step to conciliate tradition and innovation in France. Bull OIV 72: 753–761.

Castellarin SD, Matthews MA, Di Gaspero G, Gambetta GA (2007a) Water deficits accelerate ripening and induce changes in gene expression regulating flavonoid biosynthesis in grape berries. Planta 227: 101–112.

Castellarin SD, Pfeiffer A, Sivilotti P, Degan M, Peterlunger E, Di Gaspero G (2007b) Transcriptional regulation of anthocyanin biosynthesis in ripening fruits of grapevine under seasonal water deficit. Plant Cell Environ 30: 1381–1399.

Chaïb J, Torregrosa L, Mackenzie D, Corena P, Bouquet A, Thomas MR (2010) The grape microvine– a model system for rapid forward and reverse genetics of grapevines. Plant J. 62: 1083–1092.

Choquer M, Fournier E, Kunz C, Levis C, Pradier JM, Simon A, Viaud M (2007) *Botrytis cinerea* virulence factors: new insights into a necrotrophic and polyphageous pathogen. FEMS Microbiol Lett 277: 1–10.

Cimerman A, Arnaud G, Foissac X (2006) *Stolbur phytoplasma* genome survey achieved using a suppression subtractive hybridization approach with high specificity. Appl Environ Microbiol 72: 3274–3283.

Conde C, Silva P, Fontes N, Dias ACP, Tavares RM, Sousa MJ, Agasse A, Delrot S, Gerós H (2007) Biochemical changes throughout grape berry development and fruit and wine quality. Food 1: 1–22.

Conrath U, Beckers G, Flors V, Garcia-Agustin P, Jakab G, Mauch F, Newman M, Pieterse C, Poinssot B, Pozo M, Pugin A, Schaffrath U, Ton J, Wendehenne D, Zimmerli L, Mauch-Mani B (2006) Priming: getting ready for battle. Mol Plant-Microbe Interact 19: 1062–1071.

Cramer GR, Ergül A, Grimplet J, Tillett RL, Tattersall EAR, Bohlman MC, Vincent D, Sonderegger J, Evans J, Osborne C, Quilici D, Schlauch KA, Schooley DA, Cushman JC (2007) Water and salinity stress in grapevines: early and late changes in transcript and metabolite profiles. Funct Integr Genom 7: 111–134.

Dalmais M, Schmidt J, Le Signor C, Moussy F, Burstin J, Savois V, Aubert G, Brunaud V, de Oliveira Y, Guichard C, Thompson R, Bendahmane A (2008) UTILLdb, a *Pisum sativum* in silico forward and reverse genetics tool. Genome Biol 9: R43.

Daniel R (2005) The metagenomics of soil. Nat Rev Microbiol 3: 470–478.

Deluc LG, Grimplet J, Wheatley MD, Tillett RL, Quilici DR, Osborne C, Schooley DA, Schlauch KA, Cushman JC, Cramer GR (2007) Transcriptomic and metabolite analyses of Cabernet Sauvignon grape berry development. BMC Genom 8: 429.

Duchene E, Huard F, Dumas V, Schneider C, Merdinoglu D (2010) The challenge of adapting grapevine varieties to climate change. Climate Res 41: 193–204.

Espinoza C, Medina C, Somerville S, Arce-Johnson P (2007) Senescence-associated genes induced during compatible viral interactions with grapevine and Arabidopsis. J Exp Bot 58: 3197–3212.

Fernandez L, Romieu C, Moing A, Bouquet A, Maucourt M, Thomas MR, Torregrosa L (2006) The grapevine fleshless berry Mutation. A unique genotype to investigate differences between fleshy and nonfleshy fruit. Plant Physiol 140: 537–547.

Fernandez L, Torregrosa L, Segura V, Bouquet A, Martínez Zapater JM (2010) Transposon induced gene activation as a mechanism generating cluster shape somatic variation in grapevine. Plant J 61: 545–557.

Figueiredo A, Fortes AM, Ferreira S, Sebastiana M, Choi YH, Sousa L, Acioli-Santos B, Pessoa F, Verpoorte R, Pais MS (2008) Transcriptional and metabolic profiling of grape (*Vitis vinifera* L.) leaves unravel possible innate resistance against pathogenic fungi. J Exp Bot 59: 3371–3381.

Fillinger S, Amselem J, Artiguenave F, Billaut A, Choquer M, Couloux A, Cuomo C, Dickman MB, Fournier E, Gioti A, Giraud C, Kodira C, Kohn L, Legeai F, Levis C, Mauceli E, Pommier C, Pradier JM, Quevillon E, Rollins J, Ségurens B, Simon A, Viaud M, Weissenbach J, Wincker P, Lebrun M-H (2007) The genome projects of the plant pathogenic fungi *Botrytis cinerea* and Sclerotinia sclerotiorum. In: P Jeandet, C Clément, A Conreux (eds) Macromolecules of Grape and Wines. Lavoisier, London, Paris, New York, pp 125–133.

Fournier-Level A, Le Cunff L, Gomez C, Doligez A, Ageorges A, Roux C, Bertrand Y, Souquet J-M, Cheynier V, This P (2009) Quantitative genetic bases of anthocyanin variation in grape (*Vitis vinifera* L. ssp. *sativa*) berry: a quantitative trait locus to quantitative trait nucleotide integrated study. Genetics 183: 1127–1139.

Fung R.W, Gonzalo M, Fekete C, Kovacs LG, He Y, Marsh E, McIntyre LM, Schachtman DP, Qiu W (2008) Powdery mildew induces defense-oriented reprogramming of the transcriptome in a susceptible but not in a resistant grapevine. Plant Physiol 146: 236–49.

Gonçalves B, Falco V, Moutinho-Pereira J, Bacelar E, Peixoto F, Correia C (2009) Effects of elevated CO2 on grapevine (*Vitis vinifera* L.): volatile composition, phenolic content, and *in vitro* antioxidant activity of red wine. J Agric Food Chem 57: 265–273.

Iandolino A, Nobuta K, Goes da Silva F, Cook DR, Meyers BC (2008) Comparative expression profiling in grape (*Vitis vinifera*) berries derived from frequency analysis of ESTs and MPSS signatures. BMC Plant Biol 8: 53.

Jiao JY, Wang HX, Zeng Y, Shen YM (2006) Enrichment for microbes living in association with plant tissues. J Appl Microbiol 100: 830–837.

Joly P-B, Rip A (2007) A timely harvest. Nature 450: 174.

Jones GV (2007) Climate change: Observations, projections, and general implications for viticulture and wine production. Proc of the Int Symp Global Warming, which Potential Impacts on the Vineyards? March 28–30, 2007, Dijon-Beaune, France: *http://www.u-bourgogne.fr/chaireunesco-vinetculture/Actes%20clima/Accueil.htm*

Karlsson EK, Baranowska I, Wade CM, Salmon Hillbertz NH, Zody MC, Anderson N, Biagi TM, Patterson N, Pielberg GR, Kulbokas EJ 3rd, Comstock KE, Keller ET, Mesirov JP, von Euler H, Kämpe O, Hedhammar A, Lander ES, Andersson G, Andersson L, Lindblad-Toh K (2007) Efficient mapping of mendelian traits in dogs through genome-wide association. Nat Genet 39: 1321–1328.

Kobayashi S, Goto-Yamamoto N, Hirochika H (2004) Retrotransposon-induced mutations in grape skin color. Science 304: 982.

Laimer M, Lemaire O, Herrbach E, Goldschmidt V, Minafra A, Bianco O, Wetzel T (2009) Resistance to viruses, phytoplasmas and their vectors in the grapevine in Europe: a review. J Plant Pathol 91: 7–23.

Lambais MR, Goldman MH, Camargo LE, Goldman GH (2000) A genomic approach to the understanding of *Xylella fastidiosa* pathogenicity. Curr Opin Microbiol 3: 459–462.

Lin H, Doddapaneni H, Takahashi Y, Walker MA (2007) Comparative analysis of ESTs involved in grape responses to *Xylella fastidiosa* infection. BMC Plant Biol 7: 8.

Lund S, Peng F, Nayar T, Reid K, Schlosser J (2008) Gene expression analyses in individual grape (*Vitis vinifera* L.) berries during ripening initiation reveal that pigmentation intensity is a valid indicator of developmental staging within the cluster. Plant Mol Biol 68: 301–315.

Malembic-Maher S, Constable F, Cimerman A, Arnaud G, Carle P, Foissac X, Boudon-Padieu E (2008) A chromosome map of the Flavescence dorée phytoplasma. Microbiology 154: 1454–1463.

McCarthy MI, Abecasis GR, Cardon LR, Goldstein DB, Little J, Ioannidis JPA, Hirschhorn JN (2008) Genome-wide association studies for complex traits: consensus, uncertainty and challenges. Nat Rev Genet 9: 356–369.

McDonald BA, Linde C (2002) Pathogen population genetics, evolutionary potential, and durable resistance. Annu Rev Phytopathol 40: 349–379.

McMahan G, Yeh W, Marshall MN, Olsen M, Sananikone S, Wu JY, Block DE, VanderGheynst JS (2001) Characterizing the production of a wild-type and benomyl-resistant *Fusarium lateritium* for biocontrol of *Eutypa lata* on grapevine. J Ind Microbiol. Biotechnol 26: 151–155.

Menda N, Semel Y, Peled D, Eshed Y, Zamir D (2004) *In silico* screening of a saturated mutation library of tomato. Plant J 38: 861–872.

Mohamed N, Lherminier J, Farmer MJ, Fromentin J, Béno N, Houot V, Milat ML, Blein JP (2007) Defense responses in grapevine leaves against *Botrytis cinerea* induced by application of a *Pythium oligandrum* strain or its elicitin, oligandrin, to roots. Phytopathology 97: 611–620.

Mori K, Goto-Yamamoto N, Kitayama M, Hashizume K (2007) Loss of anthocyanins in red-wine grape under high temperature. J. Exp. Bot. 58: 1935–1945.

Myles S, Chia J-M, Hurwitz B, Simon C, Zhong GY, Buckler E, Ware D (2010) Rapid genomic characterization of the genus *Vitis*. PLOS One 5: e8219.

Naraghi-Arani P, Daubert S, Rowhani A (2001) Quasispecies nature of the genome of grapevine fanleaf virus. J Gen Virol 82: 1791–1795.

Ollat N, Diakou-Verdin P, Carde J-P, Barrieu F, Gaudillère J-P, Moing A (2002) Grape berry development: a review. J Int Sci Vigne Vin 36: 109–131.

Pilati S, Perazzolli M, Malossini A, Cestaro A, Demattè L, Fontana P, Dal Ri A, Viola R, Velasco R, Moser C (2007) Genome-wide transcriptional analysis of grapevine berry ripening reveals a set of genes similarly modulated during three seasons and the occurrence of an oxidative burst at véraison. BMC Genom 8: 428.

Polesani M, Desario F, Ferrarini A, Zamboni A, Pezzotti M, Kortekamp A, Polverani A (2008) cDNA-AFLP analysis of plant and pathogen genes expressed in grapevine infected with *Plasmopara viticola*. BMC Genom 9: 142.

Polesani M, Bortesi L, Ferrarini A, Zamboni A, Fasoli M, Zadra C, Lovato A, Pezzotti M, Delledonne M , Polverari A (2010) General and species-specific transcriptional responses to downy mildew infection in a susceptible (*Vitis vinifera*) and a resistant (*V. riparia*) grapevine species. BMC Genomics 11: 117.

Rabosto X, Carrau M, Paz A, Boido E, Dellacassa E, Carrau FM (2006) Grapes and vineyard soils as sources of microorganisms for biological control of *Botrytis cinerea*. Am J Enol Vitic 57: 332–338.

Roper MC, Greve LC, Warren JG, Labavitch JM, Kirkpatrick BC (2007) *Xylella fastidiosa* requires polygalacturonase for colonization and pathogenicity in *Vitis vinifera* grapevines. Mol Plant-Microbe Interact 20: 411–419.

Schmidt CS, Lorenz D, Wolf GA (2001) Biological control of the grapevine dieback fungus *Eutypa lata* I: screening of bacterial antagonists. J Phytopathol 149: 427–435.

Slatkin M (2008) Linkage disequilibrium: understanding the evolutionary past and mapping the medical future. Nat Rev Genet 9: 477–485.

Van Elsas JD, Costa R, Jansson J, Sjöling S, Bailey M, Nalin R, Vogel TM, van Overbeek L (2008) The metagenomics of disease-suppressive soils—experiences from the METACONTROL project. Trends Biotechnol 26: 591–601.

Van Sluys MA, de Oliveira MC, Monteiro-Vitorello CB, Miyaki CY, Furlan LR, Camargo LE, da Silva AC, Moon DH, Takita MA, Lemos EG, Machado MA, Ferro MI, da Silva FR, Goldman MH, Goldman GH, Lemos MV, El-Dorry H, Tsai SM, Carrer H, Carraro DM, de Oliveira RC, Nunes LR, Siqueira WJ, Coutinho LL, Kimura ET, Ferro ES, Harakava R, Kuramae EE, Marino CL, Giglioti E, Abreu IL, Alves LM, do Amaral AM, Baia GS, Blanco SR, Brito MS, Cannavan FS, Celestino AV, da Cunha AF, Fenille RC, Ferro JA, Formighieri EF, Kishi LT, Leoni SG, Oliveira AR, Rosa VE Jr, Sassaki FT, Sena JA, de Souza AA, Truffi D, Tsukumo F, Yanai GM, Zaros LG, Civerolo EL, Simpson AJ, Almeida NF Jr,

Setubal JC, Kitajima JP (2003) Comparative analyses of the complete genome sequences of Pierce's disease and citrus variegated chlorosis strains of *Xylella fastidiosa*. J Bacteriol 185: 1018–1026.
Van Wees SC, Van der Ent S, Pieterse CM (2008) Plant immune responses triggered by beneficial microbes. Curr Opin Plant Biol 11: 443–448.
Ventura M, Turroni F, Canchaya C, Vaughan EE, O'Toole PW, Sinderen D (2009) Microbial diversity in the human intestine and novel insights from metagenomics. Front Biosci 14: 3214–3221.
Viaud M, Legeai F, Pradier JM, Brygoo F, Bitton F, Weissenbach J, Brunet-Simon A, Duclert A, Fillinger S, Fortini D, Gioti A, Giraud C, Halary S, Lebrun I, Le Pêcheur P, Samson D, Levis C (2005) Expressed sequence tags from the phytopathogenic fungus *Botrytis cinerea*. Eur J Plant Pathol 111: 139–146.
Vincent D, Ergül A, Bohlman MC, Tattersall EAR, Tillett RL, Wheatley MD, Woolsey R, Quilici D, Joets J, Schlauch KA, Schooley DA, Cushman JC, Cramer GR (2007) Proteomic analysis reveals differences between *Vitis vinifera* L. cv. Chardonnay and cv. Cabernet Sauvignon and their responses to water deficit and salinity. J Exp Bot 58: 1873–1892.
Zenoni S, Ferrarini A, Giacomelli E, Xumerle L, Fasoli M, Malerba G, Bellin D, Pezzotti M, Delledonne M (2010) Characterization of transcriptional complexity during berry development in *Vitis vinifera* using RNA-Seq. Plant Physiol 152: 1787–1795.
Zhang J, Ma H, Feng J, Zeng L, Wang Z, Chen S (2008) Grape berry plasma membrane proteome analysis and its differential expression during ripening. J Exp Bot 59: 2979–2990.
Zimmerli L, Jakab G, Metraux J, Mauch-Mani B (2000) Potentiation of pathogen-specific defense mechanisms in Arabidopsis by beta-aminobutyric acid. Proc Natl Acad Sci USA 97: 12920–12925.

Index

D

G

W

X

Y

Z

Color Plate Section

Chapter 2

A

B

Figure 2-1 Illustration of morphological variation. A: Variation in berry shape and size in several *V. vinifera* cultivars. B. Illustration of morphological variation in *Cissus, Tetrastigma* and *Vitis* species.

Chapter 3

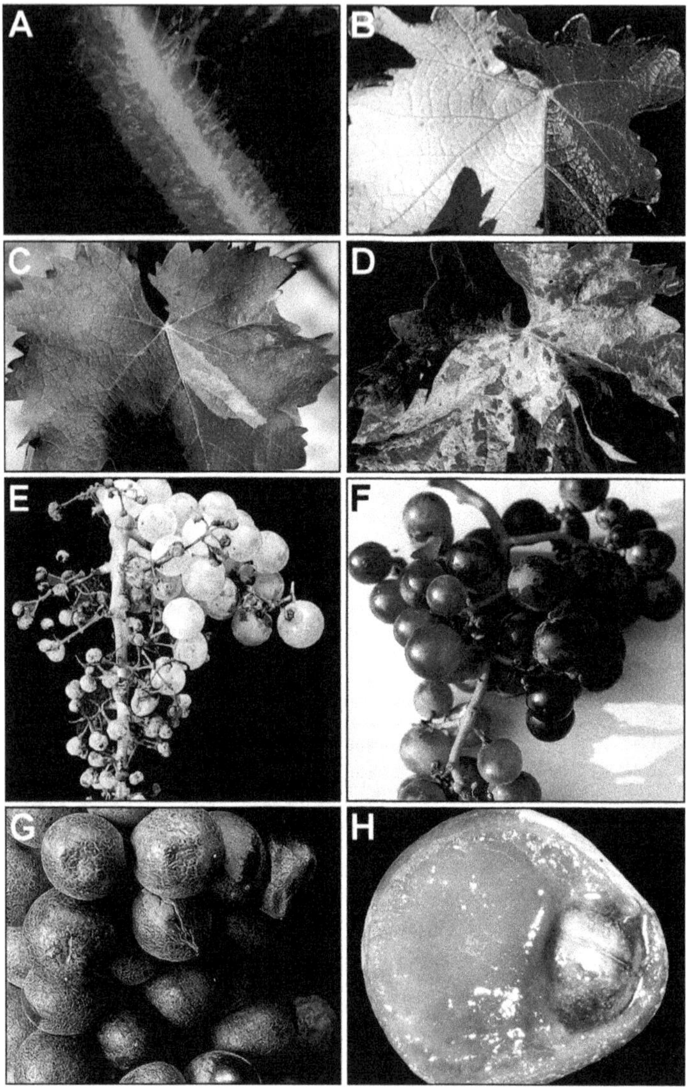

Figure 3-2 Some examples of grapevine chimeras. (A) Transgenic root expressing localized GFP ectopic expression, (B) Leaf with half of the blade showing chlorophyll deficiency, (C) Leaf blade showing three different somatic phenotypes (wild type, chlorophyll deficiency in either on cell layer, chlorophyll deficiency in L1 and L2 cell layer), (D) Patchy chlorophyll deficiency variegations, (E) Chimeric variation of FLB phenotype at bunch level, (F) Somatic variation of berry color development at bunch level, (G) Suberification of berry skin showing somatic variations at bunch and berry levels (unsuberified sector within suberified berry), (H) Ovary of the FLB somatic variant showing a reversion to normal development in one of the two carpels.

Chapter 5

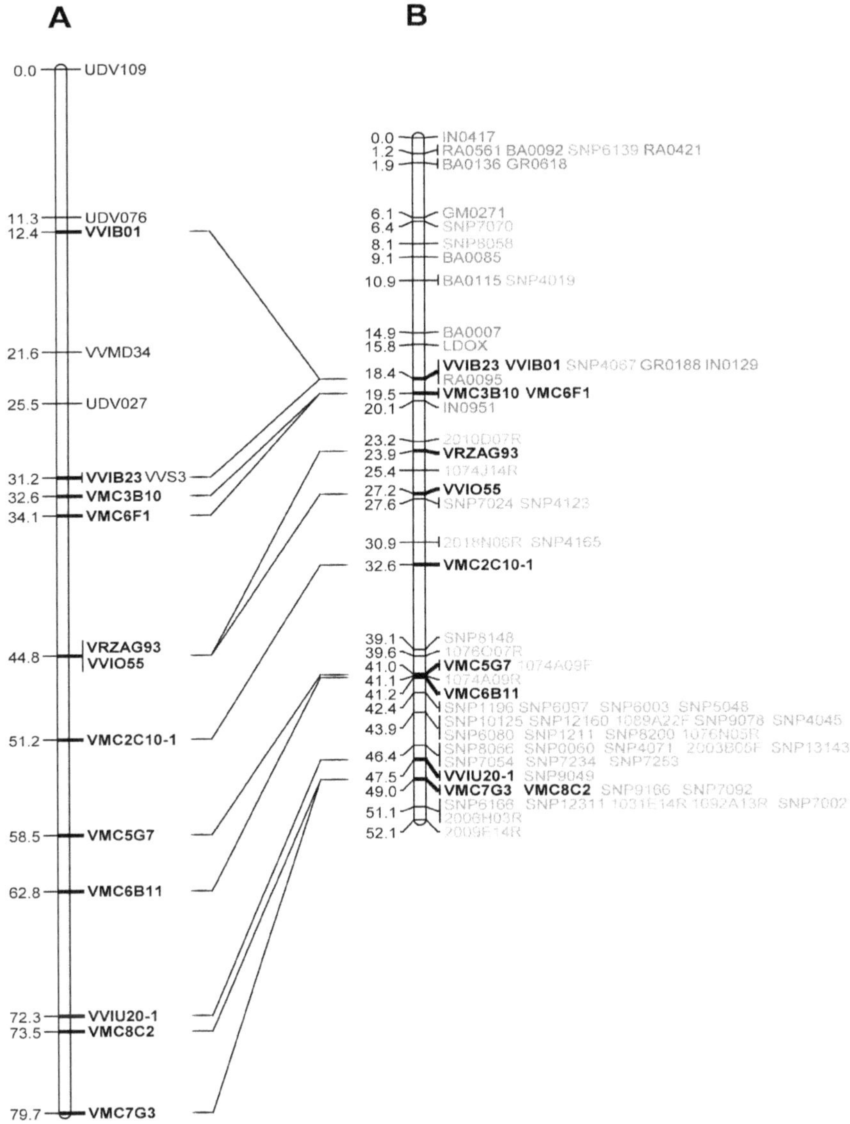

Figure 5-1 Grapevine linkage map of chromosome 2. (A) Reference map based exclusively on SSR markers (Doligez et al. 2006a); (B) the same linkage group saturated with SNP markers (Vezzulli et al. 2008a). SSR markers common to both maps are connected by lines. EST-derived SNPs are in red; BES-derived SNPs are in green (from Troggio et al. 2007); SNPs based on the heterozygous sites in the Pinot Noir sequence are in cyan (Velasco et al. 2007).

Chapter 7

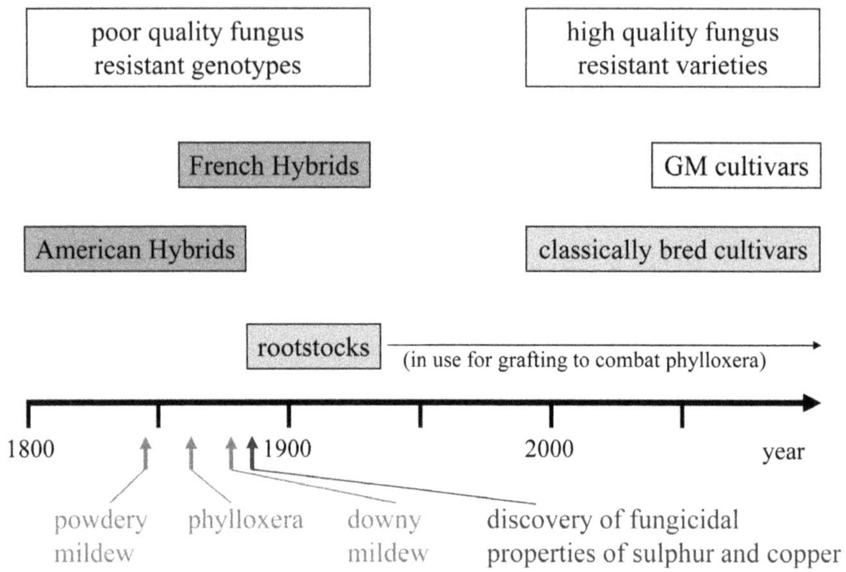

Figure 7-1 Milestones in grapevine resistance breeding on a time scale. The introduction of cultivars into the market is used as a point of reference. Red: American and French Hybrids did not succeed in the market due to poor wine quality. Even the best French Hybrids cultivated during the 20th century were almost all removed from the market. Green: Phylloxera tolerant or resistant rootstocks saved viticulture in Europe. Newly bred wine grape cultivars showing good field resistance and high wine quality entered the market around the turn of the millennium. Decoupling of resistance and quality could be proven in the 1960s but these cultivars were not accepted in the market. At the beginning of the 21st century marker-assisted selection (MAS) and marker-assisted backcrossing (MABC) became relevant for breeding programs. Yellow: Genetically modified cultivars will become available for the market in about two decades if consumer acceptance will be given. Appearance of mildew fungi and Phylloxera in Europe and the discovery of sulfur and copper as fungicides are indicated.

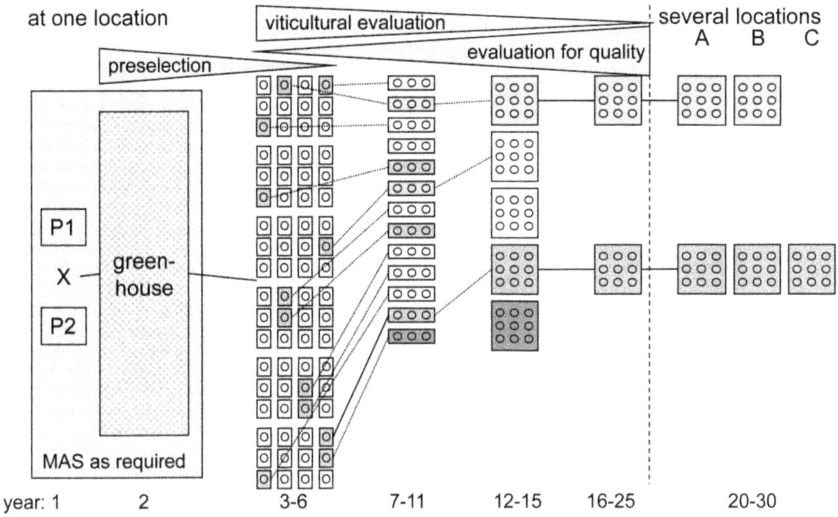

Figure 7-4 Steps in wine grape breeding: (from left to right) first selection for fungus resistance (*Plasmopara viticola* and *Erysiphe necator*) in a greenhouse, seedlings field (1 individual), pretesting (ca. 10 vines), intermediate testing (ca. 50 vines), main testing (ca. 500–1,000 vines), test planting with wine growers at different locations. A rough time scale in years is shown for each selection step.

Chapter 8

Figure 8-2 Comparison of powdery mildew (A, B) and downy mildew (C, D) susceptibility of leaves from *V. vinifera* cv. Cabernet sauvignon (A, C) and from *M. rotundifolia* x *V. vinifera* 4th generation backcross (B, D). Powdery mildew growth is completely suppressed on the BC_4 plant but a low level of downy mildew sporulation is supported on BC_4 leaves. Patches of dead cells which have undergone programmed cell death in response to downy mildew infection are visible in panel D.

Chapter 9

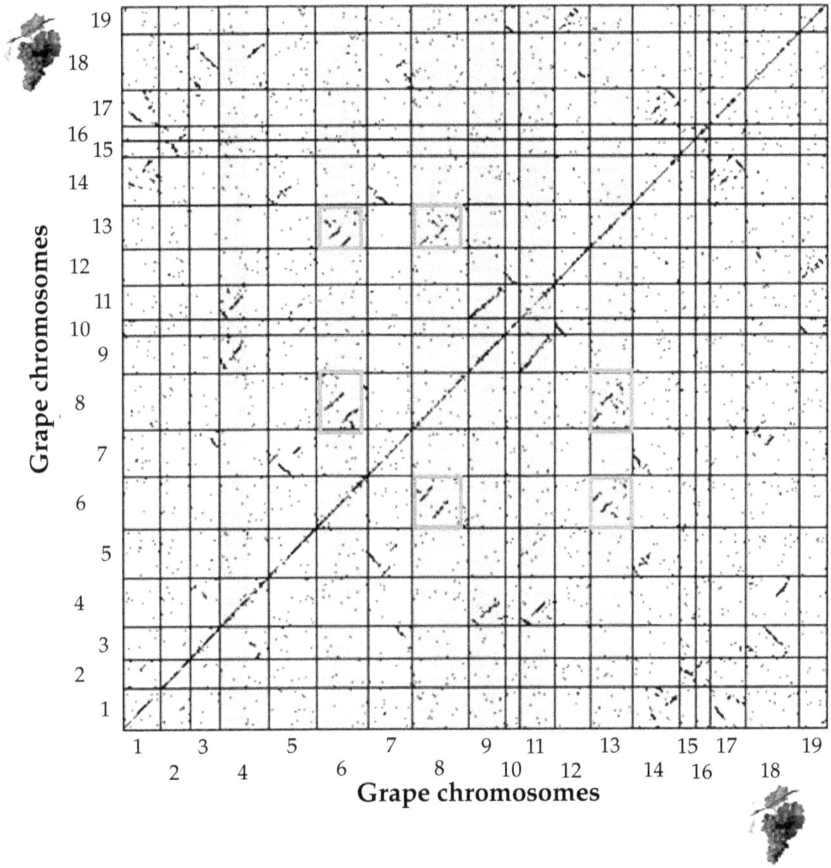

Figure 9-1 Topological distribution of paralogous genes in the grapevine genome showing a 3 to 3 relationship of blocks. The 19 chromosomes of grape are represented on both the x and y axis. Dots represent the positions of paralogous pairs of genes. Paralogous genes were computed according to a Reciprocal Best Hit (Jaillon et al. 2007). In blue is highlighted the exclusive correspondence between a half of chromosome 4 with chromosomes 9 and 11 and in red the same kind of relationships between the chromosomes 6, 8 and 13. Each and every one of the chromosomes have a relationship with two others.

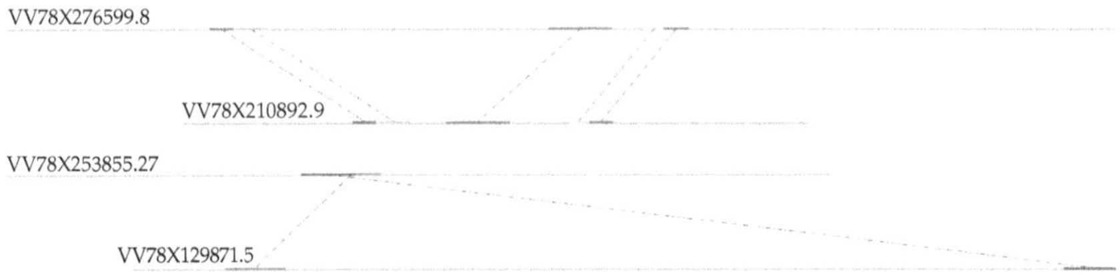

Figure 9-2 Two examples of chromosome-specific regions of the Pinot Noir genome. (a) segments with different DNA sequences (in green) flanked by orthologous regions of the two homologous chromosomes (same color for both chromosomes); (b) Gaps corresponding to a sequence missing in one chromosome but not in the other.

Chapter 14

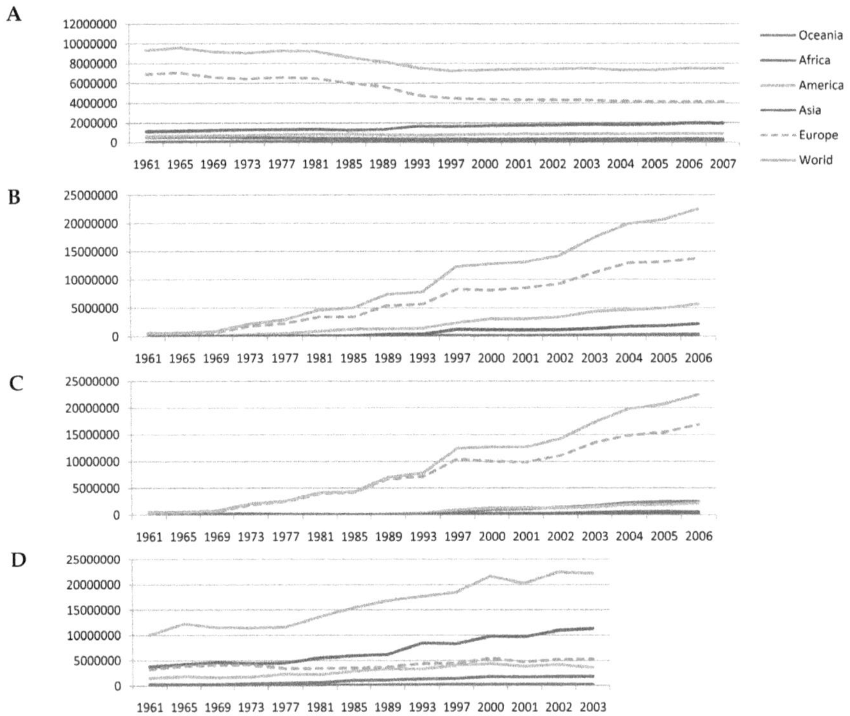

Figure 14-1 World statistics for the wine sector (from FAOSTAT; *http://faostat.fao.org/*) (A) Surface harvested (ha) (B) Wine imports (1000 $). (C) Wine exports in (1000$) (D) Wine consumption (t).

For Product Safety Concerns and Information please contact our EU representative GPSR@taylorandfrancis.com
Taylor & Francis Verlag GmbH, Kaufingerstraße 24, 80331 München, Germany

www.ingramcontent.com/pod-product-compliance
Ingram Content Group UK Ltd.
Pitfield, Milton Keynes, MK11 3LW, UK
UKHW021832260926
475702UK00049B/345
* 9 7 8 1 5 7 8 0 8 7 1 7 4 *